Noise Reduction
in Speech
Applications

THE ELECTRICAL ENGINEERING
AND APPLIED SIGNAL PROCESSING SERIES
Edited by Alexander Poularikas

The Advanced Signal Processing Handbook:
Theory and Implementation for Radar, Sonar,
and Medical Imaging Real-Time Systems
Stergios Stergiopoulos

The Transform and Data Compression Handbook
K.R. Rao and P.C. Yip

Handbook of Multisensor Data Fusion
David Hall and James Llinas

Handbook of Neural Network Signal Processing
Yu Hen Hu and Jenq-Neng Hwang

Handbook of Antennas in Wireless Communications
Lal Chand Godara

Noise Reduction in Speech Applications
Gillian M. Davis

Forthcoming Titles

Propagation Data Handbook for Wireless Communications
Robert Crane

The Digital Color Imaging Handbook
Guarav Sharma

Applications in Time Frequency Signal Processing
Antonia Papandreou-Suppappola

Signal Processing Noise
Vyacheslav P. Tuzlukov

Digital Signal Processing with Examples in MATLAB®
Samuel Stearns

Smart Antennas
Lal Chand Godara

Pattern Recognition in Speech and Language Processing
Wu Chou and Bing Huang Juang

Nonlinear Signal and Image Processing: Theory, Methods, and Applications
Kenneth Barner and Gonzalo R. Arce

Noise Reduction in Speech Applications

Edited by
Gillian M. Davis

CRC PRESS

Boca Raton London New York Washington, D.C.

Library of Congress Cataloging-in-Publication Data

Noise reduction in speech applications / edited by Gillian M. Davis.
      p.   cm. — (The electrical engineering and applied signal processing series)
      Includes bibliographical references and index.
      ISBN 0-8493-0949-2 (alk. paper)
         1. Speech processing systems. 2. Telephone systems. 3. Electronic noise—Prevention.
4. Noise control. 5. Signal processing—Digital techniques. I. Davis, Gillian M. II. Series.

      TK7882.S65 N65 2002
      621.382′8—dc21
                                                 2002017483

This book contains information obtained from authentic and highly regarded sources. Reprinted material is quoted with permission, and sources are indicated. A wide variety of references are listed. Reasonable efforts have been made to publish reliable data and information, but the author and the publisher cannot assume responsibility for the validity of all materials or for the consequences of their use.

Visit the CRC Press Web site at www.crcpress.com

Preface

A wide range of potential sources of noise and distortion can degrade the quality of the speech signal in a communication system. *Noise Reduction in Speech Applications* explores the effects of these interfering sounds on speech applications and introduces a range of techniques for reducing their influence and enhancing the acceptability, intelligibility, and speaker recognizability of the communications signal. A systems approach to noise reduction is taken that emphasizes the advantage of minimizing noise pickup and creation, in the first instance, in addition to choosing the most appropriate noise reduction technique at each stage to obtain the best overall result. This handbook aims to make the available technologies better known and to set expectations of what can actually be achieved in practice at a realistic level. Sufficient detail is given for readers to decide which, if any, of the noise reduction techniques discussed is an appropriate solution for their own systems and also to help them make the best use of these technologies.

The timing of this book is particularly appropriate. Although much of the technology required for noise reduction has existed for some time, it is only with the recent development of powerful but inexpensive digital signal processing (DSP) hardware that implementation of the technology in everyday systems has started to become practical.

Noise Reduction in Speech Applications begins with a tutorial chapter covering background material on digital signal processing and adaptive filtering. Emphasis is placed on techniques relevant to noise reduction in speech applications that are referenced by the authors of later chapters. This tutorial chapter is written at a level suitable for students studying DSP techniques as part of their electrical engineering or acoustics courses and for master's degree students taking specialist digital signal processing courses.

The remainder of the book is divided into three sections:

Systems Aspects addresses the need to consider the complete system and apply the most appropriate noise reduction technique at each stage to achieve the best overall result.

Digital Algorithms and Implementation looks at three types of digital noise reduction algorithms in detail: single-channel speech enhancers, microphone arrays, and echo cancellers. Example code and audio wavefiles illustrating the noise problems and solutions are provided to accompany these chapters. These files are available at http://www.crcpress.com/e_products/download.asp?cat_no=0949. The example code will be of particular interest to students of the subject, whereas the audio wavefiles will be of interest to a wider audience, including readers with limited technical knowledge but an interest in the noise reduction achievable with such algorithms.

Special Applications investigates the use of noise reduction techniques in eight application areas, including speech recognition, Internet telephony, and digital hearing aids. This final section is aimed at potential commercial customers of this technology and focuses on the sorts of results that can be achieved in practice. Audio wavefiles are provided to accompany these chapters at http://www.crcpress.com/e_products/download.asp?cat_no=0949.

Each chapter of this book concludes with a list of references that provide guidance for readers wishing to examine the subject of the chapter in more detail. In addition, because many of the chapters use acronyms that may be unfamiliar to the reader, a general list of acronyms is provided at the front of the book.

It has been a great pleasure working with the chapter authors on the production of this book. The willingness of these specialists to sacrifice valuable research time to prepare this review material is greatly appreciated. Without their commitment, this book would not have been possible. I have enjoyed watching this book develop, and I hope that those who read it will find it to be a valuable source of information.

Many thanks are due to NCT Group, Inc. U.S.A. for supporting my involvement in the preparation of this book. A substantial debt of gratitude is due also to James Elburn of NCT (Europe) Ltd., U.K. who provided invaluable IT support throughout this project, and to Stephen Leese, who advised on suitable subjects and contributors.

Gillian M. Davis

The Editor

Gillian M. Davis, D.Phil., is Managing Director of Noise Cancellation Technologies (Europe) Ltd. and a Vice President of the parent company, NCT Group, Inc. Previously she held research positions at Sharp Laboratories of Europe Ltd., NTT, Japan, and Rutherford Appleton Laboratory. Dr. Davis received her D.Phil. from the Clarendon Laboratory, Oxford University and her M.B.A. from the Open University, Milton Keynes.

Contributors

Victor Bray Auditory Research Department, Sonic Innovations, Inc., Salt Lake City, Utah, U.S.A.

Douglas M. Chabries College of Engineering and Technology, Brigham Young University, Provo, Utah, U.S.A.

Ingvar Claesson Department of Telecommunications and Signal Processing, Blekinge Institute of Technology, Ronneby, Sweden

John W. Cook BTexact Technologies, Martlesham Heath, Ipswich, U.K.

Mattias Dahl Department of Telecommunications and Signal Processing, Blekinge Institute of Technology, Ronneby, Sweden

Gillian M. Davis NCT (Europe) Ltd., Cambridge, U.K.

Graham P. Eatwell Adaptive Audio, Inc., Annapolis, Maryland, U.S.A.

Craig Fancourt Sarnoff Corporation, Princeton, New Jersey, U.S.A.

Lars Håkansson Department of Telecommunications and Signal Processing, Blekinge Institute of Technology, Ronneby, Sweden

Dennis Hardman Agilent Technologies, Inc., Colorado Springs, Colorado, U.S.A.

Malcolm J. Hawksford Department of Electronic Systems Engineering, University of Essex, Colchester, Essex, U.K.

Sven Johansson Department of Telecommunications and Signal Processing, Blekinge Institute of Technology, Ronneby, Sweden

Elizabeth G. Keate Texas Instruments, Santa Barbara, California, U.S.A.

George Keratiotis BTexact Technologies, Martlesham Heath, Ipswich, U.K.

Stephen J. Leese NCT (Europe) Ltd., Cambridge, U.K.

Larry Lind   Department of Electronic Systems Engineering, University of Essex, Colchester, Essex, U.K.

Robert S. Oshana   Software Development Systems, Texas Instruments, Dallas, Texas, U.S.A.

Ira L. Panzer   Dynastat, Inc., Austin, Texas, U.S.A.

Lucas Parra   Sarnoff Corporation, Princeton, New Jersey, U.S.A.

Minesh Patel   BTexact Technologies, Martlesham Heath, Ipswich, U.K.

Bhiksha Raj   Mitsubishi Electric Research Laboratories, Cambridge, Massachusetts, U.S.A.

Alan D. Sharpley   Dynastat, Inc., Austin, Texas, U.S.A.

Rita Singh   School of Computer Science, Carnegie Mellon University, Pittsburgh, Pennsylvania, U.S.A.

Per Sjösten   National Institute for Working Life, Göteborg, Sweden

Richard M. Stern   Department of Electrical and Computer Engineering and School of Computer Science, Carnegie Mellon University, Pittsburgh, Pennsylvania, U.S.A.

Robert W. Stewart   Department of Electronic and Electrical Engineering, University of Strathclyde, Glasgow, Scotland, U.K.

William D. Voiers   Dynastat, Inc., Austin, Texas, U.S.A.

Darren B. Ward   Department of Electrical and Electronic Engineering, Imperial College of Science, Technology and Medicine, London, U.K.

Stephan Weiss   Department of Electronics and Computer Science, University of Southampton, Southampton, U.K.

Pete Whelan   BTexact Technologies, Martlesham Heath, Ipswich, U.K.

Acronyms

ACRM	Absolute Category Rating Method
ADC	analog-to-digital converter
AFC	alternative forced choice
AGC	automatic gain control
ALU	arithmetic logic unit
ANC	active noise control
ANCU	adaptive noise cancellation unit
ANN	artificial neural network
API	application program interface
APLB	adaptive phase-locked buffer algorithm
AR	autoregressive
ASIC	application-specific integrated circuit
ATM	asynchronous transfer mode; automated teller machine
BB	background buzz (specific to the DAM)
BC	background chirping (specific to the DAM)
BJT	bipolar junction transistor
BNH	background noise high frequency (specific to the DAM)
BNM	background noise mid frequency (specific to the DAM)
BNL	background noise low frequency (specific to the DAM)
BS	background static (specific to the DAM)
BSS	blind source separation
CAE	composite acceptability estimate (specific to the DAM)
CBA	composite background acceptability (specific to the DAM)
CCS	crosstalk cancellation system
CDCN	codeword-dependent cepstral normalization
CE	common mode error
CER	command error rate
CIA	composite isometric acceptability (specific to the DAM)
CM	common mode
CMN	cepstral mean normalization
CNG	comfort noise generator
COTS	commercial off-the-shelf
CPA	composite perceptual acceptability (specific to the DAM)
CPU	central processing unit
CSA	composite signal acceptability (specific to the DAM)
CSU	critical speech unit

DAC	digital-to-analog converter
DACS	Digital Access Carrier Systems
DALT	Diagnostic Alliteration Test
DAM	Diagnostic Acceptability Measure
DAT	Diagnostic Alliteration Test
DCRM	Degradation Category Rating Method
DCT	discrete cosine transform
DE	differential error
DFT	discrete Fourier transform
DM	differential mode
DMA	direct memory access
DMCT	Diagnostic Medial Consonant Test
DMOS	Degradation Mean Opinion Score
DPLL	digital phase-locked loop
DRT	Diagnostic Rhyme Test
DSL	digital subscriber line
DSP	digital signal processing/processor
DSRT	Diagnostic Speaker Recognizability Test
DST	discrete sine transform
DTFT	discrete-time Fourier transform
DWT	discrete wavelet transform

EM	expectation maximization
EMAP	extended MAP
emf	electromagnetic field
EPQ	elementary perceived qualities
ERL	echo return loss
ERLE	echo return loss enhancement
ETSI	European Telecommunications Standards Institute

FEC	front-end clipping
FFT	fast Fourier transform
FIR	finite impulse response
FXLMS	filtered-x least mean squares
FXO	foreign exchange office
FXS	foreign exchange station

GCC	generalized cross correlation
GJB	Griffiths-Jim beamformer
GPP	general-purpose processor
GSC	generalized sidelobe canceller
GSD	generalized sidelobe decorrelator
GSS	geometric source separation
GUI	graphical user interface

HINT	Hearing in Noise Test
HMM	hidden Markov model
HOS	higher-order statistics
HOT	hold-over time
HPI	host port interface
HRTF	head-related transfer function
IBA	isometric background acceptability (specific to the DAM)
IIR	infinite impulse response
IMC	internal model control
INT	induction neutralizing transformers
ISA	instruction set architecture; isometric signal acceptability (specific to the DAM)
ISDN	integrated services digital network
ISR	interrupt service routine
ITU	International Telecommunication Union
JFET	junction field-effect transistor
KLT	Karhunen-Loeve transform
LAN	local area network
LCMV	linearly constrained minimum variance
LMS	least mean square(s)
LP	linear prediction
LPF	low-pass filter
LSS	linear spectral subtraction
LTI	linear and time invariant
MAC	multiply and accumulate
MAP	maximum *a posteriori*
MFCC	Mel frequency cepstral coefficients
MLLR	maximum likelihood linear regression
MMSE	minimum mean square(d) error
MNRU	modulated noise reference unit
MOS	mean opinion score
MPLS	multiprotocol label switching
MRT	Modified Rhyme Test
MSE	mean squared error
MSUB	magnitude spectrum of noise
NG	noise generator
NLP	nonlinear processing
NOC	network operations center

NOP	null operations
NSS	nonlinear spectral subtraction
OS	operating system
OSI	open system interconnection
PAC	physical acoustical correlates
PAMS	perceptual analysis measurement system
PB	phonetically balanced
PBA	perceptual background acceptability (specific to the DAM)
PCM	pulse code modulated
PCU	pipeline control unit
pdf	probability density function
PDF	probability distribution function
PESQ	perceptual evaluation of speech quality
PMC	parallel model combination
POF	probabilistic optimal filtering
PSA	perceptual signal acceptability (specific to the DAM)
PSD	power spectral density
PSTN	public-switched telephone network
PSUB	power spectrum of noise
PVT	perceived voice trait
QoS	quality of service
RASTA	relative spectral processing
RFI	radio frequency interference
RISC	reduced instruction set computer
RLS	recursive least squares (algorithm)
RMA	rate-monotonic analysis
RMS	root-mean square
RSVP	resource reservation protocol
RTCP	real-time transport control protocol
RTOS	real-time operating system
RTP	real-time transport protocol
SB	signal babble (specific to the DAM)
SD	signal distortion (specific to the DAM)
SF	signal flutter (specific to the DAM)
SH	signal high pass (specific to the DAM)
SI	signal interrupted (specific to the DAM)
SIP	Session Initiation Protocol
SIR	signal-to-interference ratio
SL	signal low pass (specific to the DAM)

SN	signal nasal (specific to the DAM)
SNR	signal-to-noise ratio; speech-to-noise ratio
SPINE	speech in noisy environments (database)
SQNR	signal-to-quantization-noise ratio
SRAM	static RAM
SS	spectral subtraction
ST	signal thin (specific to the DAM)
TCB	task control block
TCL	terminal coupling loss
TCP	transport control protocol
TF	transfer function
THD	total harmonic distortion
TIA	Telecommunications Industry Association
TOS	type of service
TRI	transformed rating intelligibility (specific to the DAM)
TRP	transformed rating pleasantness (specific to the DAM)
UDP	user datagram protocol
VAD	voice activity detector
VCA	voltage-controlled amplifier
VLIW	very long instruction word
VoFR	voice over frame relay
VoIP	Voice over Internet Protocol
VOX	voice-operated switch
VTS	vector Taylor series
WAN	wide area network
WDRC	wide dynamic range compression
WS	waveform synthesis

Contents

Section IV Special Applications

Section I:

Tutorial

1

Noise and Digital Signal Processing

Stephan Weiss, Robert W. Stewart, and Gillian M. Davis

CONTENTS

Introduction

Electronic systems in the context of audio communication perform transmission, recording, playback, analysis, or synthesis of speech signals. When designing a system for any of these purposes, noise influences must be carefully considered. Different types of noise and distortion can be characterized, and a number of signal processing concepts exist that can assist in mitigating their effect, thus enhancing the quality or intelligibility of the speech signal. Digital signal processing (DSP) offers a number of powerful tools that, depending on the circumstances, may or may not be applicable to specific types of noise corruptions. Therefore, the purpose of this introductory chapter is to create some awareness of the "noise chain" in a speech communication system and the relevant fundamental digital signal processing concepts, which can be exploited to design a system that is robust toward noise.

"Analog/Digital Interfacing and Noise Chain" gives an overview of the different stages in a general digital speech communication system and their exposure to, and inherent production of, noise. To characterize noise, which is generally assumed to be random, some background of stochastic signals and their quantities are reviewed in "Stochastic Signals and Their Characteristics." If the noise does not share the same frequency range as the speech signal, digital filtering can be a promising technique for noise reduction; this is discussed in "Digital Filtering." If noise and speech overlap in frequency but the speech signal exhibits specific features, processing in a transform domain may present a viable solution. The fundamentals of transforms and transform-domain processing are detailed in "Discrete Signal Transforms." Finally, if a probe of the corrupting noise is available or the noise is periodic, powerful adaptive DSP algorithms can be applied; these are reviewed in "Adaptive Digital Filtering."

The scope of this chapter is to provide a simple and generally intuitive overview of the DSP fundamentals upon which noise reduction methods can be based. Examples are provided that give additional insight into the working and application of the techniques discussed. For further information, the reader should consult the reference list at the end of this chapter and the specialized DSP applications discussed elsewhere in Sections III and IV of this book.

Analog/Digital Interfacing and Noise Chain

Any system involving the transmission, acquisition, or generation of speech is subject to a wide range of influences that may deteriorate the quality of the speech signal. This can include external interferences such as background

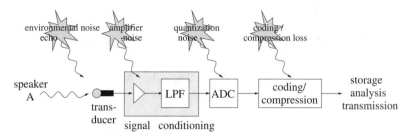

FIGURE 1.1
Stages of acquiring a speech signal, which are prone to noise corruption and distortion.

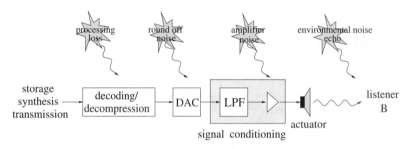

FIGURE 1.2
Reverse operation to Figure 1.1 where different stages suffer from noise and distortion.

noise in the recording, but it can also extend to echoic effects or nonlinear distortions introduced by analog electroacoustic devices or amplifiers.

Some of these adverse influences can be controlled through system parameters, such as sampling rate and word length, while others can be mitigated by DSP techniques. In the following, potential sources of noise and distortion in a speech communication system are highlighted and categorized. We then review basic signal processing techniques and discuss which particular noise influences can be addressed by them.

A Generic Digital Speech Communication System

A generic speech communication system connecting a speaker A to a listener B is shown in Figures 1.1 and 1.2. Taken on its own, Figure 1.1 represents a recording or general speech acquisition system, while Figure 1.2 implements a mechanism to synthesize speech or play back speech. Concatenated, the two block diagrams implement a simplex speech transmission system. From this simplex system, duplex communication can be established by additionally operating the same concatenated system from B back to A. In either of the two cases of Figures 1.1 and 1.2, a similar "noise chain" is encountered that will affect the quality of the signal of interest. In the case of the speech communication system in Figure 1.1, distortion includes:

- Environmental noise: The transducer picking up the voice of speaker A may be corrupted by environmental noise, e.g., in an airplane cockpit, where engine noise corrupts the pilot's voice in the microphone signal.
- Unwanted reverberation or echo: If speaker A is in an echoic environment, the recorded speech signal may be subject to severe dispersion and echo.
- Acoustic echo feedback: If the speech communication is full duplex, i.e., listener B can also communicate back to speaker A, then B's voice might be picked up by A's microphone and fed back to B. This situation arises in, e.g., hands-free telephony and results in severe intelligibility problems.
- Nonlinear amplifier distortion: Nonlinear distortion occurs if the amplifier is overdriven; in an extreme case, the signal becomes clipped.
- Amplifier noise: The amplifier may produce noise itself through, e.g., thermal influences. This noise can become audible, particularly for amplification with high gain.
- The anti-alias low-pass filter (LPF) in the signal conditioning block may distort the signal by suppressing high-frequency components of the speech.
- Jitter: This results in an irreversible signal distortion through an inaccurate clock signal for sampling. However, jitter is a problem that only affects high sampling rates that usually lie beyond the ones used for speech and audio.
- Quantization noise in the analog-to-digital converter (ADC): Distortion is introduced due to the round-off operation of sample values.
- Loss in coding and compression: Unlike lossless coders, lossy coding or compression schemes achieve a high compression of data through a trade-off against signal distortion. Although the aim of lossy coders is to refrain only from the coding of information that cannot be perceived by the human ear anyway, some audible distortion can be incurred.

When the data are finally stored, transmitted, or processed, further signal distortion can be encountered, e.g., through channel errors or interference in a transmission scenario. In the second part of the speech communication system in Figure 1.2, similar noise influences are found in reversed order:

- Noise introduced in decoding, decompression, or retrieval from storage medium.
- Inaccuracies in the signal reconstruction.

- Nonlinear distortion and amplifier noise.
- Nonlinear distortion and noise in the loudspeaker.
- Environmental noise: Interfering acoustic noise can disturb the quality or intelligibility of the signal for listener B. For example, the headphone signal is likely to be superposed with engine noise at the pilot's ear.

The noise influences listed above are not comprehensive but give an idea of the various sources of distortion to which speech communication is prone in the "noise chain." If the distortion becomes too dominant, then the signal quality quickly degrades.

In the remainder of this section, we concentrate on distortion due to the conversion of a signal from an analog waveform to its digital representation. In particular, the distortion introduced by quantization, discussed below, will provide the motivation to define a suitable "noise model." This additive noise model in turn permits the definition of a measure, the signal-to-noise ratio (SNR), to assess the severeness of noise distortion (see below). Before we treat this "classic" model, and its assumptions and characteristics, we consider sampling in the following section.

Sampling

A first step toward a digital number representation is the sampling of a signal $x(t)$, with t being the continous time variable, to obtain time-discrete values. This is generally performed by a sample-and-hold device. If we assume ideal sampling of an analog electrical signal $x(t)$ with a sampling period T_s (and hence a sampling rate of $f_s = 1/T_s$), the result of the sampling process, $x_s(t)$, can be expressed by multiplying $x(t)$ with a pulse train,

$$x_s(t) = x(t) \cdot \sum_{n=-\infty}^{\infty} \delta(t - nT_s) = \sum_{n=-\infty}^{\infty} x[nT_s] \cdot \delta(t - nT_s) \tag{1.1}$$

The pulse train consists of Dirac impulses $\delta(t)$,

$$\delta(t) = \begin{cases} 0 & \text{for} \quad t \neq 0 \\ 1 & \text{for} \quad t = 0 \end{cases} \tag{1.2}$$

which are spaced at integer multiples of the sampling period T_s. By exploiting the relation between an impulse train and its Fourier series representation, the Fourier transform $X_s(j\omega) \bullet\!\!-\!\!\circ x_s(t)$,

$$X_s(j\omega) = \int_{-\infty}^{\infty} x_s(t) e^{-j\omega t} dt \tag{1.3}$$

can be expressed as[1]

$$X_s(j\omega) = \int_{-\infty}^{\infty} x(t) \sum_{n=-\infty}^{\infty} \delta(t - nT_s)dt \tag{1.4}$$

$$= \sum_{k=-\infty}^{\infty} \int_{-\infty}^{\infty} x(t)e^{-jk\omega_s t}e^{-j\omega t}dt = \sum_{k=-\infty}^{\infty} X(j(\omega + k\omega_s)) \tag{1.5}$$

As an example, consider the spectrum $X(j\omega)$ of the continuous time signal $x(t)$ as given in Figure 1.3(a). Then according to Equation (1.5), the spectrum $X_s(j\omega)$ consists of a superposition of spectra of the continuous signal, $X(j\omega)$, shifted by integer multiples of the (angular) sampling rate $\omega_s = 2\pi f_s$, as shown in Figure 1.3(b). Therefore, unless the continuous signal $x(t)$ is sufficiently bandlimited, an overlap of spectral contributions occurs in the sampling process. This spectral overlap, known as aliasing and visualized in Figure 1.3(c), can be avoided by selecting the sampling rate appropriately as $\omega_s > 2\omega_{max}$, where ω_{max} is the maximum frequency component of the continuous time signal $x(t)$. The rate $2\omega_{max}$ is commonly labeled as the Nyquist rate $\omega_{Nyquist}$, which permits us to formulate the sampling theorem as

$$\omega_s > \omega_{Nyquist} \tag{1.6}$$

According to Equation (1.6), sampling has to occur at least at the Nyquist rate to ensure the correct and faithful representation of an analog signal in the digital domain, such that, e.g., the underlying analog signal can be reconstructed. As the exact spectral content of the signal supplied to an ADC

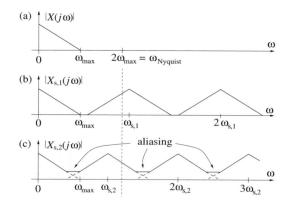

FIGURE 1.3
(a) Spectrum of analog signal $x(t)$ O—● $X(j\omega)$, and the spectra of sampled signals with (b) $\omega_{s,1} > 2\omega_{max}$ and (c) $\omega_{s,2} < 2\omega_{max}$; note the periodicity ω_s of the sampled spectra.

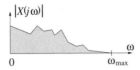

FIGURE 1.4
Spectrum $X(j\omega)$ corresponding to an analog continuous time signal $x(t)$.

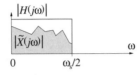

FIGURE 1.5
Spectrum of continuous time signal $x(t)$ after anti-alias filtering with a filter $H(j\omega)$ to remove spectral components above $\omega_s/2$.

is *a priori* unknown, the signal has to be appropriately bandlimited to avoid potential aliasing. This is performed by an anti-alias low-pass filter with a stopband frequency of $\omega_s/2$. The spectra of a signal prior to and after anti-alias filtering are depicted in Figures 1.4 and 1.5. Note that after filtering in Figure 1.5, the signal possesses no more components at or above $\omega_s/2$, and hence sampling can be performed perfectly satisfying the Nyquist theorem in Equation (1.6).

An alternative representation of the spectrum of $x_s(t)$ considers the sampled signal, $x[nT_s]$, as sifted by the pulse train. This forms a discrete time signal, in shorthand denoted as $x[n]$, with a discrete time index $n \in \mathbb{Z}$ of integer values. The Fourier transform $X_s(j\omega) \bullet\!\!-\!\!\circ x_s(t)$ based on $x[nT_s]$ can be written as

$$X_s(j\omega) = \int_{-\infty}^{\infty} \sum_{n=-\infty}^{\infty} x[n] \cdot \delta(t - nT_s)dt = \sum_{n=-\infty}^{\infty} x[n] \cdot e^{j\omega nT_s} \tag{1.7}$$

and is known as the Fourier transform of a discrete time sequence $x[n]$. The periodicity of $X_s(j\omega)$, which may also be noted from Equation (1.5), can be reflected in the notation by writing $X_s(e^{j\Omega})$ instead of $X_s(j\omega)$, whereby $\Omega = \omega/T_s$ is the normalized angular frequency.

Reconstruction of the underlying continuous-time analog signal $x(t)$ can be accomplished from the sampled signal values $x[n]$ by filtering the impulse train $x_s(t)$ of Equation (1.1) by a reconstruction filter — an ideal LPF with cut-off frequency $\omega_s/2$. In the time domain, such an ideal reconstruction filter is given by the sinc-function $\operatorname{sinc}(\frac{\omega_s}{2}t)$ defined as

$$\operatorname{sinc}(t) = \begin{cases} \frac{\sin t}{t} & t \neq 0 \\ 1 & t = 0 \end{cases} \tag{1.8}$$

Hence the reconstructed signal $\hat{x}(t)$ is

$$\hat{x}(t) = \sin c\left(\frac{\omega_s}{2}\right) * \sum_{n=-\infty}^{\infty} x[n] \cdot \delta(t - nT_s) = \sum_{n=-\infty}^{\infty} x[n] \cdot \sin c\left(\frac{\omega_s}{2}(t - nT_s)\right) \quad (1.9)$$

where $*$ denotes convolution. Ideal reconstruction $\hat{x}(t) = x(t)$ from the sampled signal values is theoretically possible, if $x(t)$ is suitably bandlimited and the Nyquist criterion (Equation [1.6]) is satisfied. In practice, some inaccuracies have to be permitted, as neither the anti-alias nor the reconstruction filter is ideal but must be approximated by a realizable system, which can introduce amplitude and phase distortions and only provide a finite stopband attenuation.

Quantization

After a sample-and-hold device has created discrete-time (usually voltage) signal values, $x[n]$, these values need to be rounded such that representation in a given number format is possible. DSP number formats can be fixed-point or floating-point. We consider here only fixed-point systems, where N bits are dedicated to represent a signal value within a manifold of 2^N possible levels. To represent both positive and negative values, 2's complement notation is employed. This allows the following range of integer values to be represented by an N bit number:

Word Length, N	Number Range
8 bits	$-128 \dots 127$
16 bits	$-32768 \dots 32767$
24 bits	$-8388608 \dots 8388607$
N bits	$-2^{N-1} \dots 2^{N-1} - 1$

The rounding operation ("quantization") to one of the 2^N discrete signal values is performed by a quantizer characteristic, as shown in Figure 1.6. Although quantization itself is a nonlinear operation, the quantizer characteristic as, e.g., given in Figure 1.6, a so-called mid-tread quantizer with coded zero-level,[2] at least approximates a linear function.

Two sources of error arise. First, the rounding operation of the quantizer from $x[n]$ to a quantized signal, $x_q[n]$, at its output causes a quantization error $e[n]$. As is obvious from Figure 1.6, the maximum modulus of $e[n]$ is half of the quantization step size, $q/2$. Therefore, the quantization error can be controlled by the word length N, whereby the exact relation between N and the power of the resulting quantization error will be established in the next section, "Signal-to-Noise Ratio." Second, if the amplitude of the input signal, $x[n]$, exceeds the range $[V_{min}, V_{max}]$ of the quantizer, the signal is clipped and harmonic distortion is introduced into $x_q[n]$. To avoid clipping,

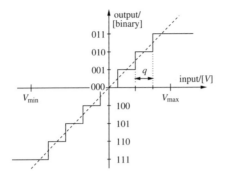

FIGURE 1.6
Quantizer characteristic with N = 3 bits.

the analog input signal requires preprocessing by an amplifier, as given in the block diagram of Figure 1.1, to limit the amplitude of $x(t)$. Also, this amplifier is to ensure that small signal amplitudes are avoided which would only excite the quantizer characterized in Figure 1.6 around the origin and hence waste resolution.

The problem of resolving small signal amplitudes while permitting larger amplitudes without clipping has led to nonlinear quantizer characteristics. These characteristics offer a smaller step size q around the origin and larger step sizes toward larger signal amplitudes. Although the optimum quantizer characteristic in this respect is determined by the probability density function (i.e., the histogram) of $x[n]$, usually standardized nonlinearities are employed. These encompass A-law, a European standard, and μ-law, which is used in the United States and Japan.[2]

Signal-to-Noise Ratio

To assess the effect that noise has on a signal, a measure known popularly as signal-to-noise ratio (SNR) is applied. This measure is based on an additive noise model, as shown in Figure 1.7, where the quantized signal $x_q[n]$ is a superposition of the unquantized, undistorted signal $x[n]$ and the additive quantization error $e[n]$. The ratio between the signal powers of $x[n]$ and $e[n]$ defines the SNR. To capture the wide range of potential SNR values and to consider the logarithmic perception of loudness in humans, SNR is generally given in a logarithmic scale, in decibels (dB),

$$\text{SNR}_{\text{dB}} = 10 \cdot \log_{10} \frac{\sigma_x^2}{\sigma_e^2} \qquad (1.10)$$

where σ_x^2 and σ_e^2 are the powers of $x[n]$ and $e[n]$, respectively. Specifically for the assessment of quantization noise, SNR is often labeled as the signal-to-quantization-noise ratio (SQNR).

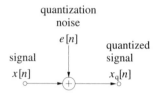

FIGURE 1.7
Additive noise model of quantizer.

For a linear quantizer characteristic as in Figure 1.6, the SNR can be quantified if the word length N is known. Let us assume a sinusoidal signal $x[n]$ with maximum amplitude V_{max} and therefore power $\sigma_x^2 = V_{max}/2$. The quantization noise is assumed to be uniformly distributed within the interval $[-q/2; q/2]$, and the resulting noise power is $\sigma_e^2 = q^2/12$, as will be derived below in "Expectation, Mean, and Variance." Hence, we can evaluate Equation (1.10) as

$$\text{SNR}_{dB} = 10 \cdot \log_{10} \frac{V_{max}/2}{q^2/12} = 10 \cdot \log_{10} \frac{6V_{max}}{[(2V_{max})/2^N]^2} = 10 \cdot \log_{10} 2^{2N} \cdot \frac{3}{2}$$

$$= N \cdot 20 \log_{10} 2 + 10 \log_{10} \frac{3}{2} = N \cdot 6.02 + 1.76$$

(1.11)

Thus, for example, a word length of $N = 16$ bits results in an SNR of 98 dB. This SNR value represents the power ratio between an input signal of maximum amplitude and the quantization noise. However, even for smaller input signal amplitudes, the noise power remains $q^2/12$ and therefore forms a "noise floor" that can only be altered by modifying the word length N. Therefore, Equation (1.11) also states the dynamic range of the digital system, i.e., the range between the minimally and maximally resolvable input signal amplitude.

Similar to the effect of the quantization noise on the signal, most other noise sources in the noise chain of Figures 1.1 and 1.2 can be described by the additive noise model of Figure 1.7. Subsequently, SNR measures can be used to characterize the severeness of the inflicted distortion.

Stochastic Signals and Their Characteristics

The description of a time domain waveform by other means such as its Fourier transform or by parameters such as the signal power or the maximum frequency was found to be useful in "Analog/Digital Interfacing and Noise Chain." Such parameters are well defined for deterministic signals for which an exact analytical formulation of the waveform is generally possible, in particular for periodic or quasiperiodic signals, which exhibit repetitiveness or repetitiveness over at least short time intervals, respectively. Signals, however, whose continued behavior is not known, are termed random or stochastic. Various quantities can be used to characterize such stochastic signals; the most important ones will be briefly reviewed in the following sections.

Probability Density Function

A very fundamental characteristic of a random signal is its probability density function (PDF), i.e., the distribution of signal values. For an intuitive understanding of the PDF, consider the random signal $x[n]$ in Figure 1.8. In a first step, a histogram is recorded by counting how often certain intervals of signal amplitudes are hit by the signal $x[n]$. By taking a large enough sample set, making the bins of the histogram infinitesimally small, and normalizing the area under the histogram curve to one, finally the PDF in Figure 1.8(bottom right) emerges.

The shape of the PDF bears important characteristics; for example, the signal for which the PDF is shown in Figure 1.8 has a normal or Gaussian distribution. A particular property of this distribution is that the PDF remains Gaussian even if the signal is subjected to a linear filtering operation. For

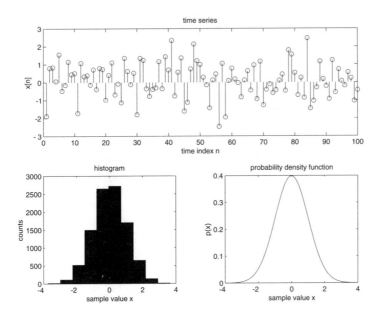

FIGURE 1.8
Description of a time series $x[n]$ (top) by a histogram (bottom left) and its probability density function (PDF, bottom right).

FIGURE 1.9
Uniform PDF.

this reason, and because many real-world noise signals are at least approximately Gaussian, the Gaussian PDF

$$p(x) = \frac{1}{\sqrt{2\pi}\sigma_x} e^{-\frac{(x-\mu_x)^2}{2\sigma_x^2}} \tag{1.12}$$

is particularly important for modeling signals. The PDF gives access to other important signal parameters, such as the variance or signal power σ_x^2 or the mean value μ_x, which will be derived in the following section, "Expectation, Mean, and Variance." Besides the Gaussian PDF, many other relevant distributions exist,[3] e.g., the uniform PDF as given in Figure 1.9, where all signal amplitudes are equally likely to occur.

Expectation, Mean, and Variance

Although the stochastic signal itself is random, some of the underlying parameters actually are deterministic, such as the mean μ_x and variance σ_x^2 in Equation (1.12) or the PDF itself. To evaluate the PDF in the example of Figure 1.8, a large amount of data had to be considered. In general, to determine or at least faithfully estimate the underlying parameters of a stochastic signal, intuitively some form of averaging over either many realizations of the signal or a large number of signal values is required to bypass randomness. For analysis purposes, this is performed by the *expectation operator* $\mathcal{E}\{\cdot\}$ by averaging over an ensemble of values. Here an ensemble represents an ideally infinite amount of realizations of random signals, which are parallel in time and are all drawn from the same statistics, i.e., have the same underlying PDF, mean, variance, and so forth.[1,4]

This expectation operator can be used to calculate the pth-order moment for a random variable x as $\mathcal{E}\{x^p\} = \int_{-\infty}^{\infty} x^p \cdot p(x)dx$. Most important are first- and second-order moments,

$$\mu_x = \mathcal{E}\{x\} \int_{-\infty}^{\infty} x \cdot p(x)dx \tag{1.13}$$

$$\sigma_x^2 = \mathcal{E}\{(x-\mu_x)^2\} = \int_{-\infty}^{\infty} (x-\mu_x)^2 \cdot p(x)dx \tag{1.14}$$

which yield the mean μ_x and variance σ_x^2, whereby a small modification was applied for the variance in Equation (1.14) to compensate for the presence of a mean μ_x.

Example: To calculate the quantization noise power in the section "Signal-to-Noise Ratio," we note that the quantization error cannot exceed half of the quantizer's step size $q/2$ and $e[n]$ is therefore uniformly distributed on the interval $[-q/2; q/2]$ as shown in Figure 1.9. As the PDF is symmetric about $e = 0$, the mean value according to Equation (1.13) is zero, and we can calculate the quantization noise power σ_e^2, using Equation (1.14),

$$P_{noise} = \sigma_e^2 = \int_{-\infty}^{\infty} e^2 p(e)\, de = \int_{-q/2}^{q/2} e^2 \frac{1}{q}\, de = \left[\frac{1}{3q} e^3 \right]_{e=-q/2}^{e=q/2} = \frac{q^2}{12} \qquad (1.15)$$

which is the result stated earlier in the section "Signal-to-Noise Ratio."

The above definitions were based on the expectation operator and ensemble averages. If the random signal $x[n]$ is *ergodic*, stochastic quantities such as μ_x and σ_x^2 can be estimated over a time interval I rather than an ensemble,

$$\hat{\mu}_x = \frac{1}{I} \sum_{n=0}^{I-1} x[n] \qquad (1.16)$$

$$\hat{\sigma}_x^2 = \frac{1}{I} \sum_{n=0}^{I-1} (x[n] - \hat{\mu}_x)^2 \qquad (1.17)$$

where ^ indicates estimates of the true quantities. The property encompassed by the term *ergodicity* implies that ensemble averages over several trials with a random signal can be replaced by time averages. As an example, let us consider the mean number of eyes when throwing a die. Ideally, the mean is calculated over an ensemble of a large number of different dice that have been thrown. If ergodicity holds, then the same result for the mean can be obtained from throwing a single die 1000 times and averaging the results. An obvious case where ergodicity is not given is if the random signal $x[n]$ is nonstationary, i.e., its mean μ_x and variance σ_x^2 vary over time.

Correlation and Power Spectral Density

The similarity between two stochastic variables can be expressed by means of their *correlation*. Specifically, a measure of similarity between two random signals $x[n]$ and $y[n]$ as a function of the relative shift, or lag, κ between the two sequences is defined by the cross-correlation

$$r_{xy}[\kappa] = \mathcal{E}\{x[n+\kappa] \cdot y^*[n]\} \tag{1.18}$$

where the superscript * denotes complex conjugate. Large values of $r_{xy}[\kappa]$ indicate strong similarity between $y[n]$ and a version of $x[n]$ shifted by κ samples, while a small value means that the similarity is weak, with $r_{xy}[\kappa]$ = 0 in the extreme case indicating no correlation between the two (shifted) signals. Also note from Equation (1.18) that

$$r_{xy}[\kappa] = r_{yx}^*[-\kappa] \tag{1.19}$$

Example: The cross-correlation function of two independent (e.g., produced by two independent sources) Gaussian random signals, $x[n]$ and $y[n]$, is

$$\mathcal{E}\{x[n+\kappa] \cdot y[n]\} = \mathcal{E}\{x[n+\kappa]\} \cdot \mathcal{E}\{y[n]\} = \mu_x \cdot \mu_y \tag{1.20}$$

and therefore the cross-correlation is zero if at least one of the two signals has zero mean.

The autocorrelation sequence can be employed to test the self-similarity of a signal and to evaluate how predictable successive signal samples are. It can be derived from Equation (1.18) for $y[n] = x[n]$, yielding

$$r_{xx}[\kappa] = \mathcal{E}\{x[n+\kappa] \cdot x^*[n]\} \tag{1.21}$$

Obvious properties of this autocorrelation sequence are its symmetry with respect to lag zero and complex conjugation, $r_{xx}[\kappa] = r_{xx}^*[-\kappa]$, and its maximum value for zero lag ($\kappa = 0$), as the signal $x[n]$ is perfectly self-similar to itself when no shifts are applied.

The Fourier transform $X(e^{j\Omega})$ of a signal $x[n]$,

$$X(e^{j\Omega}) = \sum_{n=-\infty}^{\infty} x[n] \cdot e^{-j\Omega n} \tag{1.22}$$

and its inverse transform

$$x[n] = \frac{1}{2\pi} \int_{-\pi}^{\pi} X(e^{j\Omega}) \cdot e^{j\Omega n} \tag{1.23}$$

only exist for signal $x[n]$ of finite energy, which generally is not the case for random signals. Therefore, amendments have to be made to define the spectrum of a random signal. This is performed by power spectral densities, which are based on the above correlation sequences of the signal rather than the signal itself. The power spectral density (PSD) of $x[n]$ is computed by taking the Fourier transform of its autocorrelation function,

$$P_{xx}(\Omega) = \sum_{\kappa=-\infty}^{\infty} r_{xx}[\kappa] \cdot e^{-j\Omega\kappa} \qquad (1.24)$$

which is also known as the Wiener-Khintchine transform.[3] Although not as widely used as the PSD of Equation (1.24), a cross power spectral density (cross-PSD) between two random signals $x[n]$ and $y[n]$ can be defined as $P_{xy}(e^{j\Omega}) \circ\!\!-\!\!\circ r_{xy}[\tau]$ based on the cross-correlation function of Equation (1.18). Due to Equation (1.19), note that $P_{xy}(e^{-j\Omega}) = P_{yx}(e^{j\Omega}) \circ\!\!-\!\!\circ r_{yx}[\tau]$. Although the definitions of PSD and cross-PSD are based on correlation sequences, they can be evaluated practically using averaging procedures on Fourier-transformed segments of the random signal $x[n]$.[3]

Example: For a completely random signal $x[n]$, shifted versions bear no correlation with the unshifted signal; hence, $r_{xx}[n] = \sigma_x^2[n]$. Therefore, solving Equation (1.24) gives

$$P_{xx}(\Omega) = \sum_{n=-\infty}^{\infty} \sigma_x^2 \delta[n] \cdot e^{-j\Omega n} = \sigma_x^2 \cdot e^{-j\Omega} \qquad (1.25)$$

as the PSD of $x[n]$. The magnitude of the PSD,

$$\left| P_{xx}(\Omega) \right| = \sigma_x^2 \qquad (1.26)$$

is now constant. In analogy to the visible color spectrum, the fact that all frequencies are equally present in the PSD has led to the term *white noise* when describing an uncorrelated random signal with the above autocorrelation.

Example: The filter in Figure 1.10 with impulse response $h[n]$ is excited by Gaussian white noise with autocorrelation $r_{xx}[n] = \sigma_x^2 \cdot \delta[n]$. We are interested in finding the PSD of the filter output $y[n]$. The autocorrelation of $y[n]$, $r_{yy}[n]$ is given by Equation (1.21), and the relation between $y[n]$ and $x[n]$ is defined by the convolution, $y[n] = h[n] * x[n] = \sum_{j=-\infty}^{\infty} h[j] \cdot x[n-j]$. Inserting the convolution into Equation (1.21) yields

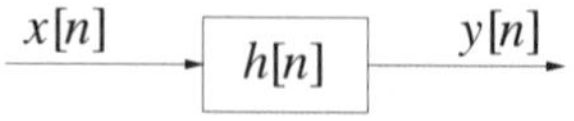

FIGURE 1.10
Random signal $y[n]$ at the output of a filter impulse response $h[n]$ excited by a Gaussian white signal $x[n]$.

$$r_{yy}[\kappa] = \mathcal{E}\{y[n+\kappa] \cdot y^*[n]\} = \mathcal{E}\left\{ \sum_{j=-\infty}^{\infty} h[j] \cdot x[n+\kappa-j] \sum_{i=-\infty}^{\infty} h^*[i] \cdot x^*[n-i] \right\}$$

$$= \sum_{j=-\infty}^{\infty} h[j] \sum_{i=-\infty}^{\infty} h^*[i] \cdot \mathcal{E}\{x[n+\kappa-j] \cdot x^*[n-i]\}$$

$$= \sum_{j=-\infty}^{\infty} h[j] \sum_{i=-\infty}^{\infty} h^*[i] \cdot \sigma_x^2 \cdot \delta[i+\kappa-j]$$

$$= \sum_{i=-\infty}^{\infty} h[j] \cdot h^*[j-\kappa] = h[\kappa] * h^*[-\kappa]$$

(1.27)

When Fourier transformed, the convolution (Equation [1.27]) corresponds to a multiplication of $H(e^{j\Omega})$ ●─○ $h[n]$ with $H^*(e^{j\Omega})$ ●─○ $h^*[-n]$, therefore

$$P_{yy}(\Omega) = H(e^{j\Omega}) \cdot H^*(e^{j\Omega}) = \left| H(e^{j\Omega}) \right|^2$$

(1.28)

represents the magnitude of the PSD. Hence, the initially white input signal $x[n]$ is "colored" by the filter $h[n]$, and according to Equation (1.28), the resulting PSD at the filter output is given by the squared magnitude of the filter.

As a generalization from the above example, the PSDs of a filter input and output are related by

$$P_{yy}(\Omega) = \left| H(e^{j\Omega}) \right|^2 \cdot P_{xx}(\Omega)$$

(1.29)

Similarly, the cross-PSDs $P_{xy}(e^{j\Omega})$ and $P_{yx}(e^{j\Omega})$ can be derived if the two random signals are the input and output of a system $h[n]$. Although not as often used as the PSDs, the cross-PSD derivations

$$P_{xy}(e^{j\Omega}) = H^*(e^{j\Omega}) \cdot P_{xx}(e^{j\Omega})$$

(1.30)

$$P_{yx}(e^{j\Omega}) = H(e^{j\Omega}) \cdot P_{xx}(e^{j\Omega})$$

(1.31)

will be useful when defining optimum filters in the section "Optimal or Wiener Filtering."

Correlation functions and spectral densities are therefore not only important characteristics of random signals themselves but also play an eminent role when the interaction with a linear time-invariant system is to be described. This will be further elaborated in the following section, which focuses mainly on filtering by linear time-varying systems, their characterization, and design.

Digital Filtering

Linear filtering of digital signals is an essential technique to either enhance signal components of interest or to attenuate noise components within the signal. The filter operation, i.e., the relation between the filter input and output, is in the time domain represented by a difference equation implementating a discrete convolution. This mathematical description can be brought into other forms and representations, which permit simple realization or analysis of the filter system.

Difference Equation and z-Domain Representation

The input-output behavior of a digital filter is characterized by a difference equation

$$y[n] = \sum_{v=0}^{L-1} b[v] \cdot x[n-v] - \sum_{v=1}^{J-1} a[v] \cdot y[n-v] \tag{1.32}$$

where the parameters $a[n]$ and $b[n]$ are the filter coefficients and $y[n]$ is the filter output in response to an excitation $x[n]$ applied to the filter input. The difference equation consists of a weighted average over the current input sample $x[n]$ and $L - 1$ past input values, whereby $L - 1$ is called the order of the feedforward section. Additionally, a $J - 1$ order section performs a feedback from $J - 1$ past filter output samples.

The filter coefficients in Equation (1.32) uniquely characterize the behavior of the filter. The difference equation (1.32) can be implemented as a signal flow graph, whereby a manifold of different implementations is possible. Figure 1.11 shows one such flow graph representation, whereby the feedforward and feedback part are separately realized ("direct form I"[1,4]). Such flow graph representations generally include basic operations such as adders, multipliers, and delay elements, whereby the latter delay their input by one sampling period and can be stacked to delay-lines — similar to a

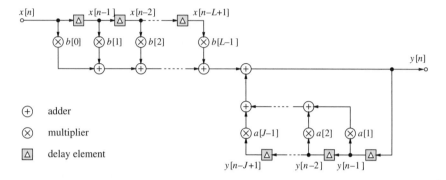

FIGURE 1.11
Flow graph of a digital filter in direct form I.

shift register — as shown in Figure 1.11 in both the feedforward and feedback part. From the signal flow graph, it is straightforward to implement a digital filter as a circuit.

An important tool to characterize both the digital filter itself, as defined through its difference equation, as well as the relation between the input and output signals to the filter is the z-transform. This transformation can be interpreted as a generalization of the Fourier transform for discrete signals given in Equation (1.7). However, unlike the Fourier transform, the z-transform is not only applicable to steady-state analysis but also can be as well employed to capture transient behavior. For the z-transform, the forward transform for a pair $h[n]$ ○——● $H(z)$ is given by

$$H(z) = \sum_{n=-\infty}^{\infty} h[n] \cdot z^{-n} \tag{1.33}$$

The inverse transform is not straightforward, and it generally is, if required, evaluated through tables for basic signal and system components into which the z-transform representation has to be decomposed.

As the transform (Equation [1.33]) is linear, applying it to Equation (1.32) results in

$$Y(z) = \sum_{n=0}^{L-1} b[n] \cdot z^{-n} \cdot X(z) + \sum_{n=1}^{J-1} a[n] \cdot z^{-n} \cdot Y(z) \tag{1.34}$$

Note that unit delays are replaced by z^{-1}. The transfer function of the digital filter in Figure 1.11 is the ratio between the output $Y(z)$ ●——○ $y[n]$ and input $X(z)$ ●——○ $x[n]$, i.e., describes the input-output behavior:

$$H(z) = \frac{Y(z)}{X(z)} = \frac{\sum_{n=0}^{L-1} b[n] \cdot z^{-n}}{1 - \sum_{n=0}^{J-1} a[n] \cdot z^{-n}} \tag{1.35}$$

$$= \frac{b[0] + b[1]z^{-1} + b[2]z^{-2} + \cdots + b[L-1]z^{-L+1}}{1 - a[1]z^{-1} - a[2]z^{-2} - \cdots - a[J-1]z^{-J+1}} = \frac{B(z)}{A(z)} \tag{1.36}$$

Similar to the Fourier transform, the z-transform has turned the (potentially awkward) convolution defined by the difference equation (1.32) in the time domain into a simple product of polynomials, $Y(z) = H(z) \cdot X(z)$, in the z-domain.

Important characteristics of a digital filter $H(z)$ can be investigated from factorizations in the z-domain. The fractional polynomial (Equation [1.36]) can be brought into the root representation

$$H(z) = b[0] \frac{(1 - \beta_0 z^{-1}) \cdot (1 - \beta_1 z^{-1}) \cdots (1 - \beta_{L-2} z^{-1})}{(1 - \alpha_0 z^{-1}) \cdot (1 - \alpha_1 z^{-1}) \cdots (1 - \alpha_{J-3} z^{-1})} \tag{1.37}$$

where α_n and β_n are the roots of the polynomials $A(z)$ and $B(z)$ in Equation (1.36), respectively. For $z = \alpha_n$ or $z = \beta_n$ singularities occur in $H(z)$, whereby the first case causes $Y(z) = 0$ and the latter case drives the output to infinity. Therefore, the parameters α_n and β_n are called the zeros and poles of the system $H(z)$. As the system exhibits poles, or in general feedback, we are particularly interested in the stability of $H(z)$, i.e., whether the filter output remains bounded for a bounded input signal. This stability can be easily observed from the *impulse response* of the system, i.e., the series $h[n]$ from Equation (1.33). The impulse response can be determined by observing the output $y[n]$ after applying an impulse $x[n] = \delta[n]$ at the filter input. For a stable system, the impulse response will converge, whereas for an unstable system it will diverge. The rare case of an oscillating, but neither decaying nor diverging system, is termed marginally stable. To address stability, we determine the roots of Equation (1.37) and define first-order systems $B_i(z)$ and $A_i(z)$,

$$H(z) = b[0] \underbrace{(1 - \beta_0 z^{-1})}_{B_0(z)} \cdots \underbrace{(1 - \beta_{L-2} z^{-1})}_{B_{L-2}(z)} \cdot \underbrace{\frac{1}{(1 - \alpha_0 z^{-1})}}_{A_0(z)} \cdots \underbrace{\frac{1}{(1 - \alpha_{J-3} z^{-1})}}_{A_{J-3}(z)} \tag{1.38}$$

Obviously, the system $H(z)$ can be brought into a cascade of subsystems $B_i(z)$ and $A_i(z)$, as shown in Figure 1.12. The overall system $H(z)$ will be stable if

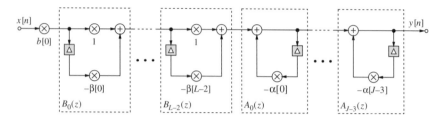

FIGURE 1.12
Factorization of the digital filter in Figure 1.11 into first-order subsystems.

and only if all subsystems are stable. The subsystems $B_i(z)$ have no feedback and are therefore so-called finite impulse response (FIR) filters, whose impulse response is identical to the filter coefficients, $h[n] = b[n]$ for $J = 1$. As the impulse response will be zero after the first L coefficients, an FIR filter is stable by default.

The impulse response of one of the feedback sections $A_i(z)$ in Figure 1.12 is given by

$$A_i(z) = 1 + \alpha_i z^{-1} + \alpha_i^2 z^{-2} + \alpha_i^3 z^{-3} + \cdots = \sum_{n=0}^{\infty} \alpha_i^n z^{-n} \qquad (1.39)$$

This impulse response converges for $|\alpha_i| < 1$. Hence, for convergence, all poles α_i must have a modulus smaller than one, i.e., when displayed in a pole-zero plot, the poles must lie within the unit circle to guarantee stability.

Another important representation for a digital filter is its frequency response, which can be useful to assess the system's behavior in the steady-state case. The frequency response $H(e^{j\Omega})$ can be calculated by evaluating $H(z) \circ\!\!\!-\!\!\!\bullet\, h[n]$ for $z = e^{j\Omega}$, i.e., on the unit circle. Alternatively, the Fourier pair relation $H(e^{j\Omega}) \bullet\!\!\!-\!\!\!\circ\, h[n]$ can be invoked by simply calculating the Fourier transform of $h[n]$ according to Equation (1.22). For display purposes, the frequency response is usually separated into

$$H(e^{j\Omega}) = \left| H(e^{j\Omega}) \right| \cdot e^{j\Phi(\Omega)} \qquad (1.40)$$

where $\left| H(e^{j\Omega}) \right|$ is the magnitude response and $\Phi(\Omega)$ the phase response. Some examples of such descriptions in the frequency domain will be given below.

Filter Design

Noise distortion of a signal of interest may be restricted to a certain frequency band, such as a jammer signal or mains hum as sources of interference. Therefore, the design of a filter with a specified frequency response $H(e^{j\Omega})$ to suppress these noise components is desirable. Four different main types

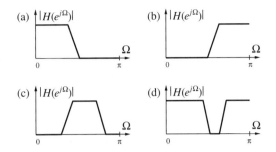

FIGURE 1.13
Four different basic filter types as defined by their magnitude response: (a) low-pass, (b) high-pass, (c) bandpass, and (d) bandstop filter.

of frequency response can be distinguished: low-pass, high-pass, bandpass, and bandstop filter, as shown in Figure 1.13. A bandstop filter with a notch in its frequency response at 50 Hz can, for example, be used to attenuate mains interference on a signal. The main characteristics of a filter are — as an example shown for an LPF in Figure 1.14 — the passband width, transition band width, the stopband edge, stopband attenuation, and passband ripple. The design of a high-quality filter that possesses a small passband ripple, a narrow transition band width, and a high stopband attenuation will generally result in a large number of coefficients, which in turn may make it difficult to implement the filter in real time.

A large number of different filter design algorithms are embedded in almost any signal processing software package. While therefore the insight into the implementation of any such algorithm is secondary, it is important to be aware of the design criteria in Figure 1.14 and the above trade-off between filter quality and filter complexity. In this respect, the differences in the design of FIR and infinite input response (IIR) filters are interesting and will be commented on briefly.

The design of FIR filters is in many cases based on the specification of a desired magnitude response, as given in Figure 1.14. After performing an

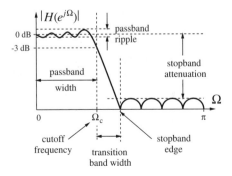

FIGURE 1.14
Criteria/quality measures in defining the magnitude response of a digital filter.

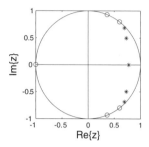

FIGURE 1.15
Pole-zero plot for the elliptic IIR $H(z)$ further characterized in Figure 1.16.

inverse Fourier transform, the time-domain response needs to be multiplied by a tapered window.[5] Sometimes several iterations between the time- and frequency-domain are performed to optimize the filter coefficients. IIR filter designs are based on well-known algorithms for analog filters, such as Butterworth or Chebychev filter designs.[5-7] The results of these analog filter designs are transferred into the digital domain by means of a so-called bilinear transform.

Some of the general advantages and disadvantages of FIR and IIR filter design are evident from the following example. At a given sampling rate of f_s = 48 kHz, the digital filter should have a cutoff frequency of f_c = 6 kHz (3 dB attenuation point), a transition band width of 1.2 kHz, and a stopband attenuation of 45 dB. In normalized angular frequency values, this yields Ω_c = $2\pi f_c/f_s$ = $2\pi \cdot$ 6 kHz/48 kHz = 0.25π and a transition band width of $\Delta\Omega$ = $2\pi \cdot$ 1.2 kHz/48 kHz = 0.05π. For the FIR filter, a Parks-McClellan design[5] is selected that fulfills the desired specification with an 83rd-order filter, i.e., L = 84 and J = 1. An IIR elliptic filter design only requires $L = J = 6$ for a comparable quality. The pole-zero plot of the elliptic IIR filter is given in Figure 1.15 with circles (o) denoting zeroes and asterisks (∗) denoting pole positions in the complex z-plane.

The characteristics for the FIR and IIR designs are compared in Figure 1.16. Although the IIR system has considerably fewer coefficients and therefore requires only $K + J - 1$ = 11 multiply-accumulate (MAC) operations per sampling period to calculate one new output sample, the FIR filter is with 84 MACs almost eight times more complex. Due to the feedback, an IIR filter can model a longer impulse response with fewer coefficients compared to the FIR filter, as evident from Figure 1.16(a) and (b). However, despite the generally much lower number of coefficients, the IIR filter parameters usually exhibit a considerable dynamic range, particularly for higher filter orders. This can result in design deterioration up to instability when transferring the designed filter coefficients to a fixed point representation with inherent round-off errors.

The magnitude responses, $|H(e^{j\Omega})|$ — fulfilling the initial specifications — are shown in Figures 1.16(c) and (d). Also displayed in Figures 1.16(e) and (f) are the phase responses $\Phi(\Omega)$ of both designs whereby notably the FIR

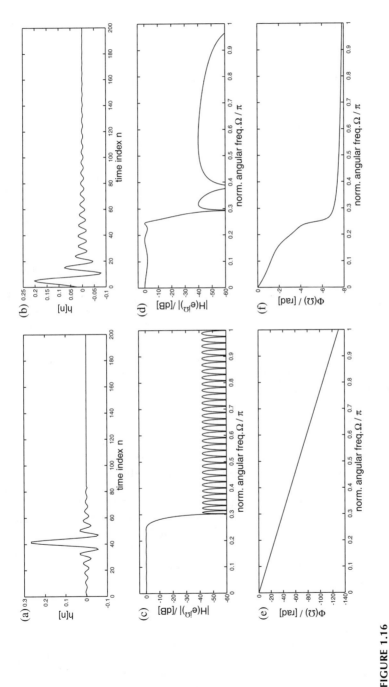

FIGURE 1.16

Filter characteristics (impulse, magnitude, and phase responses from top to bottom) for an FIR filter design (left: Parks-McClellan) and an IIR filter design (right: elliptic filter).

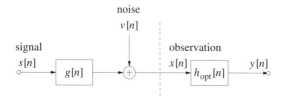

FIGURE 1.17

Reconstruction of a signal $s[n]$ distorted by a linear system $g[n]$ and corrupted by additive noise $v[n]$ by means of an optimal (Wiener) filter $h_{opt}[n]$.

has a linear phase response. This linearity is guaranteed by a symmetric impulse response, which can be embedded as a design option into the filter design algorithm. The advantage of a linear phase system is that the group delay $\gamma(\Omega) = -d\Phi(\Omega)/d\Omega$ — specifying how long different frequency components in the input signal require to propagate to the filter output — is constant and no dispersion of a signal of interest is incurred. In contrast, an IIR filter will never have a linear phase response and, therefore, may not be applicable if signals sensitive to dispersion are processed.

Optimal or Wiener Filtering

If noise corrupting a signal is restricted to a given frequency band, the previously discussed filter designs can be applied to reduce the effect of noise on the signal of interest. These designs are based on the specification of a desired filter characteristic in the frequency domain. Alternatively, the so-called "optimal" or Wiener filter can be derived if the statistics of the signal of interest and the noise are known.

The problem addressed by Wiener filtering is highlighted in Figure 1.17. There, a random input signal $s[n]$ is filtered by a system $g[n]$ and corrupted by additive noise $v[n]$, whereby $v[n]$ and $s[n]$ are uncorrelated. Based on the observable signal $x[n]$, the filter $h_{opt}[n]$ is to be found such that its output $y[n]$ is a close as possible estimate of $s[n]$ in the mean square error sense, i.e., such that the power of the error signal $e[n] = s[n] - y[n]$ is minimized. Although the exact time series is unknown, only the correlation functions, and therefore the PSDs, of the signals in Figure 1.17 are available.

The solution of the above Wiener problem is a standard exercise described in many textbooks (see, e.g., References 1, 3, and 8). Here, we only refer to the general solution based on the PSD of the observed signal $x[n]$, $P_{xx}(e^{j\Omega})$, and the cross-PSD between the signal of interest $s[n]$ and $x[n]$, $P_{sx}(e^{j\Omega})$,

$$H_{opt}(e^{j\Omega}) = \frac{P_{sx}(e^{j\Omega})}{P_{xx}(e^{j\Omega})} \tag{1.41}$$

To derive the filter coefficients $h_{opt}[n] \circ\!\!-\!\!\bullet H_{opt}(e^{j\Omega})$ in the time domain, an inverse Fourier transform has to be applied to Equation (1.41). Of the potentially noncausal result, only the causal part is realizable and is retained as the solution.

If the system $g[n] \circ\!\!-\!\!\bullet G(e^{j\Omega})$ filtering the signal of interest is known, the solution of the optimal Wiener filter in Equation (1.41) can be modified and expressed solely in terms of the PSDs of the signal of interest and the noise, $P_{ss}(e^{j\Omega})$ and $P_{vv}(e^{j\Omega})$, respectively:

$$H_{opt}(e^{j\Omega}) = \frac{P_{ss}(e^{j\Omega}) \cdot G^*(e^{j\Omega})}{P_{ss}(e^{j\Omega}) \cdot \left|G(e^{j\Omega})\right|^2 + P_{vv}(e^{j\Omega})} \tag{1.42}$$

An interesting case arises if the signal $s[n]$ is not linearly distorted but only corrupted by noise $v[n]$, i.e., $g[n] = \delta[n]$. As hence $G(e^{j\Omega}) = 1$, Equation (1.42) simplifies to

$$H_{opt}(e^{j\Omega}) = \frac{P_{ss}(e^{j\Omega})}{P_{ss}(e^{j\Omega}) + P_{vv}(e^{j\Omega})} \tag{1.43}$$

Intuitively, it can be seen that for frequencies where the noise PSD, $P_{vv}(e^{j\Omega})$, takes on large values, the Wiener filter will attain a small gain and therefore attenuate the noise. As in the previous case, the realizability of $h_{opt}[n] \circ\!\!-\!\!\bullet$ $H_{opt}(e^{j\Omega})$ is limited to the causal part of the solutions Equations (1.42) and (1.43). Note, however, that particularly with the simplification $g[n] = \delta[n]$, Equation (1.43) addresses a very basic noise suppression problem, and for the previous filter design example in the previous section, "Filter Design," to suppress mains interference the Wiener filter would similarly provide a bandstop filter at 50 Hz.

Discrete Signal Transforms

Transforms such as the Fourier and z-transforms have so far been encountered for the purpose of signal or system analysis. In this section, we will consider discrete transforms, such as, e.g., the discrete Fourier transform (DFT), which is discussed in the next section, "Discrete Fourier Transform." Such transforms fulfill two purposes. First, they can be used as an analysis tool, such as the DFT approximating the Fourier transform. We will characterize this ability and shortcomings in "Spectral Analysis with the DFT." Second, discrete transforms, among which the DFT is but one, can be utilized to parameterize signals, which bears importance for many practical applications such as coding and compression of speech signals. The section "Other Discrete Transforms" will provide a brief overview of a number of transforms. The parameterization property of such discrete transforms can be further exploited for noise reduction, as discussed in the section "Noise Reduction Based on Signal Transforms."

Discrete Fourier Transform

Fourier techniques analyze a signal or system with respect to sinusoids. The rationale behind this is that many processes produce by default a sinusoidal behavior, such as, e.g., rotating machinery, and that sinusoids are eigenfunctions of linear time-invariant (LTI) systems. The latter property is most important, and means that an LTI system with sinusoidal input will in the steady state produce a sinusoidal output. This does not hold for any other function, as, e.g., a square wave input will generally not result in a square wave output. The only change for a fixed sinusoid is with respect to the amplitude and phase of the sinusoid, which can therefore be used to uniquely describe an LTI system's behavior at that specific frequency.

Fourier analysis therefore tries to find the sinusoidal content in a signal or system response, which is performed by fitting sinusoids of different frequencies to the signal. For a fixed frequency, amplitude and phase of the sinusoid are varied until a "best fit" — mathematically in the least square error sense — is achieved. The resulting parameter pair of adjusted amplitude and phase is the Fourier coefficient at the specified frequency. Mathematically, this best fit is performed by the scalar product in Equation (1.22), which on its left-hand side yields the Fourier coefficient at frequency Ω.

To obtain a practically realizable discrete Fourier transform, we have to (1) limit the time index, n, to a causal finite interval $0 \leq n < N$ and (2) evaluate only discrete frequency points ("bins") $\Omega = 0, \Omega_0, 2\Omega_0, 3\Omega_0$, etc., yielding

$$X(e^{j\Omega})\Big|_{\Omega=\Omega_0 k} = \sum_{n=0}^{N-1} x[n]e^{-j\Omega_0 kn} \tag{1.44}$$

as the transform. As a standard, in the DFT N such equispaced frequency bins are evaluated, such that $\Omega_0 = 2\pi/N$. An example of this is given in Figure 1.18. Considering the sampling period T_s, the bin separation in terms of absolute frequency is

$$f_0 = \frac{1}{NT_s} \tag{1.45}$$

which determines the *frequency resolution* of the DFT. Therefore, the higher N, the higher is the frequency resolution of the DFT.

If we write Equation (1.44) in terms of the index k into the frequency bins,

$$X[k] = \sum_{n=0}^{N-1} x[n]e^{-j\Omega_0 kn} \tag{1.46}$$

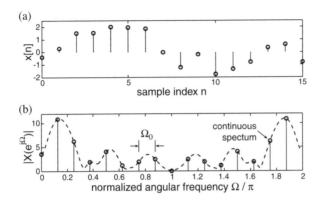

FIGURE 1.18
Example of transforming a sequence $x[n]$ using a DFT with $N = 16$ points: (a) time-domain waveform $x[n]$; (b) magnitude of DFT with underlying spectrum $X(e^{j\Omega})$ (dashed).

a matrix representation $\underline{X} = T \cdot \underline{x}$ of an N-point DFT is obtained, where $\underline{x}$ and $\underline{X}$ are vectors holding the time-domain samples and the Fourier coefficients, respectively:

$$
\underbrace{\begin{bmatrix} X[0] \\ X[1] \\ X[2] \\ \vdots \\ X[N-1] \end{bmatrix}}_{\underline{X}} = \underbrace{\begin{bmatrix} 1 & 1 & \cdots & 1 \\ 1 & e^{-j\Omega_0} & & e^{-j\Omega_0(N-1)} \\ 1 & e^{-j\Omega_0 2} & \ddots & e^{-j\Omega_0 2(N-1)} \\ \vdots & \vdots & & \vdots \\ 1 & e^{-j\Omega_0(N-1)} & \cdots & e^{-j\Omega_0(N-1)(N-1)} \end{bmatrix}}_{T} \cdot \underbrace{\begin{bmatrix} x[0] \\ x[1] \\ x[2] \\ \vdots \\ x[N-1] \end{bmatrix}}_{\underline{x}} \qquad (1.47)
$$

The matrix T is known as the DFT matrix. Due to its structure, the DFT matrix is symmetric, $T = T^T$, where $(\cdot)^T$ indicates the transpose operator.[9-11] The DFT matrix is not only invertible, but takes the very simple form $T^{-1} = \frac{1}{N} T^H$, such that the operation

$$
\underline{x} = \frac{1}{N} \cdot T^H \cdot \underline{X} \qquad (1.48)
$$

defines the inverse DFT (IDFT). The operator $(\cdot)^H$ denotes Hermitian transpose, performing a transposition and complex conjugation of its argument.

It is clear from Equations (1.47) and (1.48) that a DFT or IDFT operation requires N^2 multiplications and additions (multiply-accumulates, MACs). In general, if N is not prime, this complexity can be reduced by exploiting common arithmetic steps in the transform calculation. These redundancies can be removed by following the so-called butterfly operations of an FFT, an example of which is given in Figure 1.19. A number of implementations exist whereby here a so-called decimation-in-time FFT with permutated time

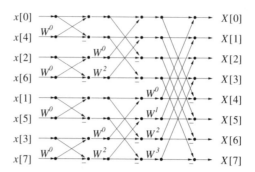

FIGURE 1.19
Flow graph of an eight-point fast Fourier transform calculation with decimation-in-time (with a factor $W = e^{j/N}$).

domain samples is shown.[7] If N is a power of two, the complexity of an FFT is given by $N \log_2 N$ MACs. Particularly for large N, this significantly reduces the transform complexity.

Spectral Analysis with the DFT

The classic use of the DFT is that of a numerical tool to calculate the Fourier transform of some time series $x[n]$. This has an important relation to noise reduction, if, for example, a harmonic signal is buried in noise and frequency domain information is required in order to design a digital filter for retrieving this signal or for determining the PSDs in case of the Wiener filter in the section "Optimal or Wiener Filtering." It is, however, important to understand the limitations of the DFT for this analysis, which will be briefly outlined in this section.

As noted in Equation (1.7), the spectrum of a discrete signal, $x[n]$, results in the Fourier domain in a periodicity of $2\pi/T_s$, where T_s is the sampling period, i.e., the distance between adjacent discrete sample values. Due to the duality of the Fourier transform, this holds vice versa:[1,4] a discrete spectrum corresponds to a periodic time domain, which is a well-known fact exploited in the Fourier series. As in Equation (1.46) both domains are discrete, periodicity is enforced in either domain. For the time domain, this results in an enforced periodicity of N, where N is the number of samples considered in a DFT, also known as the window length. Thus, the DFT "periodizes" the data, which is likely to create aberrations.

Example: Consider spectral analysis of a sampled sinusoid with a fundamental period of $N_0 = 16$ samples, as shown in Figure 1.20(a). If we apply a DFT with $N = 2\,N_0 = 32$ points, we obtain the Fourier domain in Figure 1.20(b) with a single nonzero coefficient, as may be expected for the sinusoidal signal under analysis. In Figure 1.20(d), an $N = 27$ point DFT is applied

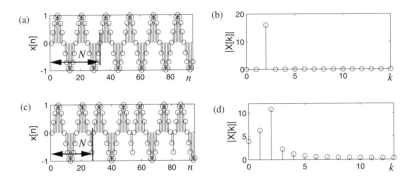

FIGURE 1.20
(a) Sinusoidal signal and (b) its magnitude response via $N = 32$ point DFT; with $N = 27$, (c) shows the periodized data with discontinuities and hence (d) spectral leakage in the Fourier domain.

to the data, and the previous single peak has now "leaked" into neighboring frequency bins. This can be explained by the time series shown in Figure 1.20(c), which the DFT inherently creates by periodizing a data segment of $N = 27$ samples. The discontinuities arising at the repetitions of this fundamental period N are responsible for introducing spurious frequencies into the spectrum that have not been present in the original data in Figure 1.20(a).

The effect noticed in the previous example is known as spectral leakage, and it arises whenever $N \cdot T_s$, with N the number of points in the DFT, is not an integer multiple of the fundamental period present in the signal to be analyzed. The resulting discontinuities at the margins of the window repetitions, as seen in Figure 1.20(c), can however be alleviated by applying a tapered window to the N-sample data segment prior to executing the DFT. A simple window based on a raised cosine, $w[n] = 1 - \cos 2\pi(n/N)$, fulfilling this purpose is shown in Figure 1.21.

A number of popular windows exist to control spectral leakage, such as Hamming, Hann, Blackman-Harris, or Bartlett,[5,7] which are part of most signal processing toolboxes and come with a specific side effect. This effect generally consists of reducing spectral leakage at the cost of a loss in spectral resolution, i.e., the main lobe of a sinusoid in the Fourier domain is widened. An example is given in Figure 1.22, where in (a) a sinusoidal data segment is subjected to a DFT with a noninteger multiple N of the sinusoid's fundamental period. The result in the Fourier domain is given in (b). The graphs in (c) and (d) are with a Hamming window applied, and clearly show the reduction in spectral leakage, but also the widened main lobe at the frequency of the sinusoid.

The widening of the main lobe can be reasoned as follows. In the frequency domain, the multiplication of the data segment with a window corresponds to a convolution between the true Fourier spectrum with the Fourier transform of the window function. As window functions have a low-pass characteristic, the true spectrum is blurred, as outlined in Figure 1.23.

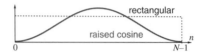

FIGURE 1.21
Tapered raised cosine window to de-emphasize the margins of the data segment obtained by rectangular windowing in order to avoid discontinuities in the periodization and therefore spectral leakage.

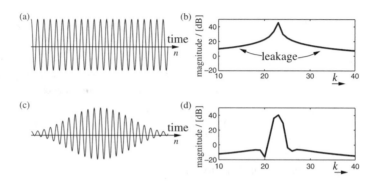

FIGURE 1.22
(a) Sinusoidal signal with a rectangular window and (b) the magnitude of its DFT coefficients; (c) sinusoidal signal with a Hamming window and (d) the magnitude of its DFT coefficients.

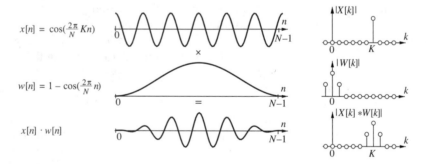

FIGURE 1.23
The multiplication of a signal to be analyzed, $x[n]$, and a window function, $w[n]$, in the time domain corresponds to a convolution of the two Fourier transforms in the frequency domain, resulting in a blurred spectrum with a widened main lobe.

Other Discrete Transforms

Besides the application to frequency domain analysis, the DFT is often used as a parameterizing transform. While, for example, a time series requires the storage of all N samples, the DFT coefficient domain may be sparse and only a small amount of non-zero coefficients may be necessary to represent the contained information. For this operation, the application of a transform matrix to a data vector can be interpreted as a change of coordinate system.

The time domain representation of a time series, $x[n]$, can be separated into the basis, formed by shifted Dirac impulses $\delta[n]$,

$$\delta[n] = \begin{cases} 0 & \text{for} \quad n \neq 0 \\ 1 & \text{for} \quad n = 0 \end{cases} \tag{1.49}$$

which are weighted by the coordinates, i.e., the sample values corresponding to these specific time shifts:

coordinates

$$x[n] = x_0 \delta[n] + x_1 \delta[n-1] + \ldots + x_{N-1} \delta[n-N+1] \tag{1.50}$$

basis/coordinate system

Note that the basis functions are orthogonal, i.e., $\sum_{n=0}^{N-1}\delta[n]\delta[n-k]=0$ for $\kappa \neq 0$. Of course, the signal can be represented differently, using, e.g., a Fourier expansion whereby a number of N orthogonal complex harmonics form the basis and the DFT coefficients represent the corresponding coordinate values:

other coordinates

$$x[n] = X_0 1 + X_1 e^{j2\pi n/N} + \ldots + X_{N-1} e^{j2\pi(N-1)n/N} \tag{1.51}$$

other basis/coordinate system

Both expansions in Equations (1.50) and (1.51) have an orthogonal basis, and their coordinates can be organized in vectors $\underline{x}$ and $\underline{X}$ and related by Equation (1.47). Therefore, Equations (1.47) and (1.48) essentially perform rotations with respect to a new coordinate system.[8,10]

The motivation of subjecting data samples to a discrete transform as in Equation (1.47) lies in the potential sparseness of the resulting new coordinates. This is, e.g., exploited in coding and compression, where the amount of data can be reduced by extracting only non-zero or sufficiently large values from a sparse representation. Besides the DFT as a potential transform, many other possibilities exist. Orthogonality of a matrix T indicates that only a rotation (no deformation) of the coordinate system is performed, which is easily reversible by applying $T^{-1} = T^H$. Candidates for this include discrete sine transforms (DST), discrete cosine transforms (DCT), discrete wavelet transforms (DWT), and many others.[12,13] Any such transform can be viewed analogously to the step from Equations (1.50) to (1.51) as an expansion of the time-domain waveform with respect to a different basis, and can be brought into a matrix notation similar to Equation (1.47). Each of the above transforms rotates a time-domain waveform into a basis with distinct properties, and thus may be more or less suitable for a specific application with

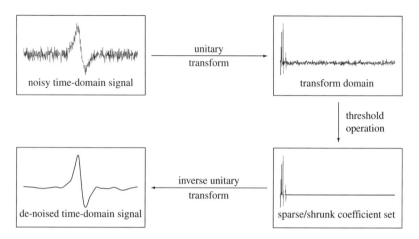

FIGURE 1.24
De-noising of a one-dimensional time series: in the transform domain, additive noise remains smeared, whereas the signal of interest is parameterized by only few coefficients. By thresholding and inverse transformation, noise reduction can be achieved.

respect to the achievable parameterization and hence sparseness of the data in the transform domain.

Noise Reduction Based on Signal Transforms

The parameterization property of discrete transforms highlighted in the previous section can be exploited for noise reduction. If a signal is corrupted by random noise in the time domain, the idea is to find a suitable discrete transform that parameterizes the signal component of interest. If the noise is white, the application of a transform does not change the whiteness, and the noise will be spread in the transform domain while coefficients corresponding to well-parameterized signal components stick out. The reconstruction of the original time series is performed only with those coefficients that are representing signal components, thus reducing the noise compared to the original time series. This procedure is known as "de-noising"[14] and its steps are demonstrated in Figure 1.24.

De-noising relies on two important steps. First, a suitable transform has to be found. Here, usually a DWT is employed, which offers a large variety of potential bases over which an optimization can be performed. Second, the decision as to which coefficients are to be retained in the transform domain is not trivial. A variety of approaches exist to determine suitable thresholds, whereby either hard thresholding, i.e., zeroing small coefficients, or soft thresholding, i.e., shrinking coefficients with an appropriate smooth nonlinear function, can be applied.

The requirements for de-noising are not as strict but are similar to Wiener filtering because a certain amount of information about the signal of interest and the corrupting noise has to be known in order to adjust the noise reduction mechanism appropriately.

Adaptive Digital Filtering

More powerful noise reduction mechanisms than Wiener filtering or de-noising are at hand if a reference probe of the corrupting noise is available. These techniques are based on adaptive digital filtering. Different from the fixed filter design in the section "Filter Design," here the filter coefficients are tunable, are adjusted in dependency of the environment that the filter is operated in, and can therefore track any potential changes in this environment.

The first section, "Structure and Architectures," presents some applications and architectures of adaptive filters, and thereafter defines an optimum filter in the mean squared error sense, given by the Wiener-Hopf solution in the section "Mean Square Error Optimization." An adaptive filter presents an iterative solution toward this optimum, and is used because either the direct optimization may be numerically costly or even unstable, or because the underlying environment has time-varying parameters, which need to be tracked by the adaptive system. As an example of the wide range of available adaptive algorithms, gradient descend methods, including the popular least mean squares (LMS) adaptive filter, are reviewed in the section "Gradient Techniques." Some exemplary characteristics regarding the convergence of the LMS filter are addressed in "LMS Convergence Characteristics," which are similarly found in other adaptive digital algorithms for which "Other Adaptive Filter Algorithms" provides an overview.

Structure and Architectures

A generic adaptive filter is shown in Figure 1.25. The aim of the filter is to produce an output signal $y[n]$ from an input $x[n]$, such that when subtracted

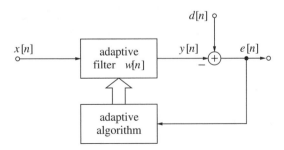

FIGURE 1.25
Generic adaptive filter.

from a desired signal $d[n]$, the error signal $e[n]$ is minimized in a suitable sense. To achieve this, the parameters of the filter are adjusted by an adaptive algorithm. In most cases, this algorithm monitors the error signal $e[n]$. This feedback intuitively works such that if the filter is adjusted well, the error signal will be small and only slight changes need to be applied to the parameters inside the filter. In contrast, if the filter is misadjusted, the error signal $e[n]$ will be large and the algorithm will subsequently apply larger changes to the filter coefficients. Exact definitions of the error criterion to be minimized and suitable adaptation algorithms will be discussed in the sections "Mean Square Error Optimization," "Gradient Techniques," and "LMS Convergence Characteristics."

The generic adaptive filter of Figure 1.25 can be applied in a number of different architectures.[15] These are shown in Figure 1.26. The functionality of these architectures, provided that the adaptive filter finds the optimum solution, is:

- **System identification:** As given in Figure 1.26(a), the adaptive filter is placed parallel to an unknown system with impulse response $c[n]$, and both systems are excited by the same input $x[n]$. If the adaptive filter $w[n]$ converges, ideally the error signal $e[n]$ should go to zero. Therefore, the input-output behavior of $c[n]$ and $w[n]$ is identical. If the excitation $x[n]$ is broadband and $c[n]$ is an LTI system, then the adapted $w[n]$ represents a model of the unknown system $c[n]$.

- **Inverse system identification:** The unknown system $c[n]$ and the adaptive filter $w[n]$ are placed in series according to Figure 1.26(b). If the error signal $e[n]$ is minimized, the convolution $c[n] * w[n]$ will approximate a Dirac impulse with a delay, depending on the delay Δ that is placed in the path of the desired signal $d[n]$. Thus, using a broadband input signal, with $w[n]$ the inverse of the unknown system, $c[n]$ will be obtained.

- **Prediction:** The signal $d[n]$ is applied delayed to the adaptive filter $w[n]$, as seen in Figure 1.26(c). From these past signal values in $x[n]$, the current sample value $d[n] = x[n + \Delta]$ has to be predicted by the adaptive filter. If the error $e[n]$ is minimized, the filter will replicate the predictable signal components at its output, $y[n]$, whereas $e[n]$ will only retain the random, uncorrelated part of the signal $d[n]$.

- **Noise cancellation:** In Figure 1.26(d) the desired signal is formed by a signal of interest, $v[n]$, corrupted by noise $\tilde{x}[n]$. A reference signal $x[n]$ of the noise — e.g., picked up from the noise source — is appropriately modified by the adaptive filter to match $\tilde{x}[n]$ once filtered. Therefore, after adaptation, the error signal $e[n]$ will ideally only contain the signal of interest $v[n]$.

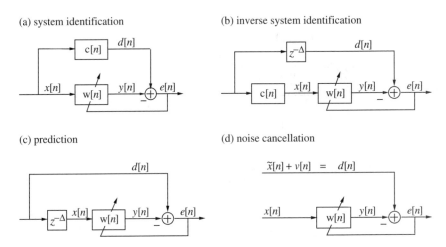

FIGURE 1.26
Adaptive filter architectures.

The noise cancellation architecture can, for example, be employed for voice communication in a helicopter cockpit. As the pilot's microphone picks up both speech and engine noise, a separate reference can be taken from the engine. Identifying the reference with $x[n]$ in Figure 1.26(d) and the microphone signal with $d[n]$, providing adaptation is good, the error signal $e[n]$ will be the pilot's voice undistorted by noise, which can be transmitted.

Although it is mostly the last of the above architectures, noise cancellation, that is immediately connected to noise reduction, other structures also find use in this respect. Examples include the removal of sinusoidal interference, e.g., through mains hum, from a speech signal. There, a prediction architecture can be employed that removes the predictable mains components, while the speech remains unaffected if the delay Δ in Figure 1.26(c) is select to be long enough to decorrelate any short-time periodicities within the speech signal.

Mean Square Error Optimization

Several minimization criteria can be employed for determining the optimum parameter setting for an adaptive filter. The most commonly used are the mean square error (MSE) criterion, i.e., the variance of the error signal $e[n]$, and the least squares criterion, i.e., the sum of squared error samples over all times n. Depending on the chosen criterion, different algorithms arise.[11] Here we concentrate on the MSE minimization.

To formulate the MSE, consider an adaptive FIR filter, as in Figure 1.27. By adopting vector notation for both the samples of the input signal in the filter's tap delay line and the coefficients at time instance n,

$$\boldsymbol{x}_n = [x[n]\, x[n-1] \cdots x[n-L+1]]^{\mathrm{T}} \qquad (1.52)$$

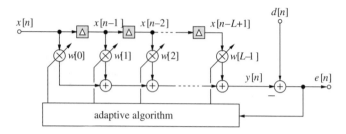

FIGURE 1.27
Adaptive FIR filter with coefficients adjusted by an adaptive algorithm.

$$w = [w_0 \ w_1 \ \cdots \ w_{L-1}]^H \qquad (1.53)$$

the filter output can be denoted by $y[n] = w^H \cdot x_n$. This can be used to formulate the mean squared error (MSE) ξ_{MSE} as

$$\xi_{MSE} = \mathcal{E}\{e[n]e^*[n]\} = \mathcal{E}\{(d_n - w^H x_n)(d_n^* - x_n^H w)\} \qquad (1.54)$$

$$= \mathcal{E}\{d_n d_n^*\} - \mathcal{E}\{w^H x_n d_n^*\} - \mathcal{E}\{d_n w_n^H w\} + \mathcal{E}\{w^H x_n x_n^H w\}$$

$$= \sigma_{dd} - w^H \mathcal{E}\{x_n d_n^*\} - w^T \mathcal{E}\{d_n x_n^*\} + w^H \mathcal{E}\{x_n x_n^H\} w \qquad (1.55)$$

$$= \sigma_{dd} - w^H p - w^T p^* + w^H R w$$

where substitutions with the cross-correlation vector p and the autocorrelation matrix (covariance matrix for zero-mean processes) R have taken place. The cross-correlation vector p is defined by

$$p = \mathcal{E}\{x_n d_n^*\} = [\mathcal{E}\{x_n d_n^*\}, \ \mathcal{E}\{x_{n-1} d_n^*\}, \dots \mathcal{E}\{x_{n-L+1} d_n^*\}]^T \qquad (1.56)$$

$$= [r_{xd}[0], r_{xd}[-1], \dots r_{xd}[-L+1]]^T \qquad (1.57)$$

where $r_{xd}[\tau]$ is the cross-correlation function between $x[n]$ and $d[n]$ according to Equation (1.18). Both $x[n]$ and $d[n]$ are assumed to be wide-sense stationary, i.e., their mean and variance as defined in Equations (1.13) and (1.14) are constant while higher-order moments may be time-varying. For the further analysis, we also make the "classic" assumption of statistical independence between w and x_n.[15] The entries of the $L \times L$ autocorrelation matrix, R

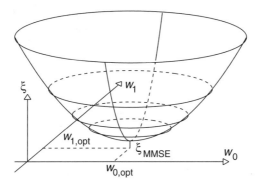

FIGURE 1.28
Mean squared error (MSE) cost function ξ for the case of a weight vector w with two coefficients.

$$
R = \mathcal{E}\{x_n x_n^H\} = \mathcal{E}\left\{
\begin{bmatrix}
x_n x_n^* & x_n x_{n-1}^* & \cdots & x_n x_{n-L+1}^* \\
x_{n-1} x_n^* & x_{n-1} x_{n-1}^* & \cdots & x_{n-1} x_{n-L+1}^* \\
\vdots & \vdots & \ddots & \vdots \\
x_{n-L+1} x_n^* & x_{n-L+1} x_{n-1}^* & \cdots & x_{n-L+1} x_{n-L+1}^*
\end{bmatrix}
\right\}
$$

$$
=
\begin{bmatrix}
r_{xx}[0] & r_{xx}[-1] & \cdots & r_{xx}[-L+1] \\
r_{xx}^*[-1] & r_{xx}[0] & \cdots & r_{xx}[-L+2] \\
\vdots & \vdots & \ddots & \vdots \\
r_{xx}^*[-L+1] & r_{xx}^*[-L+2] & \cdots & r_{xx}[0]
\end{bmatrix}
\tag{1.58}
$$

are samples of the autocorrelation function $r_{xx}[\tau]$ according to Equation (1.21). R is Töplitz, which implies that it possesses a band structure with identical elements on all diagonals, and Hermitian, i.e., $R^H = R$. By sole virtue of these structural properties,[8,10,11] R is positive semidefinite and has real-valued eigenvalues.

The cost function ξ_{MSE} in Equation (1.55) is apparently quadratic in the filter coefficients, and due to the positive semidefiniteness of R, Equation (1.55) has a minimum, which is unique for a positive definite (full-rank) autocorrelation matrix R. The cost function therefore forms an upright hyperparabola over the L-dimensional hyperplane defining all possible coefficient sets w_n. In Figure 1.28 $\xi_{MSE}(w)$ is visualized for the case $N = 2$ in dependency of the two filter coefficients.

Due to the quadratic nature of the MSE cost function, its minimum can be evaluated by differentiating ξ_{MSE} with respect to the coefficient vector and setting the derivative to zero. If vector and matrix calculus is applied (see, e.g., Reference 8), this derivative is given by

$$
\frac{\partial \xi_{MSE}}{\partial w^*} = -p + Rw
\tag{1.59}
$$

Therefore, if the autocorrelation matrix R is invertible, the optimum filter coefficients are given by

$$w_{opt} = R^{-1}p \qquad (1.60)$$

which is well known as the Wiener-Hopf solution.[11,15]

Although Equation (1.60) could be readily applied to update the filter coefficients $w[n]$, three pitfalls arise. First, R might not be invertible, or might be ill-conditioned, such that the inverse either does not exist or, more likely, is highly noisy and therefore does not present a numerically stable solution. Second, the estimation of the correlation quantities for R and p requires considerable effort. Finally, the matrix inversion is of the order $O(L^3)$ MACs, and therefore is computationally very complex and potentially unsuitable for real-time applications. Therefore, the next section reviews iterative techniques that are numerically more efficient and robust in computing w_{opt}.

Gradient Techniques

The quadratic form of the cost function ξ_{MSE} derived in the previous section allows us to search for the minimum using iterative methods. This can be performed by stepping from an initial starting point in the direction of the negative gradient of the cost function, which will eventually lead to the global minimum. Mathematically, this can be phrased as

$$w[n+1] = w[n] - \mu\nabla\xi_{MSE}[n] \qquad (1.61)$$

where $w[n]$ marks the current weight vector at time n. From this current solution, a step is taken in the direction of the negative gradient $\nabla\xi[n]$ of the cost function to yield a new improved coefficient vector $w[n + 1]$. The notation $\nabla\xi_{MSE}[n]$ is to indicate that the gradient is applied to the MSE cost function yielded by the coefficient vector w_n at time, n. The parameter μ is referred to as the *step size*, loosely defining the length of a step that the algorithm takes in each iteration toward the optimum.

The explicit term for the gradient has been derived with Equation (1.59),

$$\nabla\xi_{MSE}[n] = \frac{\partial\xi_{MSE}}{\partial w_n^*} = -p + Rw_n \qquad (1.62)$$

and insertion into Equation (1.61) leads to the update equation known as the steepest descend algorithm.[11,15] In comparison to the Wiener-Hopf solution in Equation (1.60), inversion of the autocorrelation matrix is no longer required, but both the autocorrelation matrix R and the cross-correlation

TABLE 1.1

Equations for Filter Update by
LMS Adaptive Algorithm

LMS Algorithm

1	$y_n = w_n^H x_n$
2	$e_n = d_n - y_n$
3	$w_{n+1} = w_n + \mu x_n e_n^*$

vector p have to be reliably estimated. Furthermore, the multiplication with Rw_n creates a computational cost of order $O(L^2)$ MACs.

To lower the computational complexity and statistical record of the involved signals, in a simplification step the true gradient is replaced by an estimate, which leads to so-called stochastic gradient algorithms.[11,15] In the extreme case, the gradient estimate is based only on the current samples of $x[n]$ held in the tap delay line and $d[n]$,

$$\hat{p} = x_n d_n^* \tag{1.63}$$

$$\hat{R} = x_n x_n^H \tag{1.64}$$

which is equivalent to minimizing the instantaneous squared error $e_n e_n^*$, rather than the MSE $\varepsilon\{e_n e_n^*\}$. Inserting these estimates into Equation (1.62),

$$\hat{\nabla}\xi_n = -\hat{p} + \hat{R}w_n = -x_n(d_n^* - x_n^H w_n) = -x_n e_n^* \tag{1.65}$$

gives a gradient estimate, which together with Equation (1.61) forms the basis for the least mean squares (LMS) algorithm[11,15]

$$w_{n+1} = w_n + \mu x_n e_n^* \tag{1.66}$$

The complete LMS equations are listed in Table 1.1. These steps have to be performed once in every sampling period, leading to a moderate complexity of $2L$ MAC. This complexity of $O(L)$ is considerably lower than that required for the Wiener-Hopf equation and gradient descend algorithms.

LMS Convergence Characteristics

Representative of adaptive algorithms in general, this section addresses some convergence issues of the LMS algorithm. For an exact proof of convergence, the reader is referred to standard textbooks.[11,15] To prove that the LMS (or in fact any) algorithm converges to the Wiener-Hopf solution, two steps are required: (1) convergence in the mean to show that the LMS solution is bias-free, and (2) convergence in the mean square to prove consistency.

These proofs give some insight into the functioning of an adaptive algorithm, and specifically for the LMS algorithm yield a number of interesting analytical limits. One of these limits concerns the step size parameter μ. If μ is selected within the easily obtainable bounds

$$0 < \mu < \frac{2}{L\sigma_x^2} \tag{1.67}$$

where L is the number of adaptive filter coefficients and σ_x^2 the power of the input signal, the adaptive algorithm is guaranteed to be stable.[11,15]

The LMS algorithm exhibits an exponential convergence. As a measure for the convergence speed therefore a time constant T can be derived by fitting an exponential $e^{-n/T}$ to the MSE cost function evolving over time. This time constant can be approximated analytically as[11]

$$T_i \approx \frac{1}{\mu\lambda_i} \quad \forall |\mu\lambda_i| \ll 1 \tag{1.68}$$

with λ_i being the eigenvalues of $\mathbf{R}$. Although the validity of this approximation is based on restrictions on λ such as slow convergence,[11] it can be observed that the overall convergence is governed by (1) the step size and (2) the smallest eigenvalue of $\mathbf{R}$. In particular, the smaller the step size μ, the slower the convergence of the algorithm.

Analysis of LMS convergence in the mean square reveals that the final error variance will differ from the minimum MSE (MMSE) value by an excess MSE, $\xi_{EX} = \xi_{MSE}[n] - \xi_{MMSE}$ with $n \to \infty$, which can be derived as[11]

$$\xi_{EX} = \xi_{MMSE} \cdot \frac{a}{1-a}, \quad \text{with } a = \sum_{i=0}^{L-1} \frac{\mu\lambda_i}{2-\mu\lambda_i} \tag{1.69}$$

This means that a large μ will create a large excess MSE. The influence of μ is therefore such that a trade-off is created between the convergence speed (large for large μ) and the size of the final MSE, $\xi_{MSE}[n]$ for $n \to \infty$, which is kept small if a small parameter μ is selected.

Other Adaptive Filter Algorithms

Besides the LMS adaptive filter as introduced in the section "Gradient Techniques," a large variety of other adaptive filtering algorithms exists. Algorithmically, the cost function can be modified; minimizing the sum of squared errors leads to the class of recursive least squares (RLS) adaptive algorithms, which generally tend to have a higher complexity than the LMS but exhibit faster convergence.[11] Structurally, for some problems, multichannel adaptive filters are required, or IIR filters can be employed instead of the FIR system discussed here. Even nonlinear filters can be embedded into an adaptive system, such as Volterra filters or neural networks. Algorithmically, modifications to the algorithm are possible and may be motivated by the demand for more robust, fast converging, or computationally more efficient adaptive filters.

Conclusions

This chapter has highlighted some of the noise sources that can distort speech signals and reviewed fundamental DSP techniques that allow us to combat this distortion. To describe the noise, the additive noise model was adopted, which motivated the definition of the SNR as a measure for the severeness of the noise corruption in "Analog/Digital Interfacing and Noise Chain." "Stochastic Signals and Their Characterization" characterized the noise by means of stochastic signals and their parameters, such as variance or signal power, autocorrelation and cross-correlation sequences and power spectral densities. In order to control or mitigate the influence of noise, digital filters were reviewed in "Digital Filtering." Digital filters can be best employed if the corrupting noise occupies well-defined frequency bands, as the filter can be designed accordingly. The filter design can be performed explicitly by standard design software or by means of a Wiener filter if the signal and noise statistics are known. "Discrete Signal Transforms" addressed the discrete Fourier transform (DFT) for frequency domain analysis, or general transforms in order to parameterize a signal in the transform domain. The latter culminated in de-noising techniques by thresholding of noisy transform coefficients prior to an inverse transform. More powerful noise reduction techniques were introduced in "Adaptive Digital Filtering," which can be applied if, e.g., a reference probe of the corrupting noise is available.

The techniques for noise reduction addressed in this chapter should provide a strong motivation to study the finer details of DSP, for which several excellent books are recommended as a starting point.[1,4,13,15] It should transpire from the noise reduction examples provided in this chapter that the specific DSP technique chosen for noise reduction strongly depends on the circumstances of the application, such as the availability of a reference signal and/or additional information on the signal of interest and the noise. This book covers many specialized techniques for speech enhancement and noise reduction, which are customized by exploiting as much knowledge of the involved signals and systems as possible. However, despite the potentially high specialization of these techniques, the reader will find that many of the methods introduced in later chapters of this book, such as the ones dedicated to single-channel speech enhancement (Chapter 6) or acoustic echo cancellation (Chapter 8), make strong use of the basic concepts that have been outlined in this first chapter.

References

1. Girod, B., Rabenstein, R., and Stenger, A., *Signals and Systems,* John Wiley & Sons, Chichester, U.K., 2001.
2. Jayant, N. and Noll, P., *Digital Coding of Waveforms,* Prentice-Hall, Englewood Cliffs, NJ, 1984.
3. Papoulis, A., *Probability, Random Variables, and Stochastic Processes,* 3rd ed., McGraw-Hill, New York, 1991.
4. Oppenheim, A.V. and Schafer, R.W., *Discrete-Time Signal Processing,* Prentice-Hall, Englewood Cliffs, NJ, 1989.
5. Parks, T.W. and Burrus, C.S., *Digital Filter Design,* John Wiley & Sons, New York, 1987.
6. Papoulis, A., *Signal Analysis,* McGraw-Hill, New York, 1984.
7. Press, W.H. et al., *Numerical Recipes in C,* 2nd ed., Cambridge University Press, Cambridge, U.K., 1992.
8. Moon, T.K. and Stirling, W.C., *Mathematical Methods and Algorithms,* Prentice-Hall, Upper Saddle River, NJ, 1999.
9. Strang, G., *Linear Algebra and Its Applications,* 2nd ed., Academic Press, New York, 1980.
10. Golub, G.H. and Van Loan, C.F., *Matrix Computations,* 3rd ed., Johns Hopkins University Press, Baltimore, 1996.
11. Haykin, S., *Adaptive Filter Theory,* 3rd ed., Prentice-Hall, Englewood Cliffs, NJ, 1996.
12. Vaidyanathan, P.P., *Multirate Systems and Filter Banks,* Prentice-Hall, Englewood Cliffs, NJ, 1993.
13. Strang, G. and Nguyen, T., *Wavelets and Filter Banks,* Wellesley–Cambridge Press, Wellesley, MA, 1996.
14. Donoho, D.L., De-noising by soft-thresholding, *IEEE Transactions on Information Theory,* 41(3), 613–627, 1995.
15. Widrow, B. and Stearns, S.D., *Adaptive Signal Processing,* Prentice-Hall, Englewood Cliffs, NJ, 1985.

Section II:

System Aspects

2

Analog Techniques

Malcolm Hawksford

CONTENTS

Introduction

This chapter considers a number of performance-critical applications of audio signal processing to demonstrate how modern circuit techniques are used to complement the advances being made in digital audio systems. The performance of a digital audio system (e.g., DVD-audio,* includes in its specification 24-bit at 96-kHz sampling[1]) demands a high degree of accuracy within the analog circuitry to maximize transparency and to prevent compromise in terms of distortion, noise, and bandwidth. Although much signal processing can be performed in the digital domain, certain key processes ultimately require analog techniques. Prominent examples are the anti-aliasing and signal recovery filters (referred to collectively in this text as *gateway filters*) and the transresistance amplifier required to perform current-

---

* DVD-audio refers to the audio-specific format of digital versatile disk.

0-8493-0949-2/02/$0.00+$1.50
© 2002 by CRC Press LLC

to-voltage conversion in association with current output digital-to-analog converters (DAC).

Other analog processes to be reviewed include the important class of voltage-controlled amplifiers (VCAs) that exploits the logarithmic characteristics of the bipolar junction transistor (BJT) to perform analog multiplication. Two generic applications of the VCA are then described to demonstrate dynamic range control and complementary noise reduction. Specifically, as an example of a classic analog system Dolby* A-type noise reduction is outlined, where it is characterized both by its topology and semiconductor device profiles.

Analog Filter Requirements at the Analog-to-Digital Gateway

Gateway filters are fundamental to the successful operation of a digital audio system, where the performance of the analog circuitry should approach, if not exceed, the resolution of the digital channel. However, if the digital system operates at the compact disc (CD) Nyquist sampling** rate of 44.1 kHz, then the frequency response specification for each system filter is particularly demanding. Specifically, the anti-aliasing filter requires a flat and very low ripple amplitude response extending to about 20 kHz with a rapid attenuation region over the frequency band 20 to 22.05 kHz, a frequency space designated as the filter *transition band*. Ideally, the filter should also have a linear phase characteristic to eliminate group delay distortion from degrading the time domain response. Also, for signal recovery filtering immediately following the DAC, a similarly specified analog *recovery filter* is required to eliminate ultrasonic frequency components resulting from signal reflection about the sampling frequency and its harmonics.

It is well known that for an ideal filter with a brick wall frequency domain response, the time-domain impulse response $h(t)$ of the filter is given by Equation (2.1) as

$$h(t) = \frac{\sin(\pi f_x t)}{\pi f_x t} = \text{sinc}(\pi f_x t) \tag{2.1}$$

where f_x is the filter's cut-off frequency. In practice, this impulse response is impossible to match because it extends infinitely either side of its center. Consequently, to reproduce the precursive response correctly infinite delay is required before the main peak, implying that a practical filter must be constrained in its time-domain response that in turn limits the rate of fre-

---

* Dolby is a trade name for Dolby Laboratories, San Francisco, CA.
** Nyquist sampling theorem determines the minimum sampling rate for a given signal bandwidth.

quency domain attenuation. However, the precursive response cannot be chosen arbitrarily as the impulse response $h(t)$ should have even symmetry such that if the response center is located at $t = \tau$, then $h(t) = h(2\tau - t)$. This condition is necessary and sufficient for the phase response to be a linear function of frequency $\phi = -2\pi f \tau$ that implies a constant time delay τ with no group delay distortion. From this discussion it can be concluded that if the gateway filters are to exhibit zero group delay distortion, then their time-domain impulse responses should be even symmetric.

If the gateway filters use purely analog topologies, then even symmetry imposes a formidable challenge, because upon closer inspection a fundamental problem emerges. Analog filters use resistance, capacitance, and inductance together with active amplification at their core and may also include gyrators to synthesize inductors from resistor–capacitor networks. Following the principles of electromagnetic theory, both capacitors and inductors can be modeled as first-order differential equations relating current and voltage. This implies that each reactive element has a frequency-dependent relationship where the reactance is either proportional or inversely proportional to frequency with a corresponding phase of either $\pi/2$ or $-\pi/2$. Hence, the building blocks normally incorporated in analog filters are not well matched to synthesizing a specific time-domain response, especially where even symmetry is required. It follows that where a linear phase response is required, an analog circuit can only ever approximate the required response. The better this approximation, the more elaborate becomes the filter in terms of component count, including amplifier stages if an active synthesis is required. Also, the ultimate rate of attenuation is determined by the number of reactive elements (capacitors and inductors); thus, if the circuit has N reactive elements, then the ultimate attenuation rate cannot exceed $-6.01\ N$ dB/octave*. It is evident that as the complexity of the filter increases, problems of tuning and of component tolerances become more severe.

Hence, there must be compromise in designing a high-performance analog filter that is able to match the requirement of an anti-aliasing filter. For audio applications, circuit complexity is the enemy of transparent performance. Each active amplifier stage introduces noise and distortion, and each component inevitably has a tolerance value that introduces a random error into the design and makes repeatability and matching between multiple audio channels problematic. There is considerable literature on the synthesis of analog filters, but a specification that calls for a 2 kHz transition band with >100 dB attenuation and good phase linearity becomes a complicated structure that is difficult to tune and relatively expensive to implement.

Rather than pursue high complexity in the analog circuitry, it is prudent to consider alternative solutions, now widely practiced in audio systems whereby a minimalist solution to the analog filter problem can be obtained that can lead ultimately to a more accurate and transparent performance.

---

* A slope of –6.01-dB/octave refers to a filter where gain is inversely proportional to frequency.

The key to solving this problem is to distribute the filter structure between the analog and digital domains. Within an analog-to-digital converter (ADC) this requires the sampling rate to be increased significantly above the Nyquist rate where *oversampling ratios* of between 4 and 64 are not uncommon. Increasing the sampling rate obviously enables a much wider signal bandwidth without incurring aliasing distortion; however, more significantly, it enables a wider transition band that relaxes the analog filter design. A digital filter can then be used to band-limit the signal to the required bandwidth, for example, 20 kHz. Digital filters are perfectly matched to engineering a specific impulse response, because they use discrete delays rather than circuit elements that model a first-order time derivative. As such, the so-called *finite impulse response (FIR) filter* can be designed with even symmetric coefficients, thus achieving an exact linear phase response without incurring a significant cost penalty. Such filter responses are readily synthesized in the digital domain. Once the information bandwidth is reduced, then decimation can be used to discard samples and reduce the sampling rate, for example, to 44.1 kHz. In essence, this process allows the majority of filtering to be performed in the digital domain, whereas a mild second- or third-order filter is used in the analog domain. However, the overall filter should be seen as a cascade of both the digital and the analog filters. This strategy also implies that any imperfections in the analog filter, such as mild amplitude response errors together with mild phase distortion, can in part be corrected for within the digital domain. A similar technique can be used in DAC systems to reduce the complexity of the analog recovery filter. However, in this case, up-sampling is used to increase the sampling rate, which is a process that also requires a digital low-pass filter. Finally, signal recovery is completed using a low-order low-pass filter that is located after the DAC to attenuate the highest frequency components. It will be shown that the filter requirements for anti-aliasing and signal recovery are similar and that when either decimation or interpolation is combined with oversampling, similar advantages are achieved by employing part digital and part analog processing. In this way, the problem areas associated with high-order analog filters and circuit complexity are avoided.

Anti-Aliasing and Recovery Filters in Digital Audio Systems

Consider an ADC that is oversampled by a modest factor of four. The ADC is to be used with a high-resolution audio system with a nominal sampling rate of 96 kHz and a bit depth of 24 bit. Four-times oversampling implies that the converter is sampled at 384 kHz. To meet the digital audio specification in full, the audio band must extend almost to 48 kHz, although in practice a transition band of about 2 kHz should be provided, so the workable bandwidth is reduced to 46 kHz. It is within this narrow transition band that the digital filter must provide appropriate attenuation to prevent aliased signal components reflected about 96 kHz from entering the ultrasonic region of the audio band. However, because the ADC is sampled at 384 kHz, the audio signal prior to decimation (i.e., sampling rate conversion down to 96 kHz) can extend up to 336 kHz before reflected signal components fall below 48 kHz. Consequently, an extreme specification for the analog filter would require an attenuation of about 140 dB over a frequency band extending from 46 to 336 kHz, which is just under 3 octaves, and represents about −46 dB/octave slope, roughly equivalent to an eighth-order low-pass filter. If an analog filter were to be implemented with a cascade of second-order, low-pass sections, then four such sections would be required. However, this is an extreme case and neglects to take into account specific features of the sampled audio waveform and the interaction with the digital low-pass filter that will be used in the decimation process when the sampling rate is reduced to 96 kHz. Also, the nature of most audio signals is the low ultrasonic content above 40 to 50 kHz. Consequently, a third-order filter can be a fair balance between out-of-band attenuation and complexity, although it is expedient to introduce additional attenuation in the frequency band where the first sampling side bands appear, namely, 384 ± 48 kHz.

A circuit example is shown in Figure 2.1(a) together with its corresponding frequency response in Figure 2.1(b). This circuit incorporates simple but very high-performance discrete unity gain buffers. The buffers are derived from a two-stage complementary emitter follower and yield a voltage gain close to unity. Alternatively, operational amplifiers can be used, although the discrete circuit or its integrated equivalent can offer exemplary performance with low distortion, low noise, and wide bandwidth. The circuit consists of two cascaded filter sections where the first section includes transmission zeros (i.e., notch filters) located around 384 kHz to give additional attenuation in this critical region. The overall response was calculated by simulation and shows the notch filter implemented by the addition of two inductors that together with the capacitors form series resonant circuits. Low-valued inductors can readily be incorporated into a practical circuit, although screening is required to minimize the injection of interference that is often endemic in the hostile electrical environment of fast digital circuitry.

In high-performance analog systems, unlike purely digital processing, the problems associated with interference that is injected through either electromagnetic coupling or power rail and ground rail contamination can be severe. In practice, this requires special attention to the circuit layout and the use, for example, of copper-plated metal screens that together can help achieve the wide dynamic range demanded by the extreme resolution of a 24-bit system.

In a DAC application[2] there is a need for a similar filter characteristic and configuration, especially where up-sampling is used to enable the rapid attenuation requirements demanded of the recovery filter to be performed in the digital domain. However, a major difference is that this filter only has to suppress high-frequency spuriae associated with sampling. Even if the analog recovery filtering is not performed completely, there is not a problem

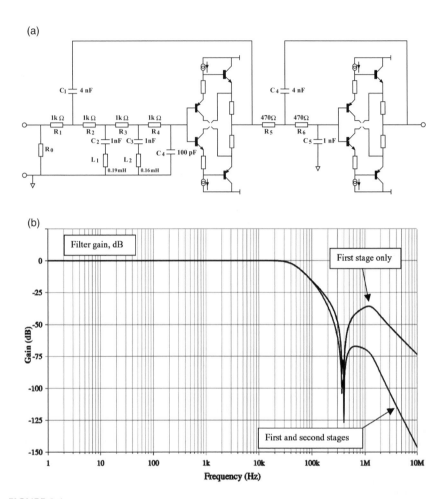

FIGURE 2.1
(a) Anti-aliasing filter for use with an ADC sampled at 384 kHz. (b) Magnitude frequency response of anti-aliasing filter.

of aliasing distortion. The only potential problem that may then occur is increased intermodulation distortion that arises in subsequent stages of analog processing, such as in a pre-amplifier or a power amplifier. Consequently, a simpler filter can be used where an example is the first stage of Figure 2.1(a) but where the notch filters (i.e., inductors) have been omitted. Inevitably there are limitations in the degree of high-frequency attenuation; however, this is a typical analog system compromise where a balance has to be achieved between out-of-band attenuation, circuit complexity, and cost. Note, however, that with proper digital filter design and the use of up-sampling, together with good linearity DACs, there should not be significant signal energy until the up-sampled, sampling frequency and its associated sidebands are encountered.[3]

Current-to-Voltage Conversion with Embedded Signal Recovery Filter

Providing the primary function of an analog system is met to an appropriate degree of accuracy, then additional and possibly unnecessary circuit complexity can often lead to deterioration in performance; usually in analog audio *less is more*. Hence, considering the analog circuitry associated with a DAC, it is expedient to identify techniques that reduce complexity yet retain a high level of performance. A typical multibit DAC is a current output device[3] and as such requires a *transresistance amplifier stage* with very low input impedance (e.g., < 1 Ω) to transform the output current to a voltage. Such a circuit is then cascaded typically with one or more filter stages, each using a buffer amplifier. In practice, the transresistance stage (or I-V stage) is a particularly critical part of a digital audio system. Not only does this amplifier have to process the audio components but it also has to respond linearly to the wide band signals produced by the DAC. At a sampling instant, when the DAC output current switches from one quantization level to another, the transresistance stage must respond momentarily extremely rapidly, a period where there is higher likelihood of nonlinear operation. Consequently, the amplifier must be designed to have both a rapid response and to generate low levels of distortion under broadband excitation. Although an operational amplifier is often employed and configured in the classic shunt–feedback amplifier arrangement (i.e., *virtual earth amplifier*) as shown in Figure 2.2(a) operational amplifiers can have relatively high open-loop nonlinearity, so they are not particularly suited to this function. However, Figure 2.2(b) illustrates an example of a current-feedback[4] configuration that uses discrete circuitry with input stage error correction[5] and overall feedback to enhance linearity. Also, to craft a more efficient topology while simultaneously improving performance, the recovery filter is embedded within the feedback structure, thus eliminating the need for additional cascaded stages.

The use of local, input stage error feedback[2,5] will virtually eliminate the nonlinear modulation of transistor base–emitter slope resistance. In Figure 2.2(b) the input stage consists of two matched complementary transistors, T_1 and T_2, together with a grounded base stage, T_3. Transistors T_1 and T_3 form a cascode* stage to steer the DAC output current, i, to the current mirror formed by T_4 and T_5 and the three equal-valued resistors, R_0, such that the collector currents of T_4 and T_5 each carry a mirror of the current, i. As a result, the changes in emitter currents within T_1 and T_2 are identical, where assuming parametric and thermal matching, then $V_{BE1} = -V_{BE2}$. Consequently, the emitter potential of T_1 remains theoretically zero even though the base–emitter voltages are nonlinear functions of signal current. This

---

* A cascode is a series connection of a common-emitter transistor stage and a grounded-base transistor stage.

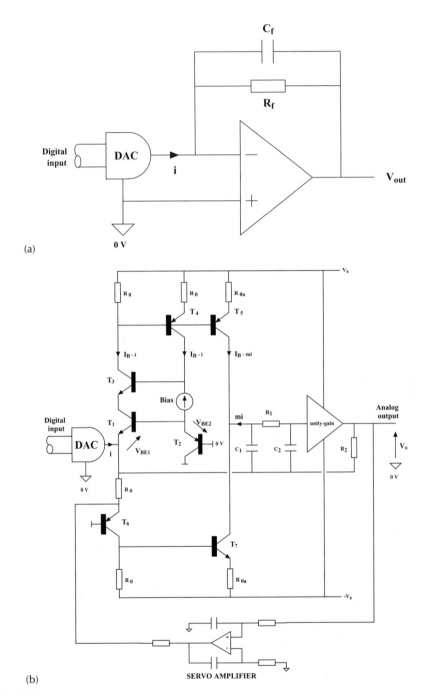

FIGURE 2.2
(a) Transresistance I-V stage using shunt feedback (*virtual earth*) amplifier. (b) Transresistance I-V stage with embedded low-pass filter and servo.

operation implies zero input impedance even under large signal conditions. A constant current generator formed by transistors T_6 and T_7 then sinks the collector bias current of T_5, thus completing the driving circuitry for the output stage.

A principal feature of this topology is the inclusion of a second-order, low-pass filter that is embedded within the output stage and is formed by the π-network R_1, C_1, and C_2, a unity-gain buffer and resistor, R_2, where the latter defines the low-frequency transresistance of the stage. However, the common signal line of this inner filter is not connected to ground but is returned to the input node where the whole structure constitutes a current-feedback path back to the emitter of T_1. At low frequency, overall feedback is derived from the output voltage via R_2 and includes the output buffer, whereas at higher frequency in the filter attenuation region, the feedback current is derived primarily from the collector current of T_6. However, under all signal conditions, the signal current produced at the collector of T_5 is fed back to the input, which helps both to improve overall linearity and to lower the input impedance.

In this respect, the high-frequency feedback path is similar to a simple dc-coupled feedback pair of transistors. Benefits derived from this configuration include reduced output impedance and enhanced linearity together with an embedded low-order reconstruction filter, where the filter behaves as an integral part of the feedback path while returning no current to ground and thus aiding ground-rail purity. Although a second-order low-pass filter is illustrated in Figure 2.2(b), higher-order filters can be accommodated without incurring a stability penalty. Also, the DAC signal current i is returned directly to the power supply and does not require transient currents to flow in the ground bus; this helps reduce circuit layout problems associated with electromagnetic compatibility. Assuming the current mirror formed by transistors T_4 and T_5 has unity current gain, then the closed-loop transimpedance $Z_{I/V}$ of the overall amplifier is given as

$$Z_{I/V} = \frac{-R_2}{1 + j0.5\omega R_2(C_1 + C_2) - 0.5\omega^2 R_1 C_1 R_2 C_2} \qquad (2.2)$$

Equation (2.2) describes the overall circuit transresistance and confirms that the transfer function is a second-order low-pass filter response.

Servo Amplifier to Establish Output dc Conditions

A technique that can be used to give precise control of the output dc conditions is to incorporate an analog servo amplifier in the feedback path of a negative feedback control loop. The transresistance amplifier shown in Figure 2.2(b) includes a noninverting servo amplifier. A servo amplifier is normally a single linear integrator (i.e., gain inversely proportional to frequency,

−90° phase shift) with a suitably large time constant so as only to influence the closed-loop gain at low frequency. At signal frequencies approaching dc, the servo amplifier should have an extremely high gain (typically 10^5 for an operational amplifier) that then forces the closed-loop gain of the transresistance amplifier to become virtually zero at dc. If the servo amplifier dc gain is assumed to be infinite, then the output voltage of the transresistance amplifier is controlled by negative feedback so as to have an average value (i.e., the quiescent dc value) that is equal to the dc input offset voltage of the servo amplifier. Normally with operational amplifiers such as a BI-FET,* the input bias current is negligible and the input offset voltage is typically under 1 mV. A servo loop can accurately maintain the average output voltage close to zero enabling dc coupling to be used at the output of the transresistance amplifier, where the closed-loop transfer function now includes a first-order high-pass filter response. The servo amplifier can achieve a low value of cut-off frequency without recourse to high-value capacitors, as the input resistors in the servo amplifier that are instrumental in determining the integrator time constant can be large.

Voltage-Controlled Amplifiers

An important class of analog amplifiers is the voltage-controlled amplifier (VCA). The VCA is a two-quadrant analog multiplier (as opposed to a modulator, which is a four-quadrant multiplier**), where the signal input is bipolar, whereas the gain control input is constrained to be zero or positive. A VCA finds use in audio system applications such as programmable analog mixing desks where, e.g., gain or filter parameters require dynamic control from signals derived within a computer or remote controller. VCAs are also widely used in phase-lock loops as the phase-sensitive detector, although in this application a four-quadrant multiplier implementation is required. The basic specification requirements of a VCA are similar to other audio amplifiers in terms of noise and distortion. However, because the gain is programmable, a method is required to embody active devices that, although nonlinear, appear to be linear from the signal's perspective. The core principle exploited by BJT-based VCAs is to use the logarithmic method of multiplication, where

$$x \cdot y \equiv e^{\{\log_e(x) + \log_e(y)\}} \tag{2.3}$$

---

* A BI-FET incorporates field-effect transistors in the input stage with the remaining circuitry using bipolar junction transistors.
** Quadrant describes effectively the permissible polarities of the two input signals to the VCA.

However, in the multiplication described by the identity in Equation (2.3), only positive and nonzero values of inputs x and y are permissible, so in its basic form, this process is limited to one-quadrant applications. To extend the technique to two-quadrant operation requires a bias level X to be added to the input designated to handle a bipolar signal, say the x input, combined with a method of differential input drive and subtraction at the output. This technique is demonstrated as follows and is described by Equation (2.4):

Let

$$(X + x) \cdot y \equiv e^{\{\log_e(X+x)+\log_e(y)\}} \quad \text{and} \quad (X - x) \cdot y \equiv e^{\{\log_e(X+x)+\log_e(y)\}}$$

whereby

$$x \cdot y \equiv \frac{(X + x) \cdot y - (X - x) \cdot y}{2} \equiv \frac{e^{\{\log_e(X+x)+\log_e(y)\}} - e^{\{\log_e(X-x)+\log_e(y)\}}}{2} \tag{2.4}$$

The device characteristic exploited in BJT-based VCAs is the logarithmic relationship between emitter current and base-emitter voltage, where for an ideal BJT the emitter current is $V_{BE} = (kT/q)\log_e[I_E/I_s]$, where I_s is the saturation current, q is the charge on electron, k is Boltzmann's constant, and T is the junction temperature in degrees Kelvin. However, two fundamental problems of using a single transistor are that it is only a one-quadrant device and that it is highly temperature sensitive, particularly with respect to the saturation current, I_s. For BJTs to operate successfully as a VCA, means must be found to compensate for the temperature dependence. This is resolved using a transistor array consisting of at least two, but more usually four, transistors in a configuration that, if all the transistors are physically identical and their junction temperatures also identical, then the temperature-dependent parameters cancel. To understand the operation of a typical gain cell, the design is divided into two parts. First, the transistor gain cell topology is identified or conceptualized and then analyzed to confirm whether it meets the requirement of a linear multiplier. Second, support or interfacing circuitry is introduced around the basic multiplier cell to establish dc biasing, to apply appropriate input signals and to extract a suitable output signal. Three gain cell topologies that can be used at the core of a VCA are illustrated in Figure 2.3(a–c). Barry Gilbert* conceived the first cell,[6] the second was by the author,[7,8] and the third was by the dbx company.** To establish that a matched transistor array can achieve multiplication over a wide dynamic range, the cell in Figure 2.3(b) is analyzed here, where the following assumptions are made:

---

* Barry Gilbert is associated with Analog Devices, Norwood, MA, U.S.A.
** dbx is a trade name of a company in the U.S.

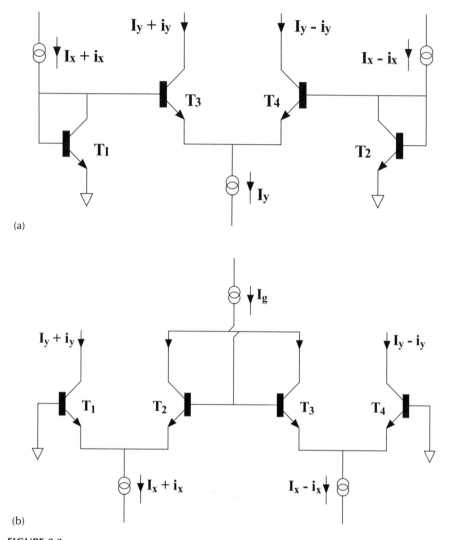

FIGURE 2.3
(a) Gilbert translinear gain cell. (b) Current-steering gain cell.

- Emitter current, base-emitter voltage exhibits exact logarithmic conformity.
- All BJTs in the array are parametrically and thermally matched.
- BJT bulk resistance is negligible.
- BJT base currents are negligible so that collector current I_C equals emitter current I_E.

If the logarithmic relationship is applied to each of the transistors T_1, T_2, T_3, and T_4, then applying Kirchhoff's voltage law to the mesh containing the

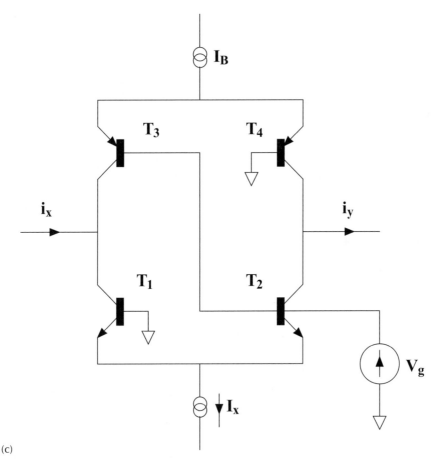

(c)

FIGURE 2.3 *Continued.*
(c) dbx-complementary transistor gain cell.

respective base-emitter voltages V_{BE1}, V_{BE2}, V_{BE3}, and V_{BE4}, then $V_{BE1} - V_{BE2} + V_{BE3} - V_{BE4} = 0$. Substituting for each base-emitter voltage, it follows that the respective transistor collector currents I_{C1}, I_{C2}, I_{C3}, and I_{C4} are related as $I_{C1}I_{C3} = I_{C4}I_{C2}$ while defining I_g as the gain control current, then the current gain of the cell is given by Equation (2.5) as

$$\frac{i_y}{i_x} = 1 - \frac{I_g}{2I_x} \tag{2.5}$$

A critical characteristic of all the cell topologies shown in Figure 2.3 is that the dominant temperature-dependent parameters cancel where providing the transistors maintain accurate logarithmic conformity over a wide current range, then the VCA cells remain linear even for large input signals. This is critical because enabling high signal levels to be used that can approach the limits set by transistor bias currents facilitates a wide dynamic range and an extended signal-to-noise ratio performance. Also, a characteristic of the current steering cell is that at maximum gain the cell operates purely as a grounded-base stage and therefore produces negligible distortion.

In practice, VCAs suffer impairment that can be grouped into linear and nonlinear errors. It is possible to visualize a three-dimensional error surface where the error is plotted against the two input functions. The error is effectively the difference between the theoretical multiplier output and the actual multiplier output. For linear distortions, the error surface remains planar, although the surface can in theory appear to be displaced and rotated. On the contrary, nonlinear multiplier errors are represented by curvature of the error surface. In practice, it is possible to trim out the linear error by applying appropriate input and output offset correction; however, the nonlinear error cannot be corrected by such means and represents a fundamental shortcoming of the multiplier. The need to adjust errors in an analog system can be problematic, because where trim controls are provided there remains the possibility of drift with time and temperature. Nevertheless, the use of properly matched devices and appropriate interface circuit design can reduce these problems and lead to cost-effective and high-performance circuit solutions.

Dynamic Range Control

It is a common requirement for an audio signal to be processed in order to limit the dynamic range, that is, the difference in level between loud and quiet sounds, where typical applications include:

- Reproduction of audio in noisy environments such as a factory or a car
- Gaining enhanced penetration in radio broadcasts where program content may be required to be produced at a near-constant level
- Hearing-impaired persons, where the suppression of loud signals and the expansion of low-level signals can improve intelligibility

At the core of a dynamic range control system is the requirement to modulate the gain of a system as a function of signal level. Consequently, in analog systems, variable gain circuitry, as discussed earlier in this section can be employed, where performance parameters such as low distortion and noise and gain control signal feed-through to output are particularly important if the quality of the input signal is to be preserved. There are many forms of dynamic range controllers with numerous characteristics where such systems may have a specific function or may be used as an effects unit where the modification is for artistic reasons. However, whatever the application, a number of key factors should be considered. First, changes in sound quality should be subjectively pleasing to the ear without obvious generation

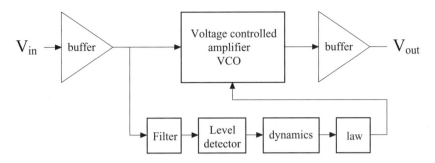

FIGURE 2.4
Basic audio compression structure.

of distortion or gain and noise-pumping effects unless, of course, such modifications are specifically required. This implies the modifications must be perceptually acceptable. It is a characteristic of a dynamic range control device that gain does not change instantaneously, otherwise gross nonlinear distortion is generated. Hence, a gain control signal is required that has metered dynamics, where typically this has a fast attack and a slower decay function. It is also desirable that some signal delay is introduced in the main signal channel so that gain changes can be predicted before their occurrence.

At the core of a dynamic range control processor are a number of subsystems that are normally located in a side chain; these influence the control range and the nonlinear characteristics of the dynamic range controller. Also, the dynamic response times of the circuit are determined by appropriate system time constants, where a basic feedforward structure is shown in Figure 2.4. The dynamic response times, particularly decay times, must be selected to avoid rapid gain changes that otherwise would be perceived as gross signal distortion. A key process is the derivation of a signal related to the input signal level. A common technique is to use a full-wave rectifier and smoothing circuit, although this can be waveform dependent in its operation. An alternative is to incorporate a true root-mean square (RMS) detector. Such detectors can use a four-quadrant multiplier as described in the previous section, but where the two inputs are connected together so that the output responds to the square of the input voltage. Following a smoothing function, a quantity related to the RMS value can be derived. The output of the detector is then processed by a time-dependent circuit to establish different attack and decay times. Finally a nonlinear system is used to shape the control function in order to produce the required overall gain control law, whether this be expansion or compression. Some of these features are addressed in the next section on complementary companding.

Dolby A-Type Noise Reduction

The family of noise reduction systems introduced by Dolby Laboratories in the 1960s[9] was designed to reduce the effects of additive noise introduced by analog tape recorders as well as other forms of analog transmission channels. The system is widely used with analog tape recorders where the Dolby A-type system finds favor with the professional studios, whereas the Dolby B-type system is widely used in domestic applications, especially with cassette tape recorders. The amount of noise reduction follows a well-defined and standardized characteristic, varying from 10 dB at 50 Hz to 10 kHz and rising to 15 dB at 15 kHz and above. This level and frequency-dependent characteristic is designed to match perceptual requirements such that the processor dynamics are inaudible, while achieving useful improvements in dynamic range. This section discusses the Dolby A-type system, which has found wide application in studios, where it can be viewed as a landmark product, although a later system, Dolby SR,[10] is now considered to be the ultimate solution to analog noise reduction.

The Dolby system works by encapsulating the channel to be protected with an encoder and decoder that operate as complementary processes in order to adapt the signal to achieve improvement in dynamic range. When considered as a composite process, the input and output signals should remain almost identical, but where the noise in the channel is no longer constant and adapts according to the signal. As such, the process can be viewed as a forerunner to the class of perceptually motivated coders, where the aim is to mask the channel noise by spectral shaping and level adaptation. A fundamental concept common to both the Dolby systems and perceptually motivated coders is masking by the human ear. It can be shown that if the noise spectrum is similar to that of a signal or signal component, then the signal will mask the noise even when the noise is only a few tens of decibel below the signal. The closer the noise spectrum matches the signal spectrum, the better can be the noise masking. However, if the noise spectrum is spectrally distant from the signal spectrum, then masking by the human hearing system is minimal.

To forge correlation between noise and signal spectra, a method of frequency discrimination is required. In the Dolby A-type system, this is achieved by dividing the audio signal into four subbands (although the two higher bands do overlap to some extent). Within a perceptually motivated coder, banks of bandpass filters are used, although here 16 to more than 50 bands are not uncommon, because this allows much tighter matching between signal and noise. However, the Dolby system is an analog process where the degree of noise reduction is relatively modest, so a four-band filter is adequate. Principal requirements of a noise reduction system are that the overall operation is complementary, that there is no discernible modulation noise (*noise pumping*), that transient overshoot is kept to a minimum, and that level matching between encoder and decoder is not too critical (typically ± 2 dB). The Dolby A-type system achieves this in part by processing only lower-level signals that in effect become amplified at the encoder and then attenuated at the decoder. However, higher-level signals are by default remote from the channel noise and are left almost unmodified; consequently, tracking errors resulting from incorrect level matching only occur on lower-level signals. In practice, the system employs a reference tone for calibration, where

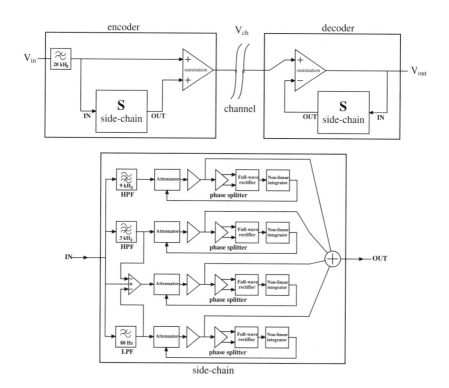

FIGURE 2.5
Dolby A-type complementary noise reduction processor.

this tone is generated at the encoder to allow the decoder to be calibrated in level prior to use. The tone is set at the standard *Dolby level*, which in tape-based applications is also related to a specific magnetic flux density on the tape (e.g., 185 nW/m Ampex NAB level and 320 nW/m DIN level).

A key system feature of the Dolby system is to locate the adaptation and filtering subsystems within a side-chain process by using a combination of feedback and feedforward, where for the special case of a noiseless channel, this technique allows, in theory, exact complementary performance. Even when channel noise is present, system tracking still has low sensitivity to this noise, because the signals are held at a much higher level and there is effective bandlimiting in the level control circuits that offer a degree of noise rejection. The complementary feedback and feedforward structures are shown in Figure 2.5, together with some of the key side-chain processes. The complementary characterization of the Dolby system can be demonstrated as follows. Let $S(V_{in})$ be the transfer function of the side-chain processor at any instant in time, noting that $S(V_{in})$ is a nonlinear function of V_{in} that depends upon the current state of the system. If V_{ch} is the output of the encoder, then if the feedforward path in the encoder is examined, it follows that $V_{ch} = V_{in}\{1 + S(V_{in})\}$. Similarly, if $S(V_{out})$ is the transfer function of the decoder side-chain processor (shown now as a function of V_{out}) where V_{ch} is also the input to the decoder, then the decoder output is calculated by applying standard analysis of a negative feedback loop as $V_{out} = V_{ch}\{1 + S(V_{out})\}^{-1}$. Hence, assuming a noiseless and unity gain channel, then the overall system transfer function is given by Equation (2.6) as

$$\frac{V_{out}}{V_{in}} = \frac{1 + S(V_{in})}{1 + S(V_{out})} \tag{2.6}$$

However, if the input and output signals of the overall process track closely as they are required to do in practice, the states of the two side-chain processors also track, implying the transfer functions $S(V_{out})$; $S(V_{in})$. Consequently, even if the side-chain process is nonlinear, providing that any additional noise in the channel does not cause the side-chain states to diverge significantly, the overall system transfer function described by Equation (2.6) is unity. This is the principal feature that enables the Dolby noise reduction to offer virtually exact complementary encode and decode functionality.

In the Dolby A-type system, the four band-pass channels are synthesized from a combination of two high-pass filters, a low-pass filter, and a matrix process that derives a band-pass response. The filters and matrix are shown in the side-chain processor of Figure 2.5, where the frequency subbands are nominally 0 to 80 Hz, 80 to 3 kHz, 3 to 9 kHz, and 9 to 20 kHz. In practice, there is significant spectral overlap between the low-order filters, as they are typically second-order Sallen and Key topologies, similar to those shown in Figure 2.1. Each subband channel uses almost identical nonlinear compressors that include variable gain, nonlinear transient limiting and nonlinear smoothing circuitry.

A compressor consists of a variable gain stage, a transient limiter, and a level detection stage to derive a gain control signal that is applied in a local feedback path back to the variable gain stage. Consequently, as the input signal level rises, the gain control signal increases and the gain of the first stage is reduced, thus realizing signal compression. Although the gain control stage could employ a translinear amplifier, as described in "Voltage-Controlled Amplifiers," in this example two cascaded junction field-effect transistors (JFETs) form a linear attenuator. The combination of specific device characteristics, nonlinear transient limiting, and nonlinear smoothing endow the Dolby noise reduction system with its idiosyncratic, yet highly effective, characterization.

The use of a dual JFET as a voltage-controlled attenuator yields a cost-effective solution to gain control, especially where the gain control range is modest. However, a JFET does not offer the large signal linearity of a translinear circuit, as described earlier in "Voltage-Controlled Amplifiers," nor does it offer a well-defined gain control characterization that would make it suitable for general analog multiplication applications. The problem of using a JFET as a variable resistor in an attenuator is that the drain current vs. drain-source voltage is nonlinear. However, if an optimum fraction of the drain-to-source voltage is fed back to the gate control voltage, a degree of

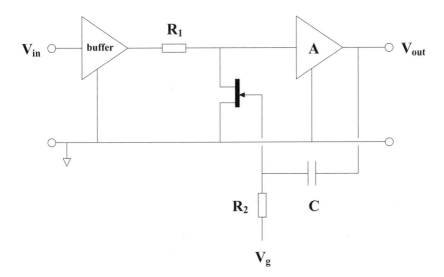

FIGURE 2.6
JFET attenuator with linearization achieved by using ac feedback.

linearization can be achieved. This technique is used in the Dolby attenuator implementation, so it is described here as an example of a subtle design change that improves the linearity up to an acceptable quality. A single JFET attenuator stage is illustrated in Figure 2.6 where an ac feedback signal is added to the dc gain control voltage V_g that in turn is applied to the gate of the JFET. In the JFET *pinch-off* or *triode region* of operation where the JFET drain-to-source voltage V_{DS} is below its saturation value,* then the drain current I_D can be expressed as $V_{DS} = V_p I_D [I_{DSS}(2 + 2V_{GS}/V_p - V_{DS}/V_p)]^{-1}$, where V_p is the pinch-off voltage and I_{DSS} the drain current when the gate-to-source voltage V_{GS} is zero. The JFET slope resistance r_d for small values of drain-to-source voltage close to zero is then calculated by differentiation as

$$r_d = \frac{\partial V_{DS}}{\partial I_D} = \frac{V_p\left(2 + 2\dfrac{V_{GS}}{V_p} - \dfrac{V_{DS}}{V_p}\right) - I_D\left(2\dfrac{\partial V_{GS}}{\partial I_D} - \dfrac{V_{DS}}{\partial I_D}\right)}{I_{DSS}\left(2 + 2\dfrac{V_{GS}}{V_p} - \dfrac{V_{DS}}{V_p}\right)^2} \tag{2.7}$$

Equation (2.7) reveals that the JFET resistance is a nonlinear function of the drain-to-source voltage. However, by adding a fraction of the drain-to-source voltage V_{DS} back to the gate-to-source control voltage, the slope resistance of the JFET is sympathetically modulated by the gate-to-source voltage so that the resistance can remain almost constant, thus substantially

---

* JFET saturation is where an increase in V_{DS} causes little change in drain current I_D, assuming V_{GS} is constant.

linearizing the attenuator circuit. Hence, by substituting $V_{GS} = V_g + 0.5V_{DS}$ in Equation (2.7) the slope resistance becomes

$$r_d = \frac{V_p}{2 I_{DSS}\left(1 + \dfrac{V_g}{V_p}\right)} \tag{2.8}$$

Equation (2.8) shows that the slope resistance has been made independent of the gate-to-source voltage. Figure 2.6 reveals this modification is simple to implement, where using a buffer amplifier, a signal equal to one half the drain-to-source voltage is derived, fed back, and superimposed onto the gate control voltage via an ac coupling network to remove any bias voltages. The ac coupling is satisfactory in this application because the mean voltage applied to the attenuator is zero and the drain is effectively biased at zero voltage.

The design of a multiband processor is relatively complicated where the performance is determined by the accuracy of the analog circuitry. It is therefore expedient to make the principal signal paths as direct as possible with the nonlinear processing placed in a parallel side-chain. A key aspect of the Dolby A-type system relates to the method by which the internal level control signals are derived, where this technique is illustrated in the simplified circuit diagram shown in Figure 2.7. Following the JFET gain control and transient limiter to prevent excessive signal overshoot, a phase-splitter circuit produces complementary, equal amplitude audio signals that are applied subsequently to a full-wave rectifier. The rectifier stage in turn drives a nonlinear smoothing circuit that has a variable but fast response time and a slow decay time where the output of this stage forms the control signal for the variable gain stage.

The dynamics of the smoothing circuits are selected on perceptual criteria and require special consideration, since short-term variations of the audio signal can introduce ripple onto the compressor gain control signal. By

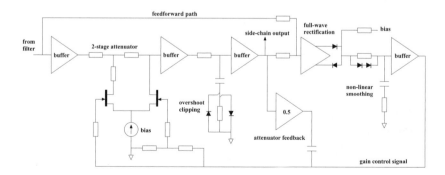

FIGURE 2.7
Dolby A-type side chain processor showing linear attenuator, nonlinear limiter, and gain control derivation using full-wave rectification and nonlinear smoothing.

introducing smoothing, this problem may be reduced, although there is a potential penalty that the response of the level detector can become too slow, leading to audible noise pumping and gain variation artefacts. As shown in Figure 2.7, this area is addressed by using a nonlinear filter that includes a diode–resistor–capacitor network, the response of which depends upon signal level. Effectively, for near-constant level signals, the time constant controlling the level detector attack response time is made relatively large, thus smoothing the control signal and reducing ripple. However, for a sudden increase in signal level, the attack time constant is reduced dynamically to enable a faster response time. This problem is common to most dynamic range controllers but is addressed in the Dolby A-type process by using a combination of four subband filters with individual side chain compressor circuits employing the circuit techniques shown in Figure 2.7, where the key features of each compressor can be summarized as

- Two-stage JFET attenuator
- Local ac-feedback loop to improve attenuator linearity
- Nonlinear transient limiter to limit the maximum signal under transient conditions
- Full-wave rectifier
- Nonlinear smoothing circuit with fast response/slow decay to derive JFET gain control signal

Conclusions

There are often significant philosophical differences in the way an analog system is conceived and designed compared to a digital system. For example, the analog designer may use a lateral approach to identify a new topology and to simplify functionality, in a way that is difficult for a formal design method to emulate. It may be argued that such differences are weaknesses, but in terms of a cost-effective design, they can also become strengths. For example, the process characterization of the Dolby noise reduction system is in part dependent not on specific algorithms but on the idiosyncratic characteristics of JFET diodes, and the implementation of the dynamic level detection circuitry. This is in strict contrast to processes that are designed specifically for digital implementation where normally more precise system definitions are made. However, this is not unusual for analog circuitry especially in the era when the Dolby noise reduction system was conceived. In the mid- to late 1960s, translinear gain circuits and low-cost digital circuits that could be used to control analog processes were not available, so in order to produce a cost-effective design, it was common practice to use the specific characteristics of available devices. Although this leads to an efficient design and considerable ingenuity in the system topology, it does require precise selection and matching of parts in order that they adhere to the required specification. In particular, the JFET devices are critical and would require careful selection to ensure proper functionality compared to a laboratory reference processor.

An interesting problem therefore emerges as to how one might characterize processes such as Dolby noise reduction in digital terms in order to emulate its performance and, e.g., enable Dolby encoded material to be decoded correctly within the digital domain. In fact, Dolby has produced an A-type digital processor for film studio applications, a task representing a formidable design challenge, especially because some of the analog processes are relatively subtle and complicated. Certain aspects map exactly; in particular, the filters are easily modeled and transformed to the z-domain (see Chapter 1). However, the attenuator and nonlinear integrators pose a greater problem. Even here analog multipliers, as discussed in "Voltage-Controlled Amplifiers," can represent the JFET attenuators and the control law emulated with an appropriate code-based lookup table. Alternatively, mathematical polynomial approximations can be made to describe in an abstract way the control laws embedded in the analog domain. For such a process, careful measurement would have to be performed, preferably on an analog laboratory reference circuit from which all other processors would be calibrated. Of course, once a digital model is created, the design process can be inverted, and a new analog design synthesized that combines modern hybrid analog–digital circuit elements, possibly augmented using digital processing and logic-based design, e.g., in the side chain processor.

This latter observation is important because it identifies a design methodology where the strengths of both analog and digital techniques can be combined to achieve high performance with accurate repeatability including self-calibration. Also, the theme of combining analog and digital filter techniques was shown also to be critical at the analog-to-digital gateways where this was highlighted at the beginning of this chapter. Elegance in analog design can often combine simplicity and functionality; the current-to-voltage conversion stage is such an example. Here, the low-pass filter is interleaved into the transresistance amplifier stage rather than performing each function as a cascade of individual processes. Thus, although it is true that many processes can be performed in the digital domain, if the input and output signals are analog signals, then the conversion and the expense of a pure digital solution may not yield the best performance for a given task. This is especially so if the gateway converters compromise performance. Finally, it is also sobering to consider that all practical electronics operate within an analog world and that the limits on speed, interconnectivity, and information communication are dictated ultimately by analog structures. It is only the information conveyed or contained that is digital data.

References

1. Funasaka, E. and Suzuki, H., DVD-audio format, *103rd AES Convention*, New York, September 26–29, 1997, preprint 4566.
2. Adams, R., Nguyen, K., and Sweetland, K., A 116 dB SNR multi-bit noise shaping DAC with 192 kHz sampling rate, *106th AES Convention*, Munich, 1999, preprint 4963 S5.
3. Hawksford, M.O.J., Introduction to digital audio (tutorial paper), *Images of Audio, Proceedings of the 10th International AES Conference*, London, September 1991.
4. Hawksford, M.O.J., Current-steering transimpedance amplifiers for high-resolution digital-to-analogue converters, *109th AES Convention*, Los Angeles, September 2000, preprint 5192.
5. Hawksford, M.J., Distortion correction circuits for audio amplifiers, *JAES*, 29(7), 8, 1981.
6. Gilbert, B., A new technique for analog multiplication, *IEEE J., Solid-State Circuits*, SC-10, 437–447, 1975.
7. Hawksford, M.O.J., Low-distortion programmable gain cell using current steering cascode topology, *JAES*, 30(6), 795–799, 1982.
8. Hawksford, M.O.J. and Mills, P.G.L., Topological enhancements of translinear two-quadrant gain cells, *JAES*, 37(6), 465–475, 1989.
9. Dolby, R., An audio noise-reduction system, *JAES*, 15(4), 383, 1967.
10. Dolby, R., The spectral recording process, *JAES*, 35(3), 99–118, 1987.

3

Hardware Design Considerations

Robert S. Oshana

CONTENTS

0-8493-0949-2/02/$0.00+$1.50
© 2002 by CRC Press LLC

Introduction

This chapter discusses digital signal processing (DSP) hardware applied to noise reduction. DSP is the method of processing signals and data in order to enhance or modify those signals or to analyze those signals to determine specific information content. It involves the processing of real-world signals that are converted and represented by sequences of numbers. These signals are then processed using mathematical techniques to extract certain information from the signal or to transform the signal in some preferably beneficial way.

There are many advantages for using a DSP solution over an analog solution, including:

- *Versatility:* It is easy to reprogram digital systems for other applications or to fine tune existing applications. A DSP solution allows for easy changes and updates to the application. A digital system can also be ported to other hardware more easily.

- *Repeatability:* As opposed to analog components whose characteristics change slightly over time, a digital solution does not depend on these tight tolerances. Digital systems can also be duplicated easily for other platforms or applications.

- *Simplicity:* It is much easier to implement a filter using simple programming than to fine tune several hardware components to achieve the same response.

- *Size, weight, and power:* A DSP solution requires mostly programming, and the DSP device itself consumes less power than a solution using all hardware components.

- *Reliability:* Analog systems are reliable to the extent the hardware devices function properly. If any of these devices fails due to physical condition, the entire system degrades or fails. A DSP solution implemented in software will function properly as long as the software is implemented correctly.

- *Expandability:* To add more functionality to the system, the engineer must add more hardware. This may not be possible due to cost or time-to-market constraints. Adding the same functionality to a DSP involves adding software, which is much easier.

- *Changeability:* A redesign or major change involves changing hardware, which is expensive and time consuming. The same redesign can be done using software in a digital system.

This chapter offers an overview of the characteristics and architectures of DSP hardware. First, a few examples of DSP hardware applied to noise reduction are presented. An overview of hardware architectures is then given followed by specific features of digital signal processors.

Noise Reduction in Hearing Devices Using DSP

Digital hearing aids convert analog sound information from the microphone into computer code, which is then processed by an amplifier consisting of microcomputers such as DSPs instead of combinations of electronic circuitry (Figure 3.1). Adjustments to the hearing aid are made by programming the DSP.[1]

Noise reduction technology is a main component applied to hearing impairment. The noise reduction operation requires frequency-dependent signal analysis in order to identify whether the signal is speech or noise. In most digital hearing aid algorithms, this basic function is the fast Fourier transform (FFT). The FFT algorithm in hearing aid applications requires an important computational load, because the time-to-frequency and frequency-to-time conversions have to be done in real time. These operations are responsible for close to 50% of the computational load as well as most power consumption. DSPs are used for digital hearing aids because of the efficiency of computation of FFTs.

Other standard noise reduction techniques are based on adaptive filtering algorithms. Adaptive filtering algorithms are also performed efficiently using DSP.

Two of the most important requirements for the modern high-performance hearing aid are low physical area consumption (it must fit inside the ear) and low-power dissipation. The processing engine of a hearing aid consists of a DSP which executes a number of signal processing routines that a DSP performs very efficiently. Examples of these algorithms are basic filter algorithms such as finite impulse response (FIR) and real-time adaptive algorithms used for noise reduction.

Digital filtering algorithms are important for hearing aids to reduce certain bands of noise. A digital filter requires a core set of components that are common on most DSPs: adders, multipliers, multiply-accumulators and memories, counters, and high-speed registers.

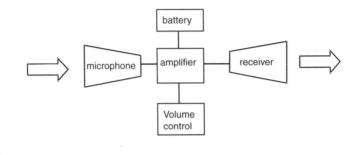

(a)

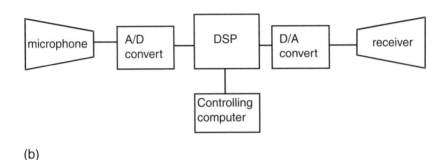

(b)

FIGURE 3.1
Block diagram of (a) a simple analog hearing aid and (b) a DSP-based hearing aid.

The main advantage of DSP is the far greater processing power over standard analog signal processing. Some of the other potential advantages of digital hearing aids include:

- Reduced feedback
- Improved sound quality
- Increased flexibility of programming or fine tuning
- More precise adjustments resulting in improved sound quality
- Elimination of circuit noise, which can be a problem for some clients, particularly those with normal or near normal low-frequency hearing

DSP-based hearing aids also have the processing power to divide the frequency response into two or more channels and adjust amplification characteristics independently for each channel. This is an important advantage for individuals who hear significantly better in one part of the frequency range than in others.

DSP provides several other advantages over an analog hardware solution. Analog systems are combinations of electronic components. Digital systems, on the other hand, have much of their functionality as software on a DSP. In order to change the parameters of an analog system, you must add or remove electronic components. This can be expensive and time consuming. To make the same change in a DSP-based system, all that needs to be changed is the software running on the DSP. This is a relatively fast and inexpensive procedure. Electronic components found in an analog hearing aid such as resistors, capacitors, and operational amplifiers will change characteristics as the temperature changes. DSP-based solutions show no variation in performance with changing temperatures because much of the processing is software based. Digital systems are also repeatable, whereas an analog circuit has variable tolerances. These tolerances make each circuit unique and more difficult to maintain. DSP-based solutions can also respond dynamically to changing conditions (e.g., noise background, etc). Analog systems cannot respond dynamically to changing conditions — they must be readjusted.

DSP Noise Reduction in Automobiles Using Electronic Mufflers

Electronic mufflers are one example of noise reduction applied to a commercial system.[2] This system receives input from a microphone and a crankshaft speed/position sensor. These sensors trap the waveform signature of the engine. A computer receives input on the pattern of pressure waves (basically sound waves) the engine is emitting at its tail pipe. These data are processed using DSP algorithms, which produce mirror-image (antinoise) pulses that are sent to speakers mounted near the exhaust outlet, creating contrawaves that cancel out the noise and remove sound energy from the environment. Although the concept can be applied to any type of unwanted sound, it works particularly well against annoying low frequencies — buzzes, hums, booms, and rumbles.

Active noise cancellation techniques have been used since the vacuum tube days of the 1930s. DSP has provided an enormous boost to the application of noise cancellation and reduction.[3] DSP performs active noise reduction and cancellation using predictive analysis and feedforward analysis techniques. Predictive analysis records rotations per minute (rpm) from a component and checks for errors with a microphone. This approach works well with the repetitive sounds caused by the rotating parts of an engine and other car parts. The feedforward approach uses a microphone input only, and it is the best technique to use for more random sounds such as wind and tire noise. An adaptive learning algorithm in the DSP is capable of storing these various characterizations so the DSP does not have to perform a full computation each time the same pattern is heard. This makes this approach to noise cancellation fast and efficient.

Active noise cancellation does not just make the car ride more enjoyable by reducing background noise. There are also performance advantages for the car itself. Active noise cancellation reduces backpressure by up to 80%, which can lead to an improvement in power output and fuel mileage. An ordinary car muffler is a passive system. There are certain bands of noise frequency where a muffler will work well, but a muffler is too simple a system to suppress all of the various noises produced by an automobile. Real-time DSP provides the ability to customize the sound control to various load conditions.

DSP-Based Digital Audio Noise Reduction

Digital audio is another example of modern noise reduction techniques. Digital audio noise reduction techniques prevent certain classes of noise and distortion from corrupting an audio signal.[4,5] These audio signals are relatively free of pops, clicks, crackles, surface noise, hums, and buzzes. Most of the noise remaining after digital noise reduction is the background noise such as that found in heating, ventilating, and air conditioning (HVAC) rumble, traffic, and other low-level noises that often creep into recordings.

Audio noise reduction techniques are divided into two main areas: the removal of ticks, pops, and other noises of an impulsive nature and removal of broad-band low-level noises. Removal of these types of noise has been done using analog electronics for quite some time. But the approaches used have been relatively expensive and inefficient. DSP techniques are able to apply greater sophistication to noise reduction. Modern digital audio noise reduction systems use hundreds of bands of multiband processes and apply psychoacoustics principles to determine when a noise band can be removed with no detrimental effects on the signal. Host side signal processing (e.g., using a desktop PC) is used to splice and rebuild damaged portions of a program segment using advanced signal processing techniques.

PC-based noise reduction systems are often based on floating point DSP boards and sophisticated DSP software that perform a number of complex operations:

- Click, pop, and scratch removal
- Splitter/combiner fine click removal (sometimes called "decrackle")
- Real-time broad-band noise reduction
- Real-time hard disk editing and assembly
- Phase/time correction

DSP is also used to reduce noise in digital cameras.[6] These DSP-based cameras provide digital performance that is far superior to conventional analog camera technology. Signal processing technologies such as three-dimensional (3D) digital noise reduction is used to improve the picture quality. 3D digital noise reduction cancels out random noise without introducing noticeable delay in the camera operation. This is done by comparing multiple frames together and applying motion-detection algorithms.

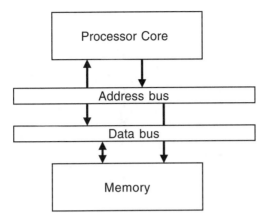

FIGURE 3.2
Von Neumann architecture.

Overview of Processor Architectures

Architecture is a term applied to both the process and the outcome of thinking out and specifying the overall structure, logical components, and logical interrelationships of a computer, its operating system, a network, or other conception. An architecture can be a reference model, or it can be a specific product architecture. Computer architecture can be divided into the fundamental components of input/output, storage, communication, control, and processing. This chapter will review DSP architectures, including the components previously mentioned.

Von Neumann Architecture

The organization of a traditional microprocessor is based on the von Neumann architecture shown in Figure 3.2. This computer architecture is characterized by two main components:

- Central processing unit (CPU) where the instructions are executed.
- Memory; most DSPs have two types of memory: slow-to-access storage area, like a hard drive, and secondary fast-access memory (RAM).

This architecture was the first model of the stored program concept. Instructions, stored as binary values, were executed sequentially. The processor would fetch instructions one at a time and process them. Today the term *von Neumann architecture* often refers to the sequential nature of computers based

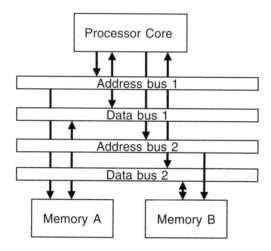

FIGURE 3.3
Harvard architecture.

on this model. There is only one memory space connected to the processor core by one system bus that is used to access address and data information. For many applications, this architecture works well. The single-system bus is able to keep the processor fed with enough data not to impact performance.

Harvard Architecture

A Harvard architecture uses two memory banks and two bus sets, as shown in Figure 3.3. Each of these two memory spaces is connected to the processor core. This allows the processor to make two simultaneous accesses to memory during each cycle. This approach basically doubles the processor bandwidth, allowing twice as much data to be fed to the processor. With the proper arrangement of data in these two memory spaces, applications such as the FIR filter can execute extremely quickly because the processor can access all the required data for each tap point from the two memory locations to perform the operation in one cycle (instead of four). The downside to this approach is the extra cost of the hardware to provide the extra bus set. This also costs in terms of power and space on the processor die.

Scalar Architectures

The scalar processor instruction cycle consists of the steps each instruction must pass through during processing. Each step in the instruction cycle requires one or more clock cycles to complete. A typical reduced instruction set computer (RISC) instruction for a scalar processor consists of the following stages:

- Fetch
- Instruction decode
- Read operands
- Execute
- Write results

A RISC processor is designed to perform a smaller number of types of computer instruction so that it can operate at a higher speed. Each instruction type that a processor must perform requires additional transistors and circuitry. Therefore, the larger the list of computer instructions, the more complicated and slower the operation of the processor. The execution time of a scalar processor is dependent on the number of instructions required and the average number of cycles per instruction. The number of instructions is dependent on the program, the compiler, and the instruction set architecture. The number of cycles per instruction is dependent on these plus the processor implementation. The execution time can be described as

Execution time = number of instructions ×
number of cycles per instruction × processor clock cycle time

For the scalar processor example described previously, the number of cycles per instruction is five.

In order to improve performance, at least one of these factors must be improved. The fact that the number of instructions, the number of cycles per instruction, and the processor clock cycle time are all interrelated results in complex trade-offs among design options.

One obvious way to increase performance is to reduce the processor clock cycle time. This is usually a technology-driven issue (over time smaller die sizes and routes result in shorter clock cycle times — this is basically Moore's Law*).

One key performance enhancement is to use a pipelined processor. Pipelining basically overlaps the processing of one instruction with the next, utilizing the different subphases of a DSP instruction (Figure 3.4). Efficient overlapping of instructions can reduce the average cycles per instruction to close to one. For a pipeline with n stages, the performance can be improved by at most a factor of n.

All instructions must pass through these phases in Figure 3.4 to be implemented. Each subphase takes one cycle. This sounds like a lot of overhead when speed is important. However, since each pipeline phase uses separate

---

* Moore's Law states that the pace of microchip technology change is such that the amount of data storage that a microchip can hold doubles every year or at least every 18 months. In 1965, when preparing a talk, Gordon Moore noticed that up to that time microchip capacity seemed to double each year. Because the pace of change has slowed down a bit over the past few years, the definition has changed (with Moore's approval) to reflect that the doubling occurs only every 18 months.

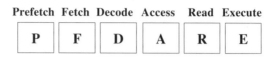

FIGURE 3.4
The subphases of a DSP instruction execution.

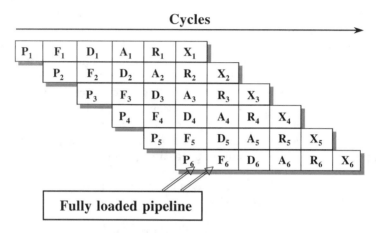

FIGURE 3.5
DSP instruction pipeline.

hardware resources from each other phase, it is possible for several instructions to be in process every cycle (Figure 3.5).

As shown in Figure 3.5, instruction 1 (P_1) first implements its Prefetch phase. The instruction then moves to the Fetch phase. The Prefetch hardware is therefore free to begin working on instruction 2 (P_2) while instruction 1 is using the Fetch hardware. In the next cycle, instruction 1 moves on to the Decode phase (D_1), allowing instruction 2 to advance to the Fetch phase (F_2), and opening the Prefetch hardware to start on instruction 3. This process continues until instruction 1 executes. Instruction 2 will now execute *one* cycle later, not six. This is what allows the high MIP rate offered in pipelined DSPs. This process is performed automatically by the DSP, and requires no awareness on the part of the programmer of its presence or actions to correctly write arithmetic code on the DSP.

Figure 3.6 shows the speed improvements to be gained by executing multiple instructions in parallel using the pipeline concept. By reducing the number of cycles required, pipelining also reduces the overall power consumed.

Adding pipeline stages increases the computational time available per stage, allowing the DSP to run at a slower clock rate. A slower clock rate minimizes data switching. Electronic circuits consume power when the transistors that they are composed of are switched on and off. So reducing the clock rate leads to lower power supply voltage. The trade-off is that large

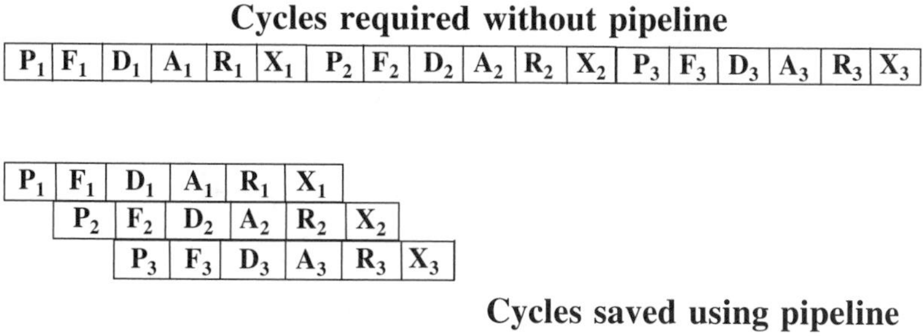

Cycles required without pipeline

P_1	F_1	D_1	A_1	R_1	X_1	P_2	F_2	D_2	A_2	R_2	X_2	P_3	F_3	D_3	A_3	R_3	X_3

P_1	F_1	D_1	A_1	R_1	X_1		
	P_2	F_2	D_2	A_2	R_2	X_2	
		P_3	F_3	D_3	A_3	R_3	X_3

Cycles saved using pipeline

FIGURE 3.6
Speed in a DSP using pipelining.

pipelines consume more silicon area. In summary, pipelining, if managed effectively, can improve the overall power efficiency of the DSP.

There are limits to increasing pipeline depth. Given that each instruction is an atomic unit of logic, there is a loss of performance due to pipeline flushes during program control flow. Another limitation to pipelines is the increased hardware/programmer complexity because of resource conflicts and data hazards. The latency increases as the depth of the pipeline increases. Sources of latency also increase, including register delays for the extra stages, register setup for the extra stages, and clock skew to the additional registers.

Superscalar Architectures

Superscalar processors are the next step beyond pipelining. In superscalar architectures, multiple pipelines operate in parallel. A superscalar processor is one that can fetch, execute, and complete more than one instruction in parallel. Superscalar designs use smart hardware controllers to manage the parallelism on chip. This approach requires less support from the software (compiler) to manage the parallel issuance of instructions. Superscalar devices can execute words of varying widths. This enhances programmer flexibility and also produces higher code densities because the less complex instructions occupy fewer bytes of memory.

An n-degree superscalar processor has n pipelines. Theoretically, the cycles per instruction for an n-degree superscalar processor can be reduced to $1/n$. If each pipeline is k stages, the potential speedup over a scalar implementation is $n \times k$. In practice, it is very difficult to keep multiple pipelines busy, and many superscalar implementations have been limited to just two or three pipeline stages. Figure 3.7 shows the structure of a three-stage superscalar pipeline.

The ZSP DSP architecture by LSI Logic Inc. is superscalar, issuing up to four instructions per cycle. The architecture is also pipelined with a five-stage pipeline and a dual multiply and accumulate (MAC) and dual arithmetic logic unit (ALU). The superscalar architecture makes the device compiler and programmer friendly, and the dynamic hardware scheduling

F	D	R	E	W		
F	D	R	E	W		
F	D	R	E	W		
	F	D	R	E	W	
	F	D	R	E	W	
	F	D	R	E	W	
		F	D	R	E	W
		F	D	R	E	W
		F	D	R	E	W

FIGURE 3.7
A three-way superscalar pipeline.

can automatically eliminate pipeline conflicts. But this same dynamic instruction scheduling also leads to more nondeterministic behavior. (The dynamic scheduling takes place in the pipeline control unit [PCU] of Figure 3.8. This is where data and resource dependencies are resolved. The PCU synchronizes the entire operation of the pipeline, arranges for operand bypass, and processes interrupt requests.)

Very Long Instruction Word Architectures

Very long instruction word (VLIW) architectures basically pack multiple opcodes into a single instruction. Each opcode describes the action of one functional unit. All of the functional units in a VLIW architecture operate in parallel.[7] The main problem with this approach is finding enough independent operations to fill the available slots in each instruction word. If enough are not found, no-ops are used instead. A no-op, for no operation, is a computer instruction that takes up a small amount of space but specifies no operation. The processor simply moves to the next sequential instruction. In order to be successful, a good compiler is required. VLIW processors work at reducing the number of instructions executed (instead of cycles per instruction).

One of the fundamental characteristics of a VLIW architecture is the ability to extract a highly parallel instruction stream from the application program. These parallel instructions are then allocated to the multiple execution units of the device (Figure 3.9). The execution units can execute multiple instructions in a single clock cycle. VLIW processors are relatively simple compared to superscalar devices. In superscalar devices, the instructions are scheduled dynamically, in real time, based on the state of the processor at the time. In

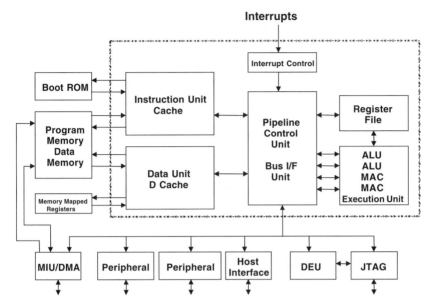

FIGURE 3.8
ZSP DSP architecture.

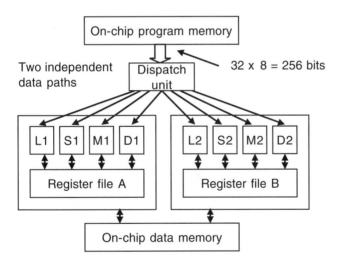

FIGURE 3.9
VLIW architecture.

VLIW processors, there is no real-time dynamic execution capability on the device. The instructions are scheduled ahead of time by the compiler.

The term *VLIW* implies a wide bit length word. Multiple different commands are packed into this wide bit word. The job of the compiler in a VLIW device is to determine which instructions can be scheduled in parallel and pack those instructions into a single word. Since this is done by the compiler,

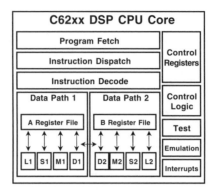

FIGURE 3.10
The Texas Instruments TMS320C6x DSP VLIW processor.

there is less overhead at run time to determine this scheduling. Also, less silicon is required to determine the proper scheduling during run time. This directly translates to simpler silicon designs for VLIW processors.

The Texas Instruments TMS320C6x DSP architecture is an example of a VLIW architecture for DSP (Figure 3.10). There are two data paths in this device, each with four instruction execution units (L, S, M, D). Up to eight instructions can be issued and executed per cycle using this VLIW approach.

Characterizing DSP Architectures

Many DSP architectures are available today. Each architecture has unique features to address specific problem domains. DSP architectures can be classified by the following attributes:

- Instruction set architecture (ISA) specialization
- Component specialization
- Processors
- Memory
- System
- Load distribution
- Component interaction

Instruction Set Architecture Specialization

The term *instruction set architecture* (ISA) defines a number of different attributes for operand storage and execution. ISA specialization determines the operand storage plan in the CPU, including the definition of the stack, accumulator, and registers data flows. The ISA also defines the number of explicit operands named per instruction (0,1,2,3), the operand location definition (memory, registers, addressing modes), the operations themselves (e.g., shift, add, store), and the type and size of operands (e.g., immediate, constant, signed, unsigned, byte, word).

The ISA specialization determines the concurrency of the individual operations as well as the chaining of operations. The ISA also defines the instruction length. This is important when considering code size. Embedded applications are code size sensitive and require a small program footprint. The ISA determines the branch control structure and the number of instructions required to express a function. The raw peak performance of a DSP ISA is measured in instructions/cycle, operations/cycle, or floating point operations/cycle. Finally, the ISA definition defines the power contribution due to operand/instruction widths. This is obviously important for power-sensitive applications.

Component Functional Unit/Data Path Specialization

The component functional unit and data path specialization defines the implementation of the internal data flow in the DSP. This specialization defines the implementation of the operations in the DSP as well as the widths of the internal paths and storage capability.

This type of specialization determines the frequency of the device (with the pipeline model), the throughput of the device, and the precision per cycle that the device can produce. This type of specialization also determines the power contribution due to the circuit style and the number of circuit transitions per operation.

Component Local Memory Specialization

The local memory of a DSP is specialized as well. The local memory specialization defines the number and size of memory. It also defines the organization of memory (e.g., RAM, ROM, cache, FIFO) and the number and size of the I/O ports. The structure of memory is also defined by this type of specialization, including how many memory banks, whether they are interleaved, and so forth).

Local memory specialization determines the efficiency of use of temporal locality in both data and program storage as well as the efficiency of use of spatial locality in data and program memory. Local memory specialization determines efficiency in data structure sizes and data flow strategies.

Component Interconnect Specialization

Another type of specialization in DSPs is the component interconnect. The component interconnect defines the physical connections and protocols between the subunits of the processor. The component interconnect scheme implements the chaining and concurrency functionality of the ISA.

This type of specialization determines the communication bandwidth in the processor. This is important for high-performance I/O applications. The component interconnect also impacts cost and power. The more buses that have to be switched on and off, the more power is consumed in the device, and the more silicon area is required.

Component Control Specialization

Component control specialization in a DSP defines such things as the control path width of the processor. The control implementation (finite state machine, controller) is also defined by this type of specialization. Component control specialization determines the pipeline model used in the DSP and the latency in the device.

System Specialization

System specialization focuses on load distribution. Load distribution is the mapping of functions to resources in a processor. This involves such techniques as data decomposition, control decomposition, and component interaction. Data decomposition, for example, considers the data flow from one computation unit to the next. Does the DSP, for example, require specialized storage to buffer data, to match formats, and to match the processing rates of the system. System specialization also includes the use of shared resources and the communication mechanisms used in the processor.

DSPs vs. General Purpose Processors

This section will describe some of the main differences between general-purpose processors (GPPs) and digital signal processors. There are several fundamental differences in these two types of processor.

Functionality

GPPs are designed to have broad functionality. These processors are designed to be used in a wide variety of applications; therefore, their architectures are designed to support a wide range of solutions. Specialized processors, on the other hand, are designed to take advantage of the limited functionality required by their applications to meet specific objectives. DSPs are specialized processors designed to focus on signal processing applications. Hence, there is little or no support to date for features such as virtual memory management, memory protection, and certain types of exceptions.[8]

Performance

The performance goal of GPPs is maximized performance over many applications. Specialized processors have customized architectures designed to achieve some required performance level in their applications. These processors have specific performance targets which may or may not be the maximum performance of the technology. DSPs, as a type of specialized processor, have customized architectures to achieve high performance in signal processing applications. Traditionally, DSPs are designed to maximize performance for inner loops containing product of sums (multiply and accumulate instructions called MACs). New commercial standards are emerging that need more than simple MACs (e.g., conditionals within inner loops).

Time to Market

GPPs meet time-to-market windows by designing the hardware with existing software used as benchmarks. Hardware development is more loosely coupled to the software development. Specialized processors meet time-to-market pressures by designing or codesigning the hardware with specific software applications in a tightly (temporal, functional) fashion. DSPs follow a general-purpose model for DSP core development. DSPs also support a specialized time-to-market model using an embedded core strategy (an example is configurable DSPs). ASICs can also be used to design specialized DSPs.

Execution Predictability

GPPs are designed to improve average performance. Predictability is a minor concern with GPPs. Many modern GPPs have complex architectures that make execution time predictability difficult (e.g., superscalar architectures dynamically select instructions for parallel execution). These processors often lack proper tool support to allow programmers to accurately predict execution time. This makes it difficult to predict execution timing of GPP code.

Specialized processors with hard time constraints such as DSPs, require designs to meet worst-case scenarios. There is little advantage in improving the average performance. Predictability is very important so that time responses can be calculated and predicted accurately. DSP processors have relatively straightforward architectures and are supported by development and analysis tools that help the programmer accurately determine execution time of applications.

The memory systems of DSPs are also designed to improve execution time predictability by limiting the use of cache. Most GPPs use cache systems to boost performance. Processor cache memory is a small amount of high-speed static RAM (SRAM) that can significantly improve CPU performance. The cache resides between the CPU and the main system memory. There are two levels of processor cache: "primary" and "secondary." Primary cache, also known as Level 1 or L1, is that cache memory built into the processor itself. Secondary cache, or Level 2 (L2), is external from the CPU. On-chip caches consume a significant portion of the silicon area of the chip, which increases the chip size as well as the cost.

If the instructions the processor executes reside in the cache, execution is much faster. If the instruction and data are not in the cache, the processor must wait while the code and data are loaded into the caches. This model of execution is probabilistic. In other words, it is not totally deterministic and predictable. This may be fine for many general computing applications, but for many real-time applications, this model of behavior can cause problems. Since many real-time applications have hard real-time constraints that must always be met, having a deterministic execution time is a requirement.

Branch prediction can also cause potential execution time predictability problems. Modern CPUs have deep pipelines that are used to exploit the parallelism in the instruction stream. This has historically been one of the most effective ways for improving performance of GPPs as well as DSPs.

Branch instructions can cause problems with pipelined machines. A branch instruction is the implementation of an if-then-else construct. If a condition is true, then jump to some other location; if false, then continue with the next instruction. This conditional forces a break in the flow of instructions through the pipeline and introduces irregularities instead of the natural, steady progression. The processor does not know which instruction comes next until it has finished executing the branch instruction. This behavior can force the processor to stall while the target location to jump to is resolved. The deeper the execution pipeline of the processor, the longer the processor will have to wait until it knows which instruction to feed next into the pipeline. This is one of the largest limiting factors in microprocessor instruction execution throughput and execution time predictability. These effects are referred to as "branch effects." These branch effects are probably the single biggest performance inhibitor of modern processors.

Dynamic instruction scheduling can also lead to execution time predictability problems. Dynamic instruction scheduling means that the processor dynamically selects sequential instructions for parallel execution, depending on the available execution units and on dependencies between instructions. Instructions may be issued out of order when dependencies allow for it. Superscalar processor architectures use dynamic instruction scheduling. Many (although not all) DSPs on the market today are not supserscalar for this reason — they need to have high execution predictability.

Power

GPPs exist in three typical environments: mainframe computers, desktop computers, and notebook computers. In these applications, the system environment is designed to the processor. Desktop computers, for example, can use large enclosures and powerful fans to cool a power-hungry processor. Specialized processors such as DSPs are often embedded in the environment. Many such environments have poor heat-dissipation characteristics (e.g., the DSP processor that performs speech processing inside a doll). Long battery life is also an important requirement. DSPs are designed to meet these important system requirements.

Cost

Specialized processors and DSPs have functionality at a cost point that drives the market. Applications are dominated by relatively low-price consumer and commercial markets. GPPs attempt to maximize performance at a cost point that drives the market. Applications are dominated by relatively high-price consumer and commercial markets.

Safety

In GPPs, reliability is the main concern. Consumers should be able to turn their desktop computers on and off several times a day without the processor failing. Specialized processors often include additional features for self-checking, fault checking, radiation hardness, error detection and correction, and electromagnetic interference. DSPs, for example, focus on applications containing error detection and correction.

Run Time Kernels and Operating Systems

One of the key elements driving DSP solutions to higher and higher levels of performance has been the evolution of real-time operating systems (RTOSs). DSP RTOs have evolved to the point that developing code for multiprocessor DSP applications is a simple extension to just programming a single processor. It is now becoming advantageous to purchase a commercial off-the-shelf (COTS) RTOS instead of developing an operating system in house. Real-time operating systems are now being built specifically for DSPs. The main features of these operating systems include:

- Preemptive priority-based real-time multitasking
- Deterministic critical times
- Time-out parameters on blocking primitives
- Memory management
- Synchronization mechanisms
- Interprocess communication mechanisms
- Special memory allocation for DSP (on-chip)

- Low interrupt latency
- Asynchronous, device-independent, low-overhead I/O

Tools

DSP processors, like many of the GPPs, come with a standard set of tools provided by the chip manufacturer. Most DSP vendors supply enhanced tool suites which are, generally, the standard code generation tool suite with an interactive graphical user interface (GUI) wrapped around them and a set of advanced analysis and profiling tools. Software and hardware development tools are necessary for rapid application development and overall time to market. Real-time applications require an enhanced set of tools to help the developer analyze real-time constraints, improve the efficiency of software, and make power, memory size, and performance trade-offs.

There are other tools that can be useful for developing DSP-based systems including simulators and emulators. Software simulators are available for many common DSPs. These tools let the engineer begin development and integration of software without the DSP and associated hardware being available. Simulators are more common in DSP applications because the algorithms typically run on a DSP are complex and mathematically oriented. This leads to many areas in which errors in design and implementation can be made. Simulators also allow the engineer to examine the device operation easily and without having to buy the device ahead of time.

Another very useful tool for DSP developers is the emulator. The purpose of an emulator is to provide the engineer access to the DSP(s) and its peripherals in a nonintrusive way to aid in debugging operations and hardware/software integration. Emulators allow engineers easy access to hardware registers and memory, allowing reading and writing to these locations. Other common functions such as breakpoints, single stepping, and benchmarking are also supported. Most emulators are nonintrusive both spatially and temporally. Spatially nonintrusive means the emulator does not require any additional hardware or software in the target environment. Temporally nonintrusive means the emulator does not prevent the processor or system from executing at full speed. These two requirements are very important when performing hardware/software integration.

Single-Cycle Multiply

DSPs were designed to process certain mathematical operations very quickly. Consider the filtering operation discussed earlier. There are actually two popular forms of filtering: finite impulse response (FIR) filters and infinite impulse response (IIR) filters (the difference in these two filters will be discussed later). The FIR filter contains a series of delay components. These delay components produce a copy of the input sample which has been delayed by one or more sample periods. These delay components are stored in some type of storage element such as memory.

To perform this type of operation the DSP will be performing many multiply operations. DSP vendors recognized this bottleneck and have added specialized hardware to compute a multiply operation in only one cycle. This dramatically improves the throughput for these types of algorithms.

Large Accumulator

Another distinguishing feature of a DSP is its large accumulator register or registers. The accumulator is used to hold the summation of several multiplication operations without overflowing. The accumulator is several bits wider than a normal processor register, which can help avoid overflow during the accumulation of many operations.

Multiply-Accumulate Instructions

DSP processors very often contain specialized instructions that take advantage of the fact that many DSP algorithms require both a multiply (which has been optimized) and an accumulate (which is handled with a larger register) to complete its operation. Specialized instructions called multiply and accumulate (MAC) were developed to perform these operations very quickly.

Characteristics of DSP Applications

As an example of a typical DSP application, consider the application of signal filtering. Signal filtering involves the manipulating of a signal in order to improve some characteristic of that signal. One goal may be to remove noise from the signal. Another goal may be to retain one small subset of signal frequencies and throw away (attenuate) the rest.

Problems with Analog Circuitry

Signal filtering used to be performed using analog circuitry. A mixture of capacitors, inductors, and resistors were used to "tune" the circuit to the desired characteristic. This approach is not desirable for the following reasons:

- Environmental factors such as temperature cause the characteristics of the circuit to change. The characteristics of the capacitor/inductor/resistor filter could change over time as the temperature of the environment warmed and cooled.

TABLE 3.1

Analog vs. Digital Filter Implementation Considerations

Analog Filter	Digital Filters
Behavior varies depending on temperature	Immune to temperature variations
Difficult to obtain very tight tolerances due to the combination of several components	Programmable to very tight tolerances
Difficult to change because filter is designed in hardware	Programmable and very easy to change

- Analog filters had to be designed to a specific characteristic and could not change unless the circuit was redesigned. This is expensive if a filter circuit has to be redesigned with new hardware components each time the filter characteristics changed.
- Analog filters, because of their very nature, could not be designed to very tight tolerances. This leads to filters that may not be able to attain the optimal signal to noise improvement required for some applications.

Digital Alternatives

DSP devices, on the other hand, are fully programmable, which eliminates many of the barriers seen in analog filter design. Table 3.1 contrasts the two approaches to filter design.

Common DSP Characteristics

Although each DSP application is different, they all share some common characteristics:

- Use of many mathematical operations, especially multiplies and adds. Because of this characteristic, using a DSP (or other processor) that performs these operations quickly will dramatically improve the response time of a signal processing based application.
- They must interact with signals from the real world. Therefore, the system must accept information from the real world in a timely manner and respond back to the real world in a timely manner.
- They must complete processing of a signal sample in a certain amount of time. Signals from the real world must be sampled at a rate that varies on the characteristics of the signal. The faster the sampling rate of the signal, the less time there is to complete processing of that sample before the next sample arrives.

There are many DSP processors on the market today. Each processor advertises its own unique functionality for improving DSP processing. However,

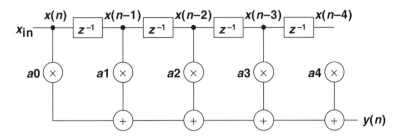

FIGURE 3.11
Block diagram of an FIR filter. The y term is the output sample, the x terms are the input samples, and the a terms are the filter weights.

as different as these processors may sound from each other, there are a few fundamental characteristics that all DSP processors share. In particular, all DSP processors must:

- Perform high-speed arithmetic
- Make multiple accesses to and from memory in a single cycle
- Be able to process data to and from the real world

Many DSP algorithms require a few specific operations to perform a majority of the computation:

- Addition
- Multiplication
- Delay
- Array indexing

DSP Example

As an example of these operations, consider the FIR filter shown in Figure 3.11. In this representation,

$$y(n) = a0 \times x(n) + a1 \times x(n-1) + a2 \times x(n-2) + a3 \times x(n-3) + a4 \times x(n-4)$$

This filter structure will produce a result that is a weighted average of the past and present inputs. For each output sample, $y(n)$, the present input, $x(n)$ and previous inputs, $x(n-2)$, $x(n-3)$, and $x(n-4)$ are all multiplied by the corresponding weights, or coefficients ($a0$, $a1$, $a3$, and $a4$), and these products are added together to produce one output sample. The FIR filter shown in Figure 3.11 requires the following mathematically based operations:

- Addition: This operation requires
  - Fetching two operands from memory

- Performing the addition on those operands
- Storing the result in memory for later use
- Multiplication: This operation requires
  - Fetching two operands from memory
  - Performing the multiplication
  - Storing the result in memory for later use
- Delay: This operation requires
  - Holding a value for later use (usually in memory)
- Array indexing: This operation requires
  - Fetching values from consecutive memory locations
  - Copying data to and from different memory locations

A DSP is designed to enable these operations to execute extremely quickly. Since many DSP algorithms are based on these fundamental concepts, if they can be made to execute quickly, the performance of the application in general will improve. The characteristics of many DSPs that optimize these operations include:

- Ability to perform a multiply and an add in parallel (at the same time). In many DSPs, there is a dedicated hardware adder and hardware multiplier. The adder and multiplier are often designed to operate in parallel so that a multiply and an add can be executed in the same clock cycle. Special MAC instructions have been designed into DSPs for this purpose. Figure 3.12 shows the high-level architecture of the multiply unit and adder unit operating in parallel.

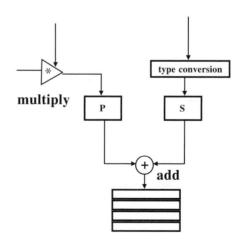

FIGURE 3.12
Adder and multiplier operating in parallel.

- Multiple accesses to memory (in the FIR example, this would include fetching two operands from memory and storing the result).
- Efficient address generation algorithms for processing arrays of data. DSPs must be able to process arrays of data quickly so generating the addresses quickly is an important feature for DSPs. Some of the on-chip DSP registers are address registers used for generating the array addresses.
- Data registers that hold data temporarily and have fast access times. Many DSP processors include a bank of internal memory with fast, single-cycle, access times. This memory is used for storing temporary values for algorithms such as the FIR filter. The single-cycle access time allows for the multiply and accumulate to complete in one cycle. The register file residing on-chip is used for storing important program information and some of the most important variables for the application. For example, the delay operation required for FIR filtering implies information must be kept around for later use. It is inefficient to store these values in external memory, because a penalty is incurred for each read or write from external memory. Therefore, these delay values are stored in the DSP registers or in the internal banks of memory for later recall by the application without incurring the wait time to access the data from external memory. The on-chip bank or memory can also be used for program and/or data and also has fast access time. External memory is the slowest to access, requiring several wait states from the DSP before the data arrive for processing. External memory is used to hold the results of processing.
- Customized features that allow for special addressing modes, delay instructions, and block data transfers.

DSP-related devices require sophisticated memory architectures. The main reason is the memory bandwidth requirement for many DSP applications. It becomes imperative in these applications to keep the processor core fed with data. A single memory interface is not good enough. For the case of a simple FIR filter, each filter tap requires up to four memory accesses:

- Fetch the instruction.
- Read the sample data value.
- Read the corresponding coefficient value.
- Write the sample value to the next memory location (shift the data).

If this is implemented using a DSP MAC instruction, four accesses to memory are required for each instruction. Using a serial von Neumann architecture, these four accesses to memory slow down the processor core to the point of causing it to wait for data. In this case, the processor is

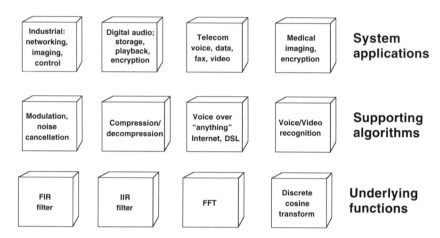

FIGURE 3.13
DSP algorithm building blocks.

"starved" for data and performance suffers. In order to overcome this performance bottleneck, DSPs have introduced a higher performance memory interface architecture.

DSP Building Blocks

Modern DSP applications are composed of very complex signal processing algorithms. These algorithms are composed of building blocks (Figure 3.13). At the heart of all of these DSP building blocks are two basic functions; the multiply and the add. These two operations are required regardless of whether the algorithm is an FFT, a filter, a correlation, or a convolution.[9] These operations can be expensive to perform (from a cycle time perspective). DSPs are designed specifically to make these operations very fast and efficient. At the heart of any DSP on the market today is the fast and efficient execution of the multiply and add instructions.

Most DSP algorithms consist of a "sum of products" type of equation. At the core of the equation is the need to multiply two numbers together. A multiplier with two inputs would serve this requirement well. Next, consider the summation of each of these products. Ideally, there should exist an adder to sum each product to create the result to store back to memory.

At the core of the DSP CPU is the MAC unit, which gets its name from the fact that the unit consists of a multiplier that feeds an accumulator. To most efficiently feed the multiplier, a "data" bus is used for the input sample array (x terms), and a "coefficient" bus is used for the filter weight array (a terms). This simultaneous use of two data buses is a fundamental characteristic of a DSP system. If the coefficients were constants, the user might want to store them in ROM. This is a common place to hold the program, and allows the user to perform the MAC operations using coefficients stored in program memory and accessed via the program bus. Results of the MAC

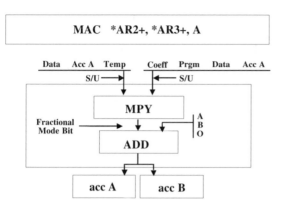

FIGURE 3.14
MAC functionality in a DSP. The MAC loads two operands (AR2 and AR3), multiplies them, and accumulates them in accumulator A (acc A).

operation are built up in either of two accumulator registers, as shown in Figure 3.14. Both of these registers can be used equally. This allows the program to maintain two chains of calculations instead of one. The MAC unit is invoked with the MAC instruction. An example of a MAC instruction is shown in Figure 3.14. The MAC instruction MAC *AR2+, *AR3+, A performs many operations in a single cycle: reading two operands from memory, incrementing two pointers, multiplying the two operands together, and accumulating this product with the running total.

There are other, more general-purpose and flexible, instructions on DSPs that allow more standard mathematical equation calculations and control loops that are not pure DSP-like. DSPs are also designed to process integer or fractional numbers efficiently as well as signed and unsigned numbers. This allows the DSP designer to adapt to a number of different environments without loss of performance.

Most DSP systems need to perform a lot of general purpose arithmetic (simple process of additions and subtractions). For this type of process, a separate arithmetic logic unit (ALU) is often added to a DSP. The results of the ALU can be basic mathematical or Boolean functions. These results are sent to either of the accumulators shown in Figure 3.14. The ALU performs standard bit manipulation and binary operations. The ALU also has an adder. The adder ensures efficient implementation of simple mathematical operations without interfering with the specialized MAC unit.

Conclusions

DSP architectures are designed to process complex signal processing algorithms efficiently, including those used to perform advanced noise reduction processing. Although much of the technology required for noise reduction has existed for some time, only with the recent introduction of powerful, but inexpensive, DSP hardware has the technology become practical. These specialized DSPs were designed for real-time numerical processing of digitized signals. These devices have enabled the low-cost implementation of powerful noise reduction algorithms and encouraged the widespread development of noise reduction systems. Filtering and transform algorithms required to perform various types of noise reduction map well to the specialized architectures of modern DSPs. Noise reduction systems that use advanced signal processing implemented on a low-cost, high-performance DSP are an emerging new technology.

References

1. Levitt, H., Digital hearing aids: a tutorial review, *J. Rehab. Res. Dev.*, 24(4), 7–20, 1992.
2. http://www.autoweb.lycos.com/garage/encyclop/ency19c.htm
3. Design of Active Noise Control Systems with the TMS320 DSP Family, Texas Instruments Application Report, Dallas, TX, 1996.
4. DSP 599zx Audio Noise Reduction Filter, Timewave Technology, Inc., Dallas, TX, 1996.
5. Giddings, P., Noise Reduction Systems, February 22, 1990, Engineering Harmonics, Inc. http://www.engineeringharmonics.com/papers/nrs.htm
6. http://www.dpreview.com/learn/Glossary/Digital_Imaging/Noise_Reduction_01.htm, Digital Photography Review, dpreview.com
7. Wolf, A., VLIW architecture emerges as embedded alternative, *Embedded Systems Programming*, February 2001.
8. Bier, J. et al., *DSP Processor Fundamentals, Architectures and Features*, BDTI, Inc., Dallas, TX, 1995.
9. Ciufo, C., DSP building blocks to handle high sample rate sensor data, *COTS J.*, March/April 2000.

4

Software Design Considerations for Real-Time DSP Systems

Elizabeth G. Keate

CONTENTS

0-8493-0949-2/02/$0.00+$1.50
© 2002 by CRC Press LLC

Understanding Real-Time DSP Embedded Systems

There are many categories of embedded systems from communication devices to home appliances to control systems. The list of applications includes car navigation, compact disk (CD) players, digital cameras, robotics, cellular phones, and base stations. These systems all contain some type of noise reduction algorithm. These systems, however, have certain design constraints that make moving from conceptual to practical application difficult. Chapter 3 discussed the hardware characteristics of digital signal processing (DSP). This chapter focuses on the software design methodologies and paradigms that are required to design these types of real-time applications.

Defining Features of Embedded Systems

Embedded systems are defined by a unique set of characteristics. Each characteristic imposes a specific set of design constraints on embedded systems designers. The challenge to designing embedded systems is to conform to the specific set of constraints for the application.

Application Specific Systems

Embedded systems are not general-purpose computers. Embedded system designs are optimized for a specific application. Many of the job characteristics are known before the hardware is designed. This allows the designer to focus on the specific design constraints of a properly defined application.

As such, there is limited user reprogrammability. Some embedded systems, however, require the flexibility of reprogrammability. Programmable DSPs are common for such applications.

Reactive Systems

A typical embedded systems model responds to the environment via sensors and controls the environment using actuators. This requires embedded systems to run at the speed of the environment. This characteristic of embedded systems is called "reactive." Reactive computation means that the system (primarily the software component) executes in response to external events. External events can be either periodic or aperiodic.* The maximum event arrival rate must be estimated in order to accommodate worst-case situations. Most embedded systems have a significant reactive component. One of the biggest challenges for embedded system designers is performing an accurate worst-case design analysis on systems with statistical performance characteristics (e.g., cache memory on a DSP or other embedded processor). Real-time system operation means that the correctness of a computation depends, in part, on the time at which it is delivered. Systems with real-time requirements must often be designed to worst-case performance to ensure that time lines are met.

Distributed Systems

A common characteristic of an embedded system is one that consists of communicating processes executing on several central processing units (CPUs) or application-specific integrated circuits (ASICs), which are connected by communication links. In this approach, multiple processors are usually required to handle multiple time-critical tasks. Devices under control of embedded systems may also be physically distributed.

Embedded Systems Examples

Figure 4.1 shows the system block diagram of a digital cell phone. This embedded system contains an analog base band section (acting as the sensor component of a typical embedded system), as well as the DSP-based processing unit and an antenna that acts as the actuator component in the embedded system.

---

* Periodic events are events that happen or appear at regular intervals such as the sampling of a signal by an analog-to-digital converter (ADC). Aperiodic events are events that occur without periodicity, irregular events include interrupts from a threshold detector.

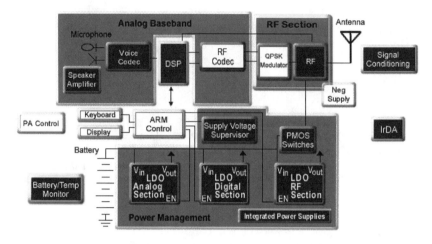

FIGURE 4.1
An embedded system example — digital cell phone.

Overview of Real-Time Systems

A real-time system is a system that is required to react to stimuli from the environment (including the passage of physical time) within time intervals dictated by the environment. The lag from input time to output time must be sufficiently small for acceptable timeliness. Another way of thinking of real-time systems is any information processing activity or system that has to respond to externally generated input stimuli within a finite and specified period. Generally, real-time systems are systems that maintain a *continuous timely* interaction with their environment such as the previous cell phone example.

There are two types of real-time systems: reactive and embedded. A reactive real-time system involves a system that has constant interaction with its environment (e.g., a pilot controlling an aircraft). An embedded real-time system is used to control specialized hardware that is installed within a larger system (e.g., a microprocessor that controls the fuel-to-air mixture for automobiles).[1]

Real-Time Event Characteristics

Real-time events fall into one of the three categories: asynchronous, synchronous, or isochronous.

- *Asynchronous events* are entirely unpredictable; for example, the event that a user makes a telephone call. As far as the telephone company is concerned, the action of making a phone call cannot be predicted.

- *Synchronous events* are predictable and occur with precise regularity if they are to occur. For example, the audio and video in a movie take place in synchronous fashion.

- *Isochronous events* occur with regularity within a given window of time. For example, audio bytes in a distributed multimedia application must appear within a window of time when the corresponding video stream arrives. Isochronous events are a subclass of asynchronous events.

Real-time systems differ from time-shared systems in several ways:

- Predictably fast response to urgent events
- High degree of schedulability; timing requirements of the system must be satisfied at high degrees of resource usage
- Stability under transient overload; when the system is overloaded by events and it is impossible to meet all deadlines, the deadlines of selected critical tasks must still be guaranteed

Characteristics of Real-Time Systems

Real-time systems have many special characteristics, which are inherent or imposed. This section discusses some of these important characteristics.

Reliable and Safe

The more society relinquishes control of its vital functions to computers, the more it becomes imperative that those computers do not fail. Failure in automated teller machines (ATMs) can result in the loss of millions of dollars. A faulty component in electricity generation could fail a life support system in a hospital intensive care unit.

Real-Time Facilities

Response time is crucial to any embedded system. It is very difficult to design and implement systems that will guarantee that the appropriate output will be generated at the appropriate times *under all possible conditions*. Accomplishing this and making use of all computing resources at all times is often impossible. Given adequate processing power, a good real-time programming language and run-time support are required to enable the programmer to:

- Specify times at which actions are to be performed
- Specify times at which actions are to be completed
- Respond to situations where *all* timing requirements cannot be met
- Respond to situations where the timing requirements are changed dynamically

Efficient Execution and the Execution Environment

Real-time systems are time critical. Therefore, the efficiency of their implementation is more important than in other systems. One of the main benefits of using a higher-level language is to allow the programmer to abstract away the details and concentrate on solving the problem. This is not always true in the embedded system world. Some higher-level languages have instruction ten times slower than assembly language. However, higher-level languages can be used in real-time systems effectively. A system operates in real time as long as its actions, which have time constraints, are performed with acceptable timeliness. *Acceptable timeliness* is defined as part of the behavioral requirements for the system. These requirements should be objectively quantifiable and measurable.

Hard Real-Time and Soft Real-Time Systems

An activity (typically, a task) is considered hard real-time if and only if it has a hard deadline for the completion of an action (typically, the execution of the whole task). This deadline must always be met; otherwise, the task has failed. The same machine having one or more hard real-time tasks may also execute other (i.e., soft real-time or nonreal-time) tasks if and when they do not interfere with any hard real-time ones. A failure to meet a deadline in a hard real-time system means the system does not deliver its output in time for its critical tasks and/or deterministic tasks (e.g., flight control laws, collision alert tasks, etc). Hard real-time systems are commonly embedded systems. In all hard real-time systems, collective timeliness is deterministic. This determinism does not imply that the actual individual task completion times or the task execution ordering are necessarily known in advance.

The feasibility and costs (e.g., in terms of system resources) of hard real-time computing depend on how well known *a priori* are the relevant future behavioral characteristics of the tasks and execution environment. These task characteristics include:

- Timeliness parameters, such as arrival periods or upper bounds
- Deadlines
- Worst-case execution times
- Ready and suspension times

- Resource utilization profiles
- Precedence and exclusion constraints

Deterministic collective task timeliness in hard (and soft) real-time computing requires that the future characteristics of the relevant tasks and execution environment be deterministic, that is, known absolutely in advance. The knowledge of these characteristics must then be used to pre-allocate resources so all deadlines will always be met. In many real-time computing applications it is common that the primary factor is dispatched on-line according to that schedule. For certain hard real-time task and environment characteristic cases, task execution eligibility indices, usually called priorities, can be assigned either off-line by application programmers or on-line by application or operating system software. For most cases of real-time systems, task and future execution environment characteristics are difficult to predict. This makes true hard real-time scheduling infeasible. In hard real-time computing, deterministic satisfaction of the collective timeliness criterion is the driving requirement. The necessary approach to meeting that requirement is static scheduling of deterministic task and execution environment characteristic cases. The requirement for advance knowledge about each of the system tasks and their future execution environment to enable off-line scheduling and resource allocation significantly restricts the applicability of hard real-time computing. Soft real-time computing is conventionally not defined except by default as "not hard real-time." It violates either, or both, of hard real-time's axioms:

- Missing some deadlines, by some amount, under some circumstances
- Not completing or even attempting the least eligible actions at all

In soft real-time systems, these violations may be acceptable rather than failures. In these systems, some tasks may have action time constraints that are multivalued (best, better, worse, and worst completion times — instead of binary-valued deadlines). In a soft real-time system, there may be soft deadlines in that they may not always have to be met. For transient or steady-state resource (e.g., computational) overload conditions, which may be routine in soft real-time computing systems, the actions whose execution priority is lowest may not be performed, completed, or even started at all. Examples of soft real-time systems include video conferencing, stock price quotation, and airline reservation systems.[2]

Introduction to Real-Time Operating Systems

The framework of real-time applications is based on fundamental infrastructure — typically an operating system. An operating system is a computer program that is initially loaded into a processor by a boot program. It then manages all the other programs in the processor. The other programs are called applications or tasks. The applications make use of the operating system by requesting services through a defined application program interface (API).

A task is a basic unit of programming that an operating system controls. Each operating system may define a task slightly differently. Fundamentally, a task also referred to as a thread, is a unit of programming that may be an entire program or each successive invocation of a program. Programs may make requests of other utility programs, and these utility programs may also be considered tasks. Many of today's widely used operating systems support multitasking. This allows multiple tasks to run concurrently. Each task takes turns using the resources of the computer.

An operating system performs these services for applications:

- In a multitasking operating system where multiple programs can be running at the same time, the operating system determines which applications should run in what order and how much time should be allowed for each application before giving another application a turn.

- It manages the sharing of internal memory among multiple applications.

- It handles input and output to and from attached hardware devices, such as hard disks, printers, and dial-up ports.

- It sends messages to each application or interactive user about the status of operation and any errors that may have occurred.

- It can offload the management of what are called batch jobs (e.g., printing) so that the initiating application is freed from this work.

- On computers that can provide parallel processing, an operating system can manage how to divide the program so that it runs on more than one processor at a time.

What Makes an Operating System a Real-Time Operating System?

An operating system (OS) must have certain properties to qualify it as a real-time operating system (RTOS). Most importantly, an RTOS must be multitasking and preemptable. The RTOS must also support task priorities. Because of the predictability and determinism requirements of real-time systems, the RTOS must support predictable task synchronization mechanisms. A system of priority inheritance must exist to limit any priority inversion conditions. Finally, the RTOS behavior should be known to allow the application developer to accurately predict performance of the system.*

An RTOS is a type of OS that guarantees a certain capability within a specified time constraint. If a certain calculation, for example, could not be performed for making a task available at a designated time, the OS would terminate with a failure. Some RTOSs are created for special applications such as DSP or even a cell phone. Others are more general-purpose operating systems. In general, RTOSs are said to require:

- Multitasking
- Process tasks that can be prioritized
- A sufficient number of interrupt levels

The purpose of an RTOS is to manage and arbitrate access to global resources such as the CPU, memory, and peripherals. The RTOS scheduler manages MIPS and real-time aspects of the processor. The memory manager allocates, frees, and protects code and memory. The RTOS drivers manage I/O devices, timers, and direct memory access units (DMAs).

Reduced functionality RTOSs, kernels, are often required in small embedded systems that are packaged as part of microdevices. Some kernels can be considered to meet the requirements of an RTOS. More than a general-purpose operating system, an RTOS should be modular and extensible. In embedded systems, the RTOS must be small because it is often in ROM and RAM space may be limited. Some systems are safety critical and require certification, including the operating system. This is why many RTOSs consist of a kernel that provides only essential services:

- Scheduling
- Synchronization
- Interrupt handling
- Multitasking

RTOS for DSP

An RTOS for DSP is somewhat specialized in itself. A typical embedded DSP application will consist of two general components: the application software and the system software. The operating system is part of the system software layer (Figure 4.2). The function of the system software is to manage the

---

* An example is the interrupt latency (i.e., time from interrupt to task run). This has to be compatible with the application requirements and has to be predictable. This value depends on the number of simultaneous pending interrupts. For every system call, this value would be the maximum time to process the interrupt. The interrupt latency should be predictable and independent from the number of objects in the system.

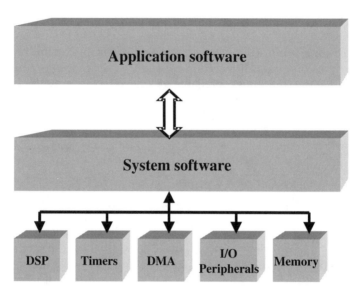

FIGURE 4.2
Embedded DSP software components.

resources for the application. Examples of system resources that must be managed are peripherals like direct memory access (DMA) units, host port interface (HPI), or on-chip memory (DSP have internal [on-chip] memory that is much faster then memory located externally). The DSP is a processing resource to be managed and scheduled like other resources.

The system software provides the infrastructure and hardware abstraction for the application software. As application complexity grows, a real-time kernel can simplify the task of managing the DSP MIPS efficiently using a multitasking design model. The developer also has access to a standard set of interfaces for performing I/O as well as handling hardware interrupts. A DSP RTOS also provides the capability to define and configure system memory efficiently. The overall structure that the multitasking design paradigm adds to the application makes it easier to scale and maintain larger applications. DSP RTOSs have very low interrupt latency. Because many DSP systems interface with the external environment, they are event or interrupt driven. Low overhead in handling interrupts is very important for DSP systems. For many of the same reasons, DSP RTOSs also ensure that the amount of time interrupts are disabled is as short as possible. When interrupts are disabled (e.g., context switching), the DSP cannot respond to the environment.

A DSP RTOS also has very high performance device independent I/O. This involves basic I/O capability for interacting with devices and other tasks. This I/O should also be asynchronous, have low overhead, and be deterministic in the sense that the completion time for an I/O transfer should not be dependent on the data size.

A DSP RTOS must also have specialized memory management. Capability to align memory allocations and multiple heaps* with very low space overhead is important. The RTOS will also have the capability to interface to the different types of memory that may be found in a DSP system, including SRAM, SDRAM, and fast on-chip memory.

Adopting a New Design Paradigm

In the early days of DSP, much of the software written was low-level assembly language that ran in a loop performing a relatively small set of functions. There are several potential problems with this approach:

- The algorithms could be running at different rates. This makes scheduling the system difficult using a "polling" approach (a technique for periodically checking on the status of an event — this is noninterrupt driven).
- Some algorithms could overshadow other algorithms, effectively starving them. With no resource management, some algorithms could never run.
- There are no guarantees of meeting real-time deadlines. Polling in the fashion described above is nondeterministic. The time it takes to go through the loop may be different each time, because the demands may change dynamically.
- Nondeterministic timing.
- No, or difficult, interrupt preemption.
- Unmanaged interrupt context switching.
- "Super loop" approach is a poor overall design approach.**

As application complexity has grown, the DSP is now required to perform very complex concurrent processing tasks at various rates. A simple polling loop to respond to these rates has become obsolete. Modern DSP applications must respond quickly to many external events, be able to prioritize processing, and perform many tasks at once. These complex applications are also changing rapidly over time, responding to ever-changing market conditions. Time to market has become more important than ever. DSP developers, like

---

* A heap is a block of memory allocated for program use, usually in conjunction with memory allocation.

** Super-loop is defined as several functions grouped in a single thread. For example, a robotic arm could have functions for moving the shoulder, the arm, and the wrist, which would in general be separate events, but implemented as a super-loop, these would be grouped in one main program loop.

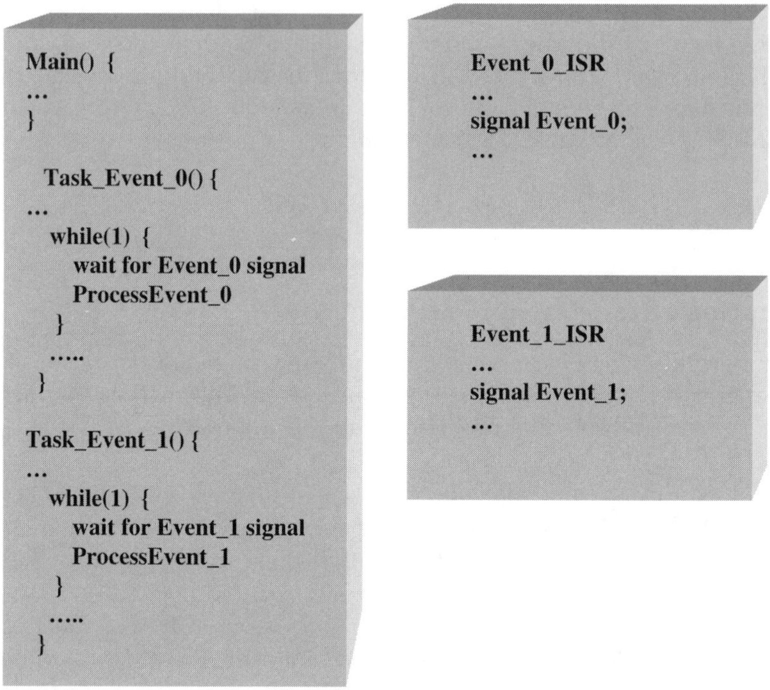

```
Main() {
...
}

   Task_Event_0() {
...
   while(1) {
       wait for Event_0 signal
       ProcessEvent_0
   }
   .....
   }

Task_Event_1() {
...
   while(1) {
       wait for Event_1 signal
       ProcessEvent_1
   }
   .....
   }
```

```
Event_0_ISR
...
signal Event_0;
...
```

```
Event_1_ISR
...
signal Event_1;
...
```

FIGURE 4.3
Multitasking application code to respond to external events.

many other software developers, must now be able to develop applications
that are maintainable, portable, reusable, and scalable.

Modern DSP systems are managed by RTOSs that manage multiple tasks,
service events from the environment based on an interrupt structure, and
effectively manage the system resource, as illustrated in Figure 4.3.

Concepts of an RTOS

An RTOS requires functionality to effectively perform its functions, that is,
to be able to execute all of its tasks without violating specified timing constraints. This section describes the major functions that RTOSs perform.

Task Based

A task implements a computation job and is the basic unit of work handled
by the scheduler. The kernel creates the task, allocates memory space to the
task, and brings the code to be executed by the task into memory. A structure

called a task control block (TCB) is created and used to manage the schedule of the task. A task is placeholder information associated with a single use of a program that can handle multiple concurrent users. From the program's point of view, a task is the information needed to serve one individual user or a particular service request.

Multitasking

Preemptive multitasking is a condition in which an operating system uses some criteria to decide how long to allocate to any one task before giving another task a turn to use the operating system. The act of taking control of the operating system from one task and giving it to another task is called preempting. A common criterion for preempting is simply elapsed time (this kind of system is sometimes called time sharing or time slicing). In some OSs, some applications are given higher priority than other applications, giving the higher priority programs control as soon as they are initiated and perhaps longer time slices.

In preemptive multitasking, each task is assigned a priority depending on its relative importance, the amount of resources it is consuming, and other factors. The OS preempts tasks with a lower priority value so that a higher priority task is given a chance to run. Cooperative multitasking is the ability of an OS to manage multiple tasks such as application programs at the same time but without the ability to necessarily preempt them. The processor operates at speeds that make it seem as though all of the user's tasks are being performed at the same time.

Multitasking is often a confusing topic for those who have not developed a multitasking system. Today's microprocessors can only execute one program instruction at a time. But because they operate so fast, they appear to run many programs and serve many users simultaneously. Each of the programs executed by the computer is viewed by the RTOS as a "task" for which certain resources are identified and kept track of in the application.

Rapid Response to Interrupts

An interrupt is a signal from a device attached to a computer or from a program within the computer that causes the RTOS to stop and figure out what to do next. Almost all DSPs and general-purpose processors are interrupt driven. The processor will begin executing a list of computer instructions in one program and keep executing the instructions until either the task is complete or cannot go any further (e.g., waiting on a system resource) or an interrupt signal is sensed. After the interrupt signal is sensed, the processor either resumes running the program it was running or begins running another program.

An RTOS has a code called an interrupt handler. The interrupt handler prioritizes the interrupts and saves them in a queue if more than one is

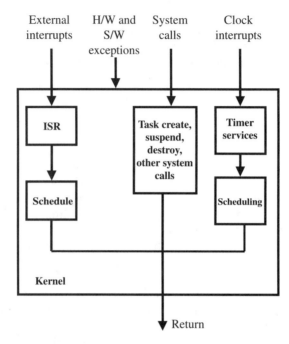

FIGURE 4.4
An RTOS must respond to and manage multiple interruptions from inside and outside the application.

waiting to be handled. The scheduler program in the RTOS then determines which program to give control to next.

In many ways, interrupts provide the "energy" for embedded real-time systems. The energy is consumed by the tasks executing in the system. Typically, in DSP systems, interrupts are generated on data buffer or event boundaries by the DMA or other equivalent hardware. In this way, every interrupt occurrence will make ready a DSP RTOS task that is iterated on full/empty data buffers. Interrupts come from many sources (Figure 4.4) and the DSP RTOS must effectively manage multiple interruptions from both inside and outside the DSP system.

RTOS Scheduling

The RTOS scheduling policy is one of the most important features of the RTOS to consider when assessing its usefulness in a real-time application. The scheduler decides which tasks are eligible to run at a given instant and which task should actually be granted the processor. The scheduler runs on the same CPU as the user tasks, and this is already the penalty in itself for using its services. There are a multitude of scheduling algorithms and scheduler characteristics, not all of which are important for a real-time system.

An RTOS for DSP requires a specific set of attributes to be effective. The task scheduling should be priority based. A task scheduler for DSP RTOS has multiple levels of interrupt priorities where the higher priority tasks run first. The task scheduler for DSP RTOS is also preemptive. If a higher priority task becomes ready to run, it will immediately preempt a lower priority running task. This is required for real-time applications. Finally, the DSP RTOS is event driven. The RTOS has the capability to respond to external events such as interrupts from the environment. DSP RTOS can also respond to internal events as required.

The major states of a typical RTOS include:

- *Sleeping:* The task is put into a sleeping state immediately after it is created and initialized. The task is released and leaves this state upon the occurrence of an event of the specified type(s). Upon the completion of a task that is to execute again, it is reinitialized and put in the sleeping state.
- *Ready:* The task enters the ready state after it is released or when it is preempted. A task in this state is in the ready queue and eligible for execution.
- *Executing:* A task is in the executing state when it executes.
- *Suspended (Blocked):* A task that has been released and is yet to complete enters the suspended or blocked state when its execution cannot proceed for some reason. The kernel puts a suspended task in the suspended queue.
- *Terminated:* A task that will not execute again enters the terminated state when it completes. A terminated task may be destroyed.

Different RTOSs will have slightly different states. Figure 4.5 shows the state model for the DSP RTOS, DSP/BIOS.[3]

The scheduler is a central part of the kernel. It executes periodically and whenever the state of a task changes. The system clock device raises (generates) interrupts periodically. This is called the clock interrupt. The period in which this interrupt is invoked is called the "tick" size. A common tick size is 10 ms. At each clock interrupt, the kernel processes the timer events. The clock device has a timer queue where pending expiration times of all timers connected to the clock are queued. The kernel uses this to determine which timer event happened. The RTOS carries out actions related to each timer event. At each clock interrupt the RTOS updates the execution budget.

The RTOS Kernel

An RTOS consists of a kernel that provides the basic OS functions. There are three reasons for the kernel to take control from the executing task and

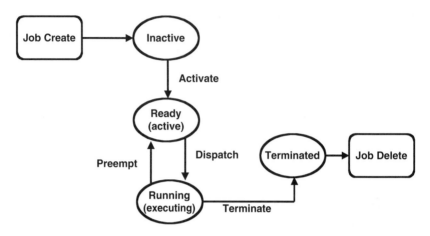

FIGURE 4.5
State model (with preemption) for the Texas Instruments DSP/BIOS RTOS.

execute itself: respond to a system call, perform scheduling and service timers, and handle external interrupts. Events are handled depending on their priority. The following lists the type of events and their relative priority levels in the system:

- Highest priority events are hardware interrupts. One hardware interrupt can interrupt another.

- Next highest priority events are software interrupts. The RTOS may also support a number of software interrupts. Software interrupts have a lower priority than hardware interrupts and are prioritized within themselves. A higher software interrupt can preempt a lower priority software interrupt. All hardware interrupts can preempt software interrupts. A software interrupt is similar to a hardware interrupt and is very useful for real-time multitasking applications. An internal RTOS clock module drives these interrupts. Software interrupts run to completion, do not block, and cannot be suspended.

- Following software interrupts in priority are tasks. Tasks can also be preempted by higher priority tasks. Tasks will run to completion or yield if blocked waiting for a resource or voluntarily to support certain scheduling algorithms such as "round robin" (a technique for scheduling tasks based on a period event to trigger a task switch; sometimes also referred to an time slicing).

System Calls

The application can access the kernel code and data via application programming interface (API) functions. An API is the specific method prescribed by a computer OS or by an application program by which a programmer writing an application program can make requests of the OS or another application. A *system call* is a call to one of the API functions. When this happens, the kernel saves the context of the calling task, switches from user mode to kernel mode (to ensure memory protection), executes the function on behalf of the calling task, and returns to user mode.

External Interrupts

Hardware interrupts provide effective means of notifying the application of the occurrence of external events. Interrupts are also used for sporadic I/O activities. The amount of time to service an interrupt varies based on the source of the interrupt. For example, handling DMA interrupts can take a significant amount of time. Interrupt handling in most processors, including DSPs, is divided into two steps: the immediate interrupt service and the scheduled interrupt service. The immediate interrupt service is executed at an interrupt priority level.

The total delay to service a DSP interrupt is the time the processor takes to complete the current instruction, do the necessary chores, jump to the interrupt handler, and interrupt the dispatcher part of the kernel. The kernel must then disable external interrupts. There may also be time required to complete the immediate service routines of higher priority interrupts, if any. The kernel must also save the context of the interrupted task, identify the interrupting device, and get the starting address of the interrupt service routine (ISR). The sum of this time is called "interrupt latency," and measures the responsiveness to external events via the interrupt mechanism. Many RTOSs provide the application the ability to control when an interrupt is enabled again. The DSP can then control the rate at which external interrupts are serviced. The flow diagram to service a nonmaskable interrupt in the TMS320C55 DSP is shown in Figure 4.6.

The second part of interrupt handling in a DSP is called the scheduled interrupt service. This is another service routine invoked to complete interrupt handling. This part is the scheduled interrupt handling routine and is typically preemptable (unless interrupts are specifically turned off by the DSP during this time).

There are various reasons for a multitasking DSP to suspend. A task may be blocked due to resource access control. A task may be waiting to synchronize execution with some other task. The task may be held waiting for some reason (I/O completion and jitter control). There may be no budget or job to execute (this is a form of bandwidth control). The RTOS maintains different queues for tasks suspended or blocked for different reasons (e.g., a queue for tasks waiting for each resource). The RTOS may also keep a number of task-ready queues. In fixed priority scheduling, there will be a queue for each priority. Rather than simply admitting the tasks to the CPU, the RTOS scheduler makes a decision based on the task state and priority.

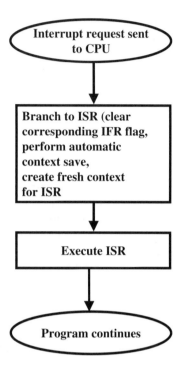

FIGURE 4.6
Interrupt flow in a TMS320C55 DSP.

There are three important multitasking algorithms:

- *Preemptive:* With this algorithm, if a high-priority task becomes ready for execution, it can immediately preempt the execution of a lower-priority task and acquire the processor without having to wait for the next regular rescheduling. In this context, "immediately" means after the scheduling latency period. This latency is one of the most important characteristics of a real-time kernel and largely defines the system responsiveness to external stimuli.

- *Cooperative:* With this algorithm, if a task is not ready to execute it voluntarily relinquishes control of the processor so that other tasks can run. This algorithm does not require much scheduling and generally is not suited for real-time applications.

- *Time sharing:* A pure time-sharing algorithm has obvious low responsiveness (limited by the length of the scheduling interval). Nevertheless, a time-sharing algorithm is always implemented in real-time operating systems, since there is almost always more than one non-real-time task in the real-time system, for example, a user interacting with a cell phone or an automated teller machine. These tasks have low priority and are scheduled with a time-sharing policy in the time when no tasks of higher priority are ready for execution.[3,4]

Synchronization and Communication

Communication can be defined as the passing of information from one task to another. Many forms of communication require synchronization. Synchronization is the satisfaction of constraints on the interleaving of the actions of different tasks. Synchronization is sometimes referred to as content less communications.

Mutual Exclusion

Mutual exclusion is the synchronization required to protect a critical section. In real-time DSP applications, there can be many of these critical sections. As an example, consider a DSP system with the following characteristics:

- Shared memory (between tasks)
- Load-store architecture system (like many of the DSP architectures today)
- Preemptive RTOS
- Two tasks running in the system, T1 and T2 (T2 has higher priority)
- Both tasks needing to increment a shared variable, X, by one

Rate-Monotonic Scheduling

Rate-monotonic scheduling is an optimal fixed priority policy where the higher the frequency (1/period) of a task, the higher is its priority. This approach can be implemented in any OS supporting the fixed priority preemptive scheme, such as DSP/BIOS. Rate-monotonic scheduling assumes the deadline of a periodic task is the same as its period.

Scheduling Periodic Tasks

Many DSP systems are multirate systems. This means there are multiple tasks in the DSP system running at different periodic rates. Multirate DSP systems can be managed using nonpreemptive as well as preemptive scheduling techniques. Nonpreemptive techniques include using state machines as well as cyclic executives.

Preemptive scheduling is an effective approach for scheduling real-time DSP systems. Its modularity simplifies the overall design. The application can be viewed as a collection of independent tasks or jobs. The complexity is reduced as the functionality becomes encapsulated into a set of well-defined tasks. Systems designed using preemptive scheduling are also more maintainable. The issue of changes to one task in the system affecting other jobs in the system is removed. New functionality can easily be added by adding a new task. A preemptive scheduling approach also makes the system more efficient, since preemptive scheduling is more efficient at utilizing time slots that may not be fully utilized.

Rate-Monotonic Analysis

Some of the scheduling strategies discussed earlier presented a means of scheduling but did not give any information on whether the deadlines would actually be met. Rate-monotonic scheduling addresses how to determine whether a group of tasks with known individual CPU utilizations will meet their deadlines. This approach assumes a priority preemption scheduling algorithm. The approach also assumes independent tasks (no communication or synchronization). (The restriction of no communication or synchronization may appear to be unrealistic, but there are techniques for dealing with this that will be discussed later.)

In this discussion, each task discussed has the following properties:

- Each task is a periodic task with a period, T, which is the frequency with which it executes.
- An execution time, C, which is the CPU time required during the period.
- A utilization, U, which is the ratio C/T.

A task is schedulable if all its deadlines are met (i.e., the task completes its execution before its period elapses). A group of tasks is considered to be schedulable if each task can meet its deadlines.

Rate-monotonic analysis (RMA) is a mathematical solution to the scheduling problem for periodic tasks with known characteristics (size, timing). The assumption with the RMA approach is that the total utilization must always be less than or equal to 100% (any more and you are exceeding the capacity of the CPU).

For a set of independent periodic tasks, the rate-monotonic algorithm assigns each task a fixed priority based on its period, such that the shorter the period of a task, the higher the priority. For three tasks T1, T2, and T3 with periods of 5, 15, and 40 ms, respectively, the highest priority is given to the task T1, as it has the shortest period; the medium priority to task T2; and the lowest priority to task T3. Note the priority assignment is independent of the applications "priority," i.e., how important meeting this deadline is to the functioning of the system or user concerns.[2]

Optimizing Real-Time DSP Applications

Many of today's DSP applications are subject to real-time constraints. Many applications will eventually grow to a point where they are stressing the available CPU and memory resources. Understanding the architecture of the DSP as well as the compiler can speed up applications, sometimes by an order of magnitude. This section summarizes some of the techniques used in practice to gain orders of magnitude speed increases from high performance DSPs.

What Is Optimization?

To start with, we must understand what optimization is and how it relates to real-time DSP systems. Optimization is a procedure used in designing systems to maximize or minimize one or more performance indices. These indices include throughput, memory usage, I/O bandwidth, power dissipation, and size.

Since many DSP systems are real-time systems, at least one (and probably more) of these indices must be optimized. It is difficult (and sometimes impossible) to optimize all these performance indices at the same time. In many cases, trade-offs will be made. For example, making the application faster may requires more memory and vice versa. The designer must weigh each of these indices and make the best trade-off. The size vs. speed trade-off is typical in embedded systems. The fastest program, e.g., is one that is perfectly linear with no branches or function calls. This is obviously an unrealistic constraint in today's sophisticated programs.

Determining which index or set of indices is important to optimize depends on the goals of the application developer. For example, optimizing for performance means that the developer can use a slow or less expensive DSP to do the same amount of work. In some embedded systems, cost savings like this can have a significant impact on the success of the product. The developer can alternatively choose to optimize the application to allow the addition of more functionality. This may be very important if the additional functionality improves the overall performance of the system, or if the developer can add more capability to the system such as an additional channel of a base station system. Optimizing for memory use can also lead to overall system cost reduction. Reducing the application size leads to a lower demand for memory, which reduces overall system cost. And finally, optimizing for power means that the application can run longer on the same amount of power. This is important for battery-powered applications.

Program optimization in some large systems is often unnecessary. In the embedded world, however, resources are still limited. Add to this the migration from assembly language to a less efficient language like C, C⁺⁺, or Java and the pressures to develop efficient code for scarce resources is a continual struggle.

Generally, DSP optimization follows the 80/20 rule. This rule states that 20% of the software in a typical application uses 80% of the processing time. This is especially true for DSP applications that spend much of their time in tight inner loops of DSP algorithms. Optimizing these loops for speed and power will make the biggest impact on overall system performance. DSP control software (best described as nonloop DSP software) consumes the majority of program memory in a typical DSP application. But this code is executed very infrequently and, therefore, consumes only a fraction of the overall system execution cycles. DSP control software is a prime target for performance-based optimizations.

The best way to determine which parts of the code should be optimized is to profile the application. This will answer the question as to which modules take the longest to execute. These will become the best candidates for performance-based optimization. Similar questions can be asked about memory usage and power consumption.

Some of the optimization techniques available are architecture and application dependent. Others are more generic approaches and can be applied to many architectures and applications. Good optimization requires a thorough understanding of the following:

- *DSP architecture:* Each target processor and compiler has different strengths and weaknesses and understanding them is critical to successful software optimization.
- *DSP compiler:* Today's DSP compilers are advanced. Many allow the developer to use a higher-order language such as C and very little, if any, assembly language. This allows for faster code development, easier debugging, and more reusable code.
- *Software application.*

Performance Improvement Guidelines

Despite a programmer's expertise and years of experience, there is always a potential for performance improvements in a DSP application. Performance-critical software like real-time embedded and DSP-based systems should always be analyzed for bottlenecks and poorly written code.

DSP application optimization requires a disciplined approach to get the best results. Here are a few guidelines:

- Performance analysis and optimization is a process of diminishing returns. Significant improvements can be found early in the process with relatively little effort. This is the "low-hanging fruit." Examples of this include accessing data from fast on-chip memory using the DMA and pipelining inner loops[5] (these techniques are discussed below).
- Change one parameter at a time. Making several optimization changes at the same time will make it difficult to determine what change led to which improvement percentage.
- Have a test plan and use it often. Optimization can be difficult. More difficult optimizations can result in subtle changes to the program behavior that leads to wrong answers.

Overview of Common Compiler Optimization Techniques

Optimizing compilers perform sophisticated program analysis including intraprocedural and interprocedural analysis. These compilers also perform data flow and control flow analysis as well as dependence analysis, and they often require provably correct methods for modifying or transforming code. Much of this analysis is to prove that the transformation or modification is correct in the general sense.

Most optimizations on code are performed while the code is in the intermediate representation (IR) stage, that is, before the application is linked. Optimizations can also be performed in the back end of the compiler where the final machine adaptations are best performed (such as allocations to the actual register files of the particular machine).[6]

Some of the common compiler optimization techniques are listed below:

- *Branch optimization:* Rearranges the program code to minimize branching logic and to combine physically separate blocks of code.
- *Code motion:* If variables used in a computation within a loop are not altered within the loop, the calculation can be performed outside of the loop and the results used within the loop.
- *Instruction scheduling:* There are various techniques of instruction scheduling that will reorder instructions to minimize execution time.
- *Interprocedural analysis:* Will uncover relationships across function calls and attempt to eliminate loads, stores, and other computations that cannot be eliminated with more straightforward optimizations.
- *Invariant IF code floating (unswitching):* Removes invariant branching code from loops.
- *Profile-driven feedback:* Uses actual results from a sample program execution. The results are used to improve optimization near conditional branches and in other frequently executed code sections.
- *Reassociation:* Rearranges the sequence of calculations in an array subscript making it more optimal for the particular machine. This technique produces more candidates for common expression elimination.
- *Store motion:* Seeks to move store instructions out of loops.
- *Strength reduction:* Replaces less efficient instructions with more efficient ones. A common example of this is in array subscripting, where the optimizer will use an "add" instruction to replace a more costly "multiply" instruction.

Last-Resort Assembly Language

Often the C code can be modified slightly to assist in optimizing the application, but it can take time and several iterations to achieve the optimal (or

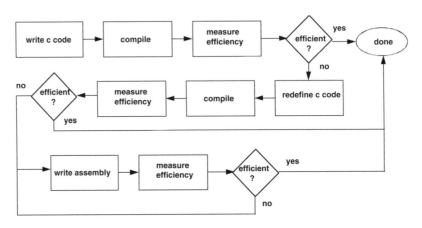

FIGURE 4.7
Code optimization process for DSP programming.

close to optimal) solution. The process of refining code in this manner is shown in Figure 4.7. The code optimization process consists of a series of iterations. In each iteration, the programmer should examine the compiler-generated code and look for optimization opportunities. The programmer will look for an abundance of null operations (NOP) in the code due to delays in accessing memory and/or another processor resource. These are the areas that become the focus of improvement. As a last resort, the programmer can consider hand-tuning the algorithms using assembly language.

Other Issues to Consider

There are other performance issues to consider when designing these types of systems. For example, when using an RTOS for task-driven systems, there is overhead to consider that increases with the number of tasks in the system. The overhead in a task switch (or semaphore pend- or postoperation) can vary based on where the operating system structures are located. If the structures are in off-chip memory, the access time to perform the operation can be much longer than if the structure was in on-chip memory. The same holds true for the task stack space. If this is in off-chip memory, the performance suffers proportionally to the number of times the stack has to be accessed. One solution is to allocate the stack in on-chip memory. If the stack is small enough, this may be a viable thing to do. But if there are many tasks in the system, there will not be enough on-chip memory to store all of the task stacks.

Compile Time Options

Many DSP optimizing compilers offer several options for code size vs. performance. Each option allows the programmer to achieve a different level of performance vs. code size. These options allow more and more aggressive

TABLE 4.1

DSP Optimizing Compiler Options

-o3	Highest performance — typically highest code size
-o3	More aggressive prologue/epilogue collapsing — minimal performance impact
-ms0	No redundant loops — required for some filter designs
-o3	Some loops not software pipelined
-ms1	Reduced unrolling — some performance degradation
-o3	No software pipelining or unrolling — code size
-ms2	Benefit with large performance degradation
-o3	No parallelism — smallest code size and lowest performance
-ms3	Typically used for control code with no performance constraints

code size reduction from the compiler. The Texas Instruments DSP compiler, e.g., has five different levels of code performance (Table 4.1). Each option allows the compiler to perform different DSP optimization techniques.[5,7]

Summary of Optimization

Application programmers have always had to develop a library of techniques to allow software to run as fast as possible. As processors continue to become more complicated, this becomes a more necessary approach. For advanced processors, managing two separate pipelines and ensuring the highest amount of parallelism requires tools support. Optimizing compilers are helping overcome many of the obstacles of these powerful new processors, but even the compilers have limitations. Application programmers should not rely on the compiler to perform all of the necessary optimizations for them.

Optimization is required in many real-time embedded DSP applications. However, optimization can also be antagonistic to other important goals for the application developer; namely, stability, maintainability, and portability. At a cursory level optimization is useful and should always be applied. But optimization can also be intrusive.

Finally, optimization of DSP applications is a process that it is part of the overall application development cycle. DSPs differ fundamentally from other more general-purpose applications. One difference is the optimization phase, which can also be iterative. Applying some of the techniques in this section will allow the developer to minimize the number of iterations required in developing DSP applications.

Conclusions

Designing noise reduction algorithms for use in real-time embedded systems requires the understanding of the application system's constraints. This chapter outlined the requirements, issues, and design methodologies required for building a real-time DSP application. The relevance of understanding the application's constraints is apparent when developing a sophisticated noise reduction algorithm.

References

1. Summerville, I., *Software Engineering,* 6th ed., Addison-Wesley, Reading, MA, 2001.
2. Liu, J.W.S., *Real-time Systems,* Prentice-Hall, Upper Saddle River, NJ, Dallas, TX, 2000.
3. The, A. and Dart, D.W., *How to Get Started with DSP/BIOS II,* Texas Instruments Applications Report SPRA697, October 2000.
4. Buttazzo, G., *Hard Real-time Computing Systems: Predictable Scheduling Algorithms and Applications,* Kluwer Academic Publishers, Boston, 1997.
5. *TMS320C62XX Programmer's Guide,* Texas Instruments, Dallas, TX, 1997.
6. Hennesey, J.L. and Patterson, D.A., *Computer Architecture, A Quantitative Approach,* Morgan Kaufmann, Palo Alto, CA, 1990.
7. *TMS320C55x_DSP_Programmer's_Guide,* Texas Instruments, Dallas, TX, 2000.

5

Evaluating the Effects of Noise on Voice Communication Systems

William D. Voiers, Alan D. Sharpley, and Ira L. Panzer

CONTENTS

Introduction

After more than a century of developments in the field of voice telecommunication, noise remains the most ubiquitous deterrent to effective voice communications. It can serve not only to reduce the intelligibility of transmitted speech, but it may also impair the aesthetic quality and obscure acoustical cues to the identity of the speaker. If we are to succeed in coping with the effects of noise on voice communications, we must understand in detail the nature of its impact on the various useful components of the speech signal. These include not only the components on which intelligibility depends but also those critical to overall acceptability and speaker recognizability.*

The need for methods of assessing the impact of noise on speech intelligibility was recognized early in the development of the field, and various methods for measuring intelligibility were developed during the first half of the 20th century. The advent of digital voice coding and communication systems obviated some noise problems but exacerbated others, particularly the problems posed by noise at the communication source. Noise in digital voice channels results in bit errors, the effects of which on intelligibility are different and more difficult to predict than effects of noise on analog systems.

Until recently, methods for removing noise from speech usually succeeded in improving acceptability only at some cost of intelligibility. Methods developed within the past few years, however, have achieved improvements in acceptability at no measurable cost of intelligibility.

Before World War II, there was little concerted effort to standardize methods of evaluating intelligibility in ways that would permit their use beyond the confines of a single laboratory. There was little concern about the effects of noise and other types of signal degradation on speech acceptability and on speaker recognizability. No generally accepted methods for evaluating these properties were available.

The difficulty of developing valid and reliable methods of evaluating intelligibility, acceptability, and speaker recognizability in speech becomes apparent when one considers the diversity of factors that may influence the results of tests designed to evaluate these properties. In many respects, the human auditory–perceptual system can perform far more complex acoustical analyses than any electro-acoustical instrument available today, but it is also characteristically susceptible to the types of error to which these devices generally have the greatest immunity. Adequate control of these errors is a formidable task. Just how formidable becomes apparent when we examine the nature and number of factors on which intelligibility, acceptability, and speaker recognizability ultimately depend both in real life and in the speech-testing laboratory. Although in the testing situation as well as in real life, intelligibility, acceptability, and speaker recognizability depend most immediately on the fidelity with which the speech signal is delivered to a communicator's ear, they also depend on a diversity of other factors, as described below.

Conceptions of Intelligibility: Implications for Test Design

In many nonspeech communication situations, a single standard waveform can serve as a test signal for a communication device or link. Depending on

---

* The term *quality* has been widely used in reference to the subjective consequences of factors such as these, but, since the ultimate effect of speech quality is *acceptability*, we propose here to replace *quality* with that term. Among other things, this convention will restore a useful term to its historical place in the vocabularies of perceptual psychology and psychophysics. *Quality* will be used here only in its traditional scientific sense, in reference to the elementary auditory qualities of pitch, loudness, timbre, and, in particular, the elementary perceptual qualities on which speech acceptability depends.

circumstances and purpose, the extent and manner to which the signal is degraded by the link can provide valuable information about the link. With speech, however, critical information is acoustically encoded in such a diversity of ways that no single test signal thus far devised can adequately reflect the fidelity with which all the intelligibility-relevant features of the signal have been transmitted. We have no practical choice but to use a variety of test signals and to use human listeners as our measuring instruments. The effect of this is to introduce a diversity of extraneous factors, which must be controlled, depending on the purposes for which testing is to be performed. These factors include:

- Linguistic content and information structure of the source signal — the predictability of the message
- Physical characteristics of the source signal as dependent on the articulatory (acoustical encoding) capabilities and performance of the speaker
- Availability of *contextual information,* i.e., intelligibility-relevant information in the communications environment and the history of the speech signal itself — the availability of *a priori* information regarding the nature or identity of a speech event
- Transmission characteristics of the *link,* i.e., the medium, channel, coding device, or combination thereof (the effects of which are our primary concern here)
- Inter- and intralistener variation in auditory-discriminative capacity and performance, i.e., "between-listeners" variation in capacity to discriminate the intelligence-bearing physical–acoustical features of the speech signal and "within-listener" variation in the availability and use of this capacity
- Interlistener variation in *situational* and *linguistic* competency — individual differences in ability to assimilate contextual information from the situation and the speech signal, respectively
- Capacity and availability of the listener's sensory–cognitive channels for the assimilation of speech intelligence as dependent on the extent of competing demands for channel space, the listener's state of alertness, level of motivation, etc.

With appropriate experimental design and procedures, tests of intelligibility can be used to evaluate the effects of any of the above factors, but rigorous control of the remaining factors is essential to the reliability and validity of such tests. This is especially true where the goal is to isolate the effects of a speech link or medium on these properties. To enable crews of human listeners to function as reliable, precise measuring instruments, we must control all factors external to the link. In particular, we must control the types of error to which humans are uniquely susceptible if we are to ensure that our human "meters" are measuring what we intend.

A crucial step in the design of tests for evaluating the effects of communication systems or links on intelligibility is the selection of appropriate "test signal(s)" or *critical speech units* (CSUs). Historically, CSUs have consisted variously of paragraphs, sentences, phrases, words, syllables, or elementary speech units (phonemes). The type of response required of listeners depends at least in part upon the nature of the CSU. At one extreme, listeners may be required to indicate their comprehension of extended prose passages. At the other, they may be asked to discriminate the state of a single phonemic feature. Most of these approaches have been used effectively in one application or another, but the advantages and limitations of each depend heavily on application and purpose.

Tests in which the CSU is a word, phrase, or larger linguistic unit have considerable intuitive appeal, but they pose some formidable problems of control when used for evaluating communications equipment or links. Almost inevitably, they allow listeners to draw on resources other than their ability to discriminate the physical properties of the received speech signal. In particular, they allow, if not invite, listeners to take advantage of the predictability of the CSU from contextual information contained in the situation or in the speech signal itself. Generally, tests in which the equivocal element or CSU is a minimal linguistic unit, or *phoneme*, permit more rigorous control of context and other extraneous factors than do tests in which the CSUs are larger linguistic units.

Intelligibility tests differ not only in the nature of the CSU they employ, but also in terms of the conception of intelligibility on which they are based. The manner in which we approach the evaluation of intelligibility in communication systems ultimately depends on whether we treat intelligibility as a unidimensional or as a multidimensional quantity, as a scalar, or as a vector. Earlier tests treated intelligibility as a scalar, yielding a single figure of merit. Most recent tests, however, are based on the conception of intelligibility as a vector, whose components are not uniformly vulnerable to the types of degradation commonly encountered in modern voice communications.

Scalar Approaches to the Evaluation of Intelligibility

Research begun at the Harvard Psycho-Acoustic Laboratory during World War II led to the development of the first widely used methods of evaluating the effects of communication media and devices on speech intelligibility. The "phonetically balanced" (PB) word lists of Egan[1] were adopted by the American National Standards Institute as the "American Standard Method for Measurement of Monosyllabic Word Intelligibility." Although still so recognized,[2] along with the Diagnostic Rhyme Test (DRT) and the Modified Rhyme Test (MRT), the PB test is rarely used for purposes of evaluating communication systems or links. However, it is still frequently used for purposes of clinical audiology.

The conventional method of scoring the PB test makes allowance only for the number, not the nature, of the errors made by the listeners. Other disadvantages of this test include the requirement that listeners undergo extensive familiarization training with the test materials. Results can be expected to vary with the amount of training to which the listeners have been exposed. Moreover, no reliable methods exist for evaluating the effects of such training on test results. In recent years, there has been a tendency to dispense with the training requirement, but numerous questions remain regarding the contribution of extraneous factors when the test is used for purposes of evaluating communication systems or links.

The Fairbanks Rhyme Test[3] is an open-response test in which the listener is presented with a word and provided a "stem" lacking the initial consonant. The listener's task is to provide the missing consonant. A purported advantage of this method is that "linguistic factors of higher order weigh lightly."[3] The Fairbanks test does provide improved control of some of the extraneous factors among those listed above, but it leaves others poorly controlled, in particular, the listener's linguistic competency.

The Modified Rhyme Test (MRT) of House et al.[4] provides somewhat improved control of contextual factors and listener variation in linguistic competence by utilizing a closed-response set. Here, in responding to a test word, the listener is required to select from among six possibilities. In half of the test items in the MRT, the six alternatives are rhyming, i.e., the alternatives differ in the initial consonant. In the remaining half, the six alternatives are alliterative, i.e., they differ only in their final consonants. One unpublished study by Voiers, Sharpley, and Panzer[5] showed that listeners' errors in the MRT tend to be highly skewed in favor of those options that differ from the correct choice by the least number of distinctive features.* In the case of many items on the MRT, some response options are virtually never exercised by listeners.

A CVC-word test (i.e., consonant–vowel–consonant) proposed by Steeneken[6] provides acceptable resolution when used to yield a gross, or overall, measure of intelligibility. However, it suffers from several disadvantages of multiple-choice and free-choice tests when used for purposes of evaluating communication systems.

In well-designed multiple-choice or free-choice tests, the incorrect choices should be equally attractive. Where this is not the case, adjustments for the effects of chance or guessing become quite complicated. "Diagnostic" scoring of results yielded by such tests is difficult if not impossible. Faced with arbitrary constraints on their response options, listeners may be forced by their perceptions of one phonemic feature to make choices that do not represent their perceptions of the states of one or more other features. Under

---

* The distinctive features of a language are the dimensions used to classify the phonemes in the language. They may have articulatory, i.e., place and manner of articulation, acoustic, or perceptual bases.

such circumstances, the nature of a particular error cannot be univocally attributed to the system or link being evaluated.

With free-choice tests of consonant recognizability, it is generally necessary to embed the critical phoneme in a CV, VCV, or VC syllable, which syllable may or may not be an actual word in the language involved. It is virtually impossible to assemble a suite of tokens similar to those used by Miller and Nicely[7] but composed entirely of words or entirely of nonsense syllables. Normally, therefore, listeners must choose among a mix of meaningful syllables and meaningless syllables in order to register their perception of a given test element. There is a wealth of experimental evidence to the effect that listeners would favor the former option. In addition, listeners must choose among familiar, frequently encountered words and less-familiar words. They must choose between euphonious or pleasant-sounding tokens and cacophonous or unpleasant-sounding tokens. Listeners must choose between tokens having pleasant connotations and tokens having unpleasant connotations, possibly between "socially acceptable" and "socially unacceptable" tokens. Finally, they must choose between tokens with familiar graphemic correlates (e.g., "b") and tokens with unfamiliar graphemic correlates (e.g., "θ").

All of the above considerations argue against the interpretability of free-choice test results even when the results are adjusted for the effects of chance.

Results described in the classic paper of Miller and Nicely[7] bear on several critical issues in the design of intelligibility tests. They reveal three important principles regarding the most appropriate means of evaluating intelligibility in voice communication systems. First, errors of speech perception are not random — phonemic confusions tend to occur in well-defined patterns. Second, error patterns vary with the type of speech degradation involved. Third, most errors can be accounted for in terms of discrimination failures with respect to a limited number of perceptual dimensions or "distinctive features" traditionally recognized by linguists, e.g., Jakobson, Fant, and Halle.[8] These principles have been validated, directly or indirectly, by other investigators (e.g., Peters,[9] Wicklegren,[10] and Singh et al.[11]), and they provided a basis for a novel approach to the evaluation of speech intelligibility: the Diagnostic Rhyme Test (DRT), developed by Voiers,[12,13] and various of its derivatives.[14,15]

Various other tests developed during the past several years have been described by Schmidt-Nielsen.[16] None of these, however, is yet to enjoy extensive use beyond the confines of a single laboratory.

Vector Approach to the Evaluation of Intelligibility

The DRT, which has undergone a succession of refinements, was the first of several tests, including the Diagnostic Medial Consonant Test (DMCT) and the Diagnostic Alliteration Test (DAT),[15] in which intelligibility is treated as a multidimensional property of the speech signal. All of these tests use a two-alternative forced choice (2AFC) paradigm in which the choices available

TABLE 5.1

Classification of 23 English Consonants by Seven Distinctive Features

Feature	M	N	V	TH	Z	ZH	DJ	B	D	G	W	R	L	Y	f	th	s	sh	ch	p	t	k	h
Voicing	+	+	+	+	+	+	+	+	+	+	+	+	+	+	−	−	−	−	−	−	−	−	−
Nasality	+	+	−	−	−	−	−	−	−	−	−	−	−	−	−	−	−	−	−	−	−	−	−
Sustention	−	−	+	+	+	+	−	−	−	−	+	+	+	+	+	+	+	+	−	−	−	−	−
Sibilation	−	−	−	−	+	+	+	−	−	−	−	−	−	−	−	−	+	+	+	−	−	−	−
Graveness	+	−	+	−	−	0	0	+	−	0	+	−	0	0	+	−	−	0	0	+	−	0	0
Compactness	−	−	−	−	−	+	+	−	−	+	−	−	0	+	−	−	−	+	+	−	−	+	+
Vowel–like*	−	−	−	−	−	−	−	−	−	−	+	+	+	+	−	−	−	−	−	−	−	−	−

* The DRT, DMCT, and DALT do not test for the discriminability of vowel-likeness, but do *not* confound the effects of this feature with those of other features.

Note: Voiced phonemes are indicated by uppercase letters, unvoiced by lowercase letters. Pluses (+) denote the nominal or positive state of the feature; minuses (−) denote the negative state; zeros (0) denote indifference or neutrality with respect to the feature.

to the listeners to each stimulus word differ only by a single distinctive feature in the critical consonant, an initial, medial, or final consonant, depending on the test involved. The classification of 23 consonants with respect to these features, shown in Table 5.1, was used in the design of these tests. Scorable to yield a gross or total intelligibility score, all of these tests yield gross diagnostic scores for each of six binary phonemic attributes or features:

- Voicing (voiced vs. unvoiced)
- Nasality (nasals vs. nonnasals)
- Sustention (sustained vs. interrupted)
- Sibilation (sibilants vs. nonsibilated)
- Graveness (grave vs. acute)
- Compactness (compact vs. diffuse)

The usefulness of such diagnostic scores depends upon knowledge of their articulatory and physical–acoustical correlates, particularly their implications regarding the acoustical characteristics of the speech signal.

The articulatory bases of the seven features are fairly well understood. All *voiced* phonemes involve some free vibration of the vocal cords; *unvoiced* phonemes do not. *Nasals* are produced by lowering of the velum, allowing air to escape through the nasal passages; *nonnasals* are produced by closing the nasal passages. *Sustained* phonemes, the continuants, are produced by incomplete constriction of the vocal tract; *interrupted* phonemes, the stops and affricates, are produced by complete constriction of the tract at some point. *Sibilants* involve extreme, but incomplete constriction of the vocal tract. *Grave* phonemes are produced by constriction toward the anterior of the vocal tract; *acute* phonemes are produced by constriction in the middle of the tract. *Compact* phonemes are produced by constriction toward the rear of the vocal tract; *diffuse* phonemes are produced by constriction near the middle. *Vowel-like* phonemes, or glides, are produced by minimal constriction of the vocal tract, with changes in the point of maximum constriction depending on the phoneme involved.

The acoustical correlates of the above features present a more complex problem. Each perceptual distinctive feature has multiple acoustical correlates, where the relative saliency of each correlate depends on the phoneme involved, its phonemic environment, and the states of one or more noncritical features. However, some rough generalizations are possible.

Voiced fricatives are distinguished from their unvoiced counterparts by the presence of periodicity and, in particular, by the time of onset in periodicity, but the acoustical correlates of voiced stops are distinguished by a more complex set of acoustical features. In voiced consonants, preceding vowels tend to be of greater duration than in unvoiced consonants. Voiers[12] and Miller and Nicely[7] showed that sufficient information for the discriminability of voicing is contained in the frequency range below 1 kHz.

Nasal phonemes are distinguished by relatively pronounced resonances at approximately 250, 800, and 2200 Hz and by the presence of nulls throughout the frequency range. The cues to nasality also vary with the height of the succeeding vowel. Several investigators, e.g., Voiers[17] and Miller and Nicely,[7] have shown that sufficient information for the reliable discrimination of nasality is contained in the frequency range below 1 kHz.

Sustained phonemes are distinguished by their gradual onset and by the presence of midfrequency noise, and interrupted phonemes are distinguished by their abrupt onset. Sustained phonemes have characteristic durational and high-frequency cues that distinguish them from their interrupted counterparts. However, as in the case of other features, the relative saliency of the various acoustical correlates of sustention varies with vowel environment and the states of other consonant features, voicing, in particular.

Sibilant consonants are characterized by higher-frequency noise and greater duration than their nonsibilant counterparts. Miller and Nicely[7] suggested that duration is the most important acoustical correlate of this feature.

Grave phonemes are distinguished by, among other properties, the origin and direction of second-formant transitions. However, the dominant acoustical cue to this feature-state depends, as in other cases, on whether the critical phoneme is voiced or unvoiced, sustained or interrupted.

In general, compact phonemes are characterized by the concentration of spectral energy in the midfrequency range, diffuse phonemes by the distribution of energy over more widely separated spectral peaks, but as in the case of the other place feature, graveness, the dominant acoustical cue depends on whether or not the phoneme is voiced or unvoiced, sustained or interrupted.

Given the complexity of the relationship between distinctive features and their acoustical correlates, the usefulness of gross feature scores for purposes of pinpointing specific system defects or deficiencies is somewhat, but not seriously, limited. However, this problem was anticipated in the design of the DRT and its various derivatives. All allow for a more detailed analysis of feature-discrimination data, in that the set of items designed to test for the discriminability of each feature is "balanced" with respect to vowel environment and to the states of various noncritical features. For example, data provided by the voicing scale of the DRT can be partitioned to yield separate discrimination scores for voicing "present" and voicing "absent," for voicing in sustained consonants and voicing in interrupted consonants, for voicing in high vowel environments vs. voicing in low vowel environments, in front vowel environments vs. back vowel environments, and all combinations thereof. The significance of these effects and their interactions can be conveniently determined by various statistical techniques, in particular, analysis of variance with factorial design. Similar analyses can be performed on data for all the other features, although the structure of the language dictates that some amount of confounding is unavoidable in some instances. For example, in half the items used to test the discriminability of sibilation, the critical phonemes are voiced; in half they are unvoiced. Orthogonal to this partitioning is one based on sustention, which, however, is completely confounded with compactness (i.e., all interrupted sibilants are compact affricates).

Although precise identification of system or channel deficiencies would in most cases require the sorts of detailed analyses described above, gross diagnostic scores can still be of considerable value in differentiating the effects of different types of speech degradation, as is shown in Figures 5.1, 5.2, and 5.3. In all three cases, the results shown are average values for three male speakers. The listening crews contained eight trained members of both sexes.

Figure 5.1 shows the effects of band-limited Gaussian noise on the major diagnostic scores of the DRT. These results are in close accord with those of Miller and Nicely.[7] *Voicing* and *nasality* appear to be very robustly encoded features. The discriminability of these features is significantly reduced only under the most extreme noise conditions. Most generally vulnerable to the effects of Gaussian noise are *sustention* and the place feature, *graveness*. The discriminability of the other place feature, *compactness*, which, in the DRT, distinguishes "back" consonants from "middle" and "front" consonants, is relatively unaffected by this type of noise.

Figure 5.2 shows the effects of the speech-modulated masking noise (modulated noise reference unit, MNRU) on diagnostic score patterns, which differ dramatically from those associated with Gaussian noise. Except under conditions of extreme degradation, the effects of this type of degradation are confined to the two place features, *graveness* and *compactness*.

Figure 5.3 shows the effects of a masking babble of eight voices on the major DRT diagnostic scores. Differences from the results of the preceding cases are evident. The discriminability of the six features is affected in a more nearly uniform manner than in those two cases.

Closer examination of data from the above studies reveals even more differences among the effects of the three types of masking noise. For example, a conspicuous effect of Gaussian noise is pronounced negative bias in

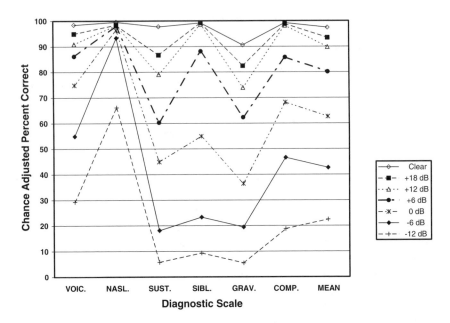

FIGURE 5.1
Effects of Gaussian noise on DRT diagnostic scores.

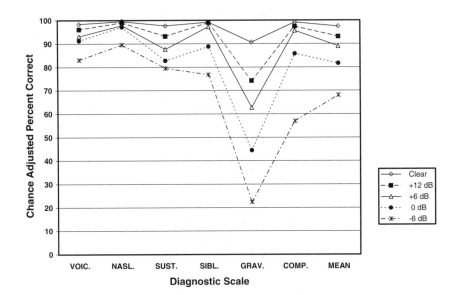

FIGURE 5.2
Effects of speech-modulated noise (MNRU) on DRT diagnostic scores.

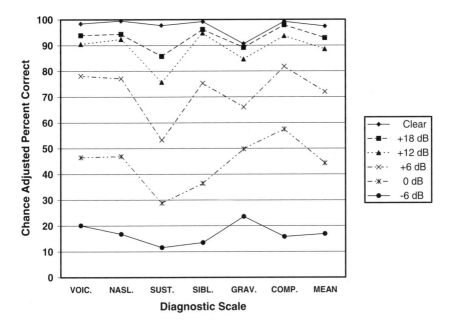

FIGURE 5.3
Effects of an eight-voice babble on DRT diagnostic scores.

listeners' perceptions of sustained and sibilated consonants — a strong tendency in listeners to perceive sustained consonants as their interrupted counterparts and sibilated consonants as their unsibilated counterparts. These biases are not evident in the case involving speech-modulated noise, but there is a strong negative bias in listeners' perception of graveness, a very pronounced tendency to perceive grave consonants as their acute counterparts.

As discussed by Voiers,[12,13] the 2AFC (alternative forced choice) paradigm provides the most effective means of controlling all of the major sources of extraneous variance in intelligibility test results, particularly those that complicate the task of diagnostic interpretation of test results.

Where an investigator's concern is only with a gross or scalar measure of intelligibility, most of the above considerations are less relevant. Moreover, there are often circumstances where such a measure will serve the investigator's purposes. Since all of the intelligibility tests in use today can be scored to yield an overall or gross intelligibility score, some interest attaches to the comparability of the various tests in use today when used for this purpose.

Figure 5.4 shows the effects of bandlimited Gaussian noise on the total or overall scores of four currently available tests. These results show that, at least for the case of Gaussian noise masking, cross predictability among these four tests is quite high. This does not, however, address the issue of relative resolving power. *Ceteris paribus*, resolving power is a function of the amount of data involved, e.g., number of replications, listening crew size. The question of resolving power is thus ultimately a question of economics. The DRT

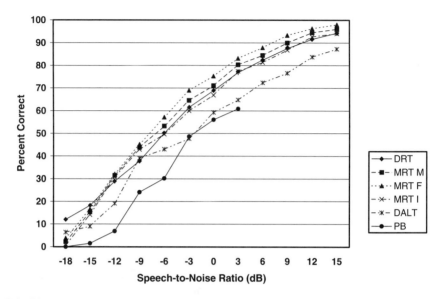

FIGURE 5.4
Effects of Gaussian noise masking on selected intelligibility tests (M = medial consonant; F = final; I = initial).

and its derivatives compare favorably in this respect with other currently available tests.

Figure 5.5 compares DRT diagnostic score profiles for the three types of noise conditions (i.e., Gaussian, MNRU, and babble) shown in Figures 5.1, 5.2, and 5.3, respectively. Here, diagnostic scores have been interpolated from the data for each of the three types of noise to yield an equivalent overall intelligibility score of 75%. The profiles illustrate how conditions can yield the same degree of overall intelligibility while exhibiting quite dramatically different feature profiles. Figure 5.5 supports the premise of intelligibility as a multidimensional rather than a unidimensional concept.

Conceptions of Speech Acceptability

It is commonly recognized that intelligibility does not alone ensure the acceptability of transmitted speech. Intelligibility is a necessary, but not sufficient, condition of acceptability. Depending on the circumstances and purposes occasioning voice communication, other factors, such as aesthetic acceptability and the recognizability of the communicator's voice, may also affect the acceptability of transmitted speech.

The ultimate criterion of acceptability is a *subjective* one, albeit *not* one based on the judgment of any single individual but rather on the judgment of a hypothetical typical, or *normative*, individual. The goal with all methods

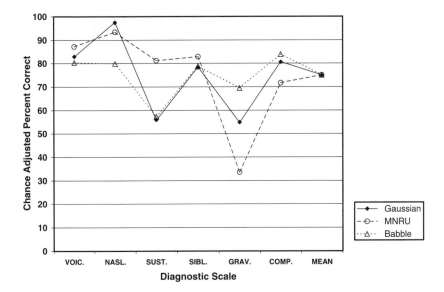

FIGURE 5.5
DRT diagnostic scores for three types of masking noise.

of evaluating the acceptability of the speech output of a communication system or device is thus to estimate the response of the normative individual by means other than exhaustive sampling of the user population. A major problem is that individuals vary widely within and among themselves in the subjective criteria of acceptability they bring to the testing situation. Our success in approximating the response of the normative user thus depends on our understanding of the nature of such variation and on our ability to control or to compensate for it.

Investigations by several researchers, e.g., Rothauser, Urbanek, and Pachl,[18] Nakatani and Dukes,[19] and Voiers,[20] have provided some important insights regarding the traditionally low reliability of subjective methods of acceptability evaluation. However, these investigators have also shown that much seemingly random variation in acceptability test results is in fact attributable to systematic, potentially controllable factors. In particular, these include:

- *Inter*individual differences in the "subjective yardsticks" listeners bring to the testing situation — differences in the subjective origins and scales to which they reference their judgments of acceptability and the perceptual qualities on which such judgments depend
- Experience-dependent *intra*individual variation in the subjective origin and scale to which listeners reference their judgments
- *Inter*individual differences in tolerance for specific types of degradation
- Experience-dependent *intra*individual variations in tolerance for specific types of degradation

- Listener uncertainty as to the nature of the task to be performed in the testing situation; uncertainty as to what aspects of the transmitted speech sample should be the basis for judgments of acceptability or other perceptual qualities
- Variation in listeners' perceptions of the actual or hypothetical purposes to be served by voice communication in a given situation

Various approaches to the control of the above factors by statistical or experimental means have been employed, but most have not been completely successful. As in the case of intelligibility, the approach to acceptability evaluation followed in any instance depends on how acceptability is conceived — whether it is conceived as a scalar or as a vector.

Diverse methods for evaluating acceptability have appeared during the past 40 years. Schmidt-Nielsen[16] describes a number of them. Only three of these, however, have been widely used during the past 25 years: the Absolute Category Rating Method (ACRM),[21] which yields a mean opinion score (MOS); the Degradation Category Rating Method (DCRM),[22] which yields a mean opinion score for perceived degradation (DMOS); and the Diagnostic Acceptability Measure (DAM) developed by Voiers and colleagues.[23]

Scalar Approaches to the Evaluation of Speech Acceptability

Both the ACRM and the DCRM treat acceptability as a scalar and yield a single figure of merit, a MOS and a DMOS, respectively. Currently, the MOS and DMOS are the standard criteria of telecommunication acceptability recommended by the Telecommunication Standards Sector of the International Telecommunications Union (ITU-T).[24,25] As the standard, the ACRM and the DCRM are routinely used by international standards organizations such as the ITU-T, the Telecommunications Industry Association (TIA), and the European Telecommunications Standards Institute (ETSI), among others. The DAM, on the other hand, treats acceptability as a vector whose components provide a variety of potentially useful information.

Although there is some variation in the procedures employed to obtain a MOS, several features of the ACRM are common to all of the most widely used versions. Speech samples (typically two sentences) for the condition or device being evaluated are presented to listening crews under controlled listening conditions. The listeners rate each sample using a five-category rating scale in which the categories are defined as "excellent," "good," "fair," " poor," or "bad." These category ratings form the basis of the five-point MOS, ranging from 1 for "bad" to 5 for "excellent."

In the case of the DCRM, each trial involves two samples, where the first sample is a reference condition and the second a test sample. The listener's task is to rate the amount of degradation in the second (test) sample compared to that of the first (reference) sample using a five-category "detectability of degradation" rating scale, in which the categories are labeled "inaudible," "audible but not annoying," "slightly annoying," "annoying," and "very annoying." The listener's category ratings form the basis of the five-point DMOS, where the scores range from "1" for "degradation is very annoying" to "5" for "degradation is inaudible." The DCRM was designed to be used in tests involving very high quality voice communication systems and in evaluating system conditions involving noisy inputs.[22] Since the DMOS provides a measure of perceived degradation of test-conditions "relative to" reference conditions, it is highly dependent on the specific reference conditions used in the evaluation. The relative nature of the DMOS precludes comparisons with "absolute" measures of acceptability yielded by the ACRM and DAM as described by Panzer, Sharpley, and Voiers.[26] The remainder of this discussion will focus on the ACRM and the DAM.

A feature of current ACRM testing practice is the inclusion of MNRU as additional test conditions. The MNRU conditions serve two purposes: (1) to "bound" the test (i.e., provide a continuous range of acceptability from "bad" to "excellent"), and (2) to provide a common reference system to be used, for example, across different laboratories, different languages, and different source materials. The error associated with the listener sample can thus be controlled to some extent by transforming raw scores into equivalent "Q-levels," each of which corresponds to a specified MNRU level. However, this practice is not universal. The repeatability or interlaboratory reliability of MOS results has not been impressive in the past, but improved experimental design, multiple speakers, and very large listening crews can ameliorate this problem to some extent.

Standard procedure with the ACRM and its derivatives calls for crews of naive listeners, and thus presents a problem where results from different experiments, and thus different listeners, are to be compared. Fortunately, the use of large listening crews (usually 32 to 64) and replication through the use of multiple talkers (usually 4 to 16), reduce listener sampling error to some extent.

Vector Approach to the Evaluation of Speech Acceptability

The use of multidimensional scaling methods by some investigators, e.g., McDermott,[27] in effect implicitly treats acceptability as a vector. However, this approach has not been widely adopted, and, in any case, is not practical in situations where the test plan requires the evaluation of a large number of systems or conditions in the same context.

Only a few investigators (e.g., McDermott,[27] Nakatani and Dukes,[19] Voiers,[20,23,28] Quackenbush, Barnwell, and Clements[29]) have attempted to understand the bases of acceptability, i.e., to identify the *perceived* speech qualities on which listeners' judgments of acceptability depend. Knowledge of the nature and number of these qualities played a crucial role in the design and development of the DAM. As described below, listener's judgments of various *elementary perceived qualities* (EPQ) of processed speech conditions provide a basis for various supplementary estimates of acceptability and also provide useful diagnostic information about the system being evaluated.

All versions of the DAM, e.g., Voiers,[20,28] Panzer, Sharpley, and Voiers,[26] share some features with earlier methods of evaluating speech acceptability, but also have some unique features. In particular, these include:

- Use of *detectability* vs. *evaluative* judgments of acceptability and of the perceptual qualities on which it depends: Listeners are asked to judge the *effects* of a system or device on the *detectability* of various simple and complex perceptual qualities.
- Provision of multiple estimates, both direct (isometric) and indirect (metametric and parametric), of speech acceptability (see below).
- Explicitly identified "end anchors" to stabilize the listeners' subjective scale and origin.
- Standard "probes" to sense shifts in listeners' subject scale or origin.
- Use of trained listeners who have been carefully screened and using the DAM itself, and the use of familiarization and training sessions to acquaint listeners with the types of degradation produced by modern speech coding systems and with the voices and the speech materials used.
- Calibration procedures to permit statistical control of interindividual and intraindividual variation in *adaptation level* and taste.
- Procedures for monitoring the performance consistency of individual listeners.

Each of these features provides some control of one or more of the major sources of extraneous variance, or error, in estimates of speech acceptability. Each source of error is in turn subject to some degree of control by one or more of these features.

Rather than being asked only to make raw subjective judgments of the acceptability of a sample of system-processed speech, listeners are also asked to judge the *detectability* of the effects of a link or system on the various EPQs on which acceptability depends. In principle, at least, such judgments should be relatively free of the effects of individual listener differences in tolerance for the qualities involved. Thus, a key feature of the DAM rests on the premise that individuals tend generally to agree better on *what they hear* rather than on *how well they like it*. For example, as noted by Nakatani and Dukes,[19] they would certainly agree more nearly on the color of a car than on how much they like it. A group of musically sophisticated individuals would be expected to agree better on how much a musical composition "sounds like Bartok" than on how well they like it, just as they might be expected to agree better on the noisiness of a sample of processed speech than on how acceptable it is. This principle has wide application and provides us with a powerful tool for controlling extraneous variation in acceptability test results.

The relationship between acceptability and various perceptual qualities, such as noisiness, varies from one individual to the next, but knowing this relationship for the *normative* individual we can, in principle, make reliable estimates of acceptability from judgments of noisiness and other perceptual qualities, as well as from "raw" judgments of acceptability.

The task of identifying the perceptual correlates of acceptability began with the compiling of a list of several hundred descriptors of undesirable speech qualities. Listening crews were asked to provide as many synonyms as they could supply for each descriptor, particularly as it might apply in an acoustical or speech context. This served to replace the original set of potential scale descriptors with a smaller, but still large, set of adjectival clusters. Rating scales defined by these clusters were then used in a succession of experimental tests to determine the nature and number of EPQs required to account for the variation in listeners' responses to the still large number of scales involved. Crews of listeners rated a wide diversity of "system conditions" — state-of-the-art speech coders under various operating conditions, simple laboratory degradations, and so forth — with respect to the various potential EPQs, as well as to the higher-order perceptual qualities, *intelligibility, pleasantness,* and *overall acceptability.* The nature and number of underlying speech qualities, the EPQs, were ultimately revealed by factor analysis cluster analysis (of variables) of data from dozens of experiments conducted over a period of 15 years. These efforts led to:

- Confirmation of the fact that listeners can and do reliably distinguish between perceptual qualities of the speech signal itself and perceptual qualities of background noises or other extraneous sounds
- The discovery that eight EPQs of the *signal* and seven EPQs of the *background* are the primary bases of listeners' judgments of the effects of state-of-the-art voice communication systems and devices on speech acceptability
- The discovery that judgments of EPQs can be used (1) to provide reliable and valid supplementary estimates of overall acceptability and (2) to diagnose specific deficiencies and malfunctions of voice communication equipment and devices
- The discovery that judgments regarding intelligibility and pleasantness can provide valuable supplementary estimates of overall speech acceptability along with useful diagnostic information

The present version of the DAM thus requires the listener to judge the *detectability* of a diversity of acceptability-related, simple and complex perceptual qualities. These fall into three categories, depending on the type of acceptability estimate they provide: *isometric* estimates, *metametric* estimates, and *parametric* estimates.

Isometric estimates are *direct* estimates of the level of an elementary or complex perceptual quality. In the case of the DAM, these qualities include *overall acceptability, signal acceptability, background acceptability, pleasantness, intelligibility,* and a variety of more elementary perceptual qualities. The distinction between isometric and other types of estimates ultimately depends on purpose. If listeners' judgments of the detectability of noise in a communication system are used to estimate speech-to-noise ratio, they provide us with isometric estimates of speech-to-noise ratio. If they are used to estimate other properties of the speech signal, such as intelligibility or acceptability, they provide us with parametric or metametric estimates, depending on the nature of the relationship between the properties involved.

The major disadvantage of isometric measures of complex speech features such as *acceptability* is their potential dependence on taste or "preference," and hence their extreme susceptibility to listener sampling error.

As used here, the term *metametric* refers to the special case of estimates that are highly, but *nonlinearly,* correlated with a given perceptual quality — estimates that appear to measure the same (or very similar) qualities but do not yield numerically interchangeable results. Ratings of pleasantness and intelligibility provide two examples. For virtually all types of degradation encountered in modern voice communications, judgments of a system's effects on intelligibility and pleasantness are highly, but curvilinearly, related to judgments of acceptability. By transforming them appropriately, we obtain two metametric estimates of acceptability.

Acceptability estimates based on EPQs are termed parametric estimates and the rating scales from which they are obtained are called parametric scales. The basis for distinguishing parametric estimates from other types of estimates of acceptability is found in the nature of the relationship between EPQs and higher-order qualities such as acceptability: Interdependencies between the various EPQs and the complex qualities of pleasantness, judged intelligibility, and acceptability are *not* symmetrical. For example, perceived raspiness in the received speech signal can reduce acceptability, but poor acceptability can occur in the absence of raspiness. Plots of acceptability vs. EPQ scores all yield triangular scattergrams consistent with the above illustration. (In the course of analyzing DAM data, all EPQ scores are transformed to approximate the acceptability level a system *would* be accorded if it were deficient only with respect to the EPQ involved.)

As used with the DAM, each of the three types of estimate described above has unique advantages and disadvantages, but the three tend to be complementary. When used in combination, they offer improved control of major types of error inherent in all subjective evaluations of speech. Additionally, metametric and parametric estimates can provide potentially valuable "diagnostic" information about the channel or link being evaluated. A list of all scores obtained with the DAM is shown in Table 5.2. Using both linear and nonlinear transformations, the various parametric, metametric, and isometric estimates are combined to yield a final *composite acceptability estimate* (CAE).

TABLE 5.2

Scores Yielded by the DAM

	Parametric Signal-Quality Scores			Parametric Background-Quality Scores	
SF	Signal flutter	Fluttering–pulsating	BNH	Background noise high	Hissing–fizzing
SH	Signal high-pass	Small–distant	BNM	Background noise mid	Rushing–roaring
SD	Signal distortion	Rasping–scratchy	BNL	Background noise low	Rumbling–rolling
SL	Signal low-pass	Dull–muffled	BB	Background buzz	Humming–buzzing
SI	Signal interruption	Interrupted–chopped	BF	Background flutter	Bubbling–percolating
SN	Signal nasality	Nasal–whining	BS	Background static	Crackling–staticky
ST	Signal thin	Thin–tinny	BC	Background chirping	Chirping–clicking
SB	Signal babble	Babbling–slobbering			
	Isometric Signal-Quality Score			**Isometric Background-Quality Score**	
ISA	Isometric signal acceptability	Unnatural–distorted	IBA	Isometric background acceptability	Conspicuous–intrusive
	Isometric Overall-Quality Scores			**Metametric Overall-Quality Scores**	
CIA	Overall acceptability		TRI	Intelligibility	
MOS	Predicted mean opinion score		TRP	Pleasantness	
	Overall Acceptability				
CAE	Composite acceptability estimate				

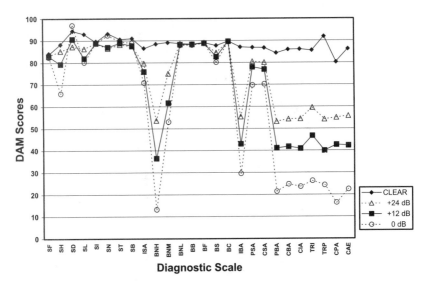

FIGURE 5.6
Effects of Gaussian noise on DAM diagnostic scores.

As with the DRT, the pattern of scores yielded by the DAM varies, depending on the type of degradation involved. Examples using an early version of the DAM are found in Barnwell and Voiers[28] and in Quackenbush, Barnwell, and Clements.[29]

Figure 5.6 shows the effects of band-limited Gaussian noise on the pattern of DAM scores. Most noteworthy are the scores for BNH (background noise high) and IBA (isometric background acceptability. Except for SH (signal highpass), the remaining parametric scores are little affected by this type of degradation. Figure 5.7 shows the effects of speech-modulated masking noise (MNRU), and Figure 5.8 the effects of an eight-voice masking babble on DAM scores. The three sets of DAM score profiles are distinct for the different forms of masking noise.

Figure 5.9 shows the relationship between the ACRM's MOS and the DAM's measure of overall acceptability, CAE, for a large number (n > 200) of common system conditions. The two measures show a substantial degree of cross predictability ($r_{MOS,CAE}$ = 0.959).

Figure 5.10 shows DAM diagnostic score profiles for the three types of noise conditions (i.e., Gaussian, MNRU, and babble) shown in Figures 5.6, 5.7, and 5.8, respectively. In Figure 5.10, the DAM diagnostic scores have been interpolated from the data for each of the three types of noise to yield an equivalent overall acceptability score, CAE, of 50. As shown above for intelligibility, conditions can yield similar scores for overall acceptability while exhibiting dramatically different diagnostic profiles. Figure 5.10, therefore, supports the premise of acceptability as a multidimensional rather than a unidimensional concept and provides a strong argument for using vector methods rather than scalar methods in the evaluation of acceptability.

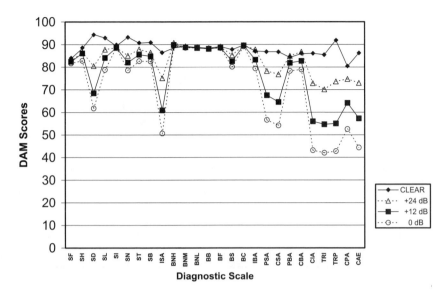

FIGURE 5.7
Effects of speech-modulated noise (MNRU) on DAM diagnostic scores.

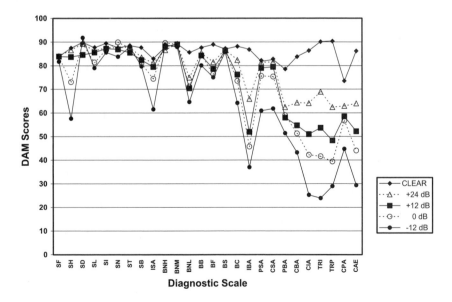

FIGURE 5.8
Effects of an eight-voice masking babble on DAM diagnostic scores.

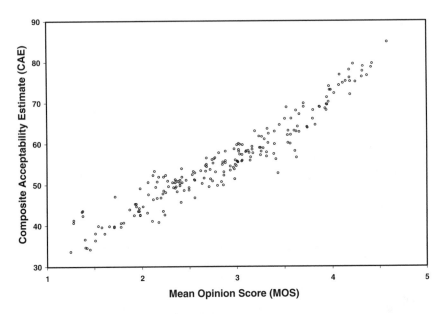

FIGURE 5.9
ACRM's mean opinion score (MOS) vs. DAM's composite acceptability estimate (CAE).

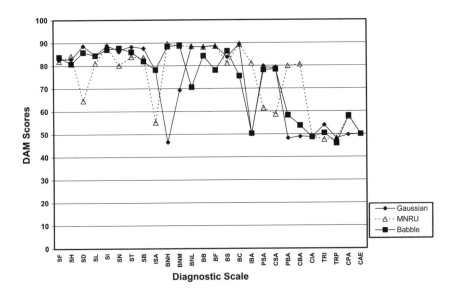

FIGURE 5.10
DAM diagnostic score profiles for three types of masking noise.

Conceptions of the Nature of Speaker Recognizability

Standardized tests for evaluating the effects of voice coding and communication systems on speech intelligibility and acceptability are presently available, but a valid, practical, and economically feasible method for evaluating the effects of such systems on the recognizability of speakers' voices has not been available heretofore. A major obstacle has been the difficulty of controlling listeners' familiarity with voices used in such tests, in particular, the difficulty of obtaining and maintaining listening crews whose members have *equal familiarity* with voices in a sample large enough (probably > 300) to provide adequate representation of the range and diversity of qualities that distinguish voices from each other.

Scalar Approaches to the Evaluation of Speaker Recognizability

Although several approaches to the evaluation of speaker recognizability in communication systems have been proposed during the past few years, none has been generally accepted for purposes of routine evaluation of communication systems. The problem noted above is prominent among the reasons for this. Under support from the Armstrong Laboratory (contract F41624-97-C-6003), Dynastat Inc. undertook the development and validation of a method, the Diagnostic Speaker Recognizability Test (DSRT), in which listeners' familiarity with voices is not a factor. The method is based on the simple premise that voice recognition by human listeners presupposes discrimination, discrimination with respect to various *perceived voice traits* (PVTs), traits that carry information as to the identity of the speaker.

The first steps in the development of the DSRT involved determining the nature and number of PVTs that carry significant amounts of speaker-identity information and the development of rating scales by means of which listeners can characterize their perceptions of these traits.

Kreiman et al.[30] studied the problem of interrater reliability in ratings made of various voice characteristics by speech professionals and by laypersons. They observed limited reliability in ratings made with respect to commonly used clinical descriptors such as *breathiness* and *roughness*. A major aspect of the development of the DSRT involved the experimental test of hundreds of potential scale descriptors to identify those specific descriptors and clusters of descriptors that provided maximum interrater reliability when used to define voice rating scales.

Appropriate multivariate analyses (factor analysis and cluster analysis in particular) of data obtained with a diversity of candidate rating scales revealed the approximate nature and number of underlying PVTs required to account for the variance (across speakers) of listeners' perceptions of voices. Listeners appeared able to discriminate 20 PVTs, 10 of which were bipolar (i.e., scales on which the modal individual falls near the middle of the range) and 10 of which were monopolar (i.e., scales on which the modal individual falls near the lower extreme of the range). The latter appear, generally, to pertain to one form or another of aberrant or dysphonic speech. Two highly correlated rating scales were developed to tap each of the above traits. Averages for each pair of scales provided the means of classifying voices in the 20-dimension trait space. Examples of the two types of rating scales are:

LOW		HIGH
DEEP	vs.	SHALLOW
BASS		TENOR

Bipolar Rating Scale

SCRATCHY
DRY
HOARSE

Monopolar Rating Scale

The 20 PVTs are labeled as follows:

Bipolar PVTs	Monopolar PVTs
Pitch	Breathiness
Rate	Shriekiness
Roughness	Stammeriness
Melodiousness	Slurpiness
Resonance	Wheeziness
Clippedness	Twanginess
Shakiness	Hoarseness
Jerkiness	Thickness
Crispness	Hissiness
Pervasiveness	Exoticness

Any reduction in speaker recognizability occasioned by signal degradation would be reflected in reduced discriminability of one or more PVTs as measured by the *interrater* reliability of voice ratings on these traits. Analysis of variance with factorial design provides a convenient means of evaluating the significance of such effects as reflected in the F-ratio for "(speakers)/(speakers × listeners)." From an F-ratio, we can also estimate the amount of *shareable* speaker identity information (SII) contained in the mean of N listeners' ratings. It also provides the means of comparing the effects of different conditions on the SII content of the received signal. Where the assumption is that both the systematic effect and the error are normally distributed, SII can be estimated as $^1/_2 \log_2(F_N)$, where F is the ratio of mean square for speakers to mean square for error (speakers × listeners) and N is the number of listeners. This statistic is quite robust across variations from normality, such as is found in the case of most monopolar traits. (For the extreme case of a rectangular distribution, this equation overestimates SII by a constant of 0.25 bits.[31])

For purposes of the DSRT, separate sets of five exemplars were selected from pools of 240 males and 240 females to represent approximately equidistant points across the range of mean ratings for each PVT. The exemplars selected for each PVT were, to the extent possible, neutral with respect to all other PVTs.

The DSRT involves having crews of 24+ listeners rate speech samples (sentences) for five exemplars for each of the 20 PVTs. For a given transmission condition, we estimate the average amount of speaker identity information in mean voice ratings (SII) for each PVT, as described above. The effect of a channel or device on the speaker identity information in mean voice ratings for each PVT is measured as SII_C-SII_E, i.e., the difference between SII for the control or clear condition and SII for an experimental condition. Separate tests are conducted for male speakers and female speakers. The results for the two sexes are then averaged.

It should be stressed that the DSRT is presently in the final stages of validation. It is still subject to minor modification, depending on the results of research in progress. Here, unlike the cases of intelligibility and acceptability, no independent criteria of speaker recognizability are available, nor ever likely will be. Hence, the results yielded by the DSRT must be taken at face value.

Figure 5.11, which shows the effects of Gaussian noise masking on the discriminability of the various PVTs (variations of 0.25 or less may be attributable to chance). It is evident that speaker identity information is more robustly encoded for some PVTs than for others. Gaussian noise masking affected the PVTs of pitch, warmth, and thickness the least and the PVTs of slurpiness and wheeziness the most.

Figure 5.12 shows the effect of speech-modulated masking noise (MNRU) on the discriminability of the 20 PVTs. Moderate levels of MNRU appear to have negligible effect on recognizability, but below an MNRU of 0 dB, the loss of SII is widespread, particularly so among the monopolar PVTs. Most affected is the PVT, *exoticness*, which is presumably sensitive to a speaker's linguistic heritage — to speakers whose native language is not American English.

Not shown here are the effects of other forms of degradation, speech coding, etc. The DSRT has been proven highly sensitive to various forms of frequency pass band restriction, to differences among speech coding algorithms, and to data rate in coders thus far tested.

The DSRT is still being validated and refined within the limitations imposed by economic considerations. The ultimate product must be available at prices comparable to those typically charged for tests of intelligibility and acceptability.

Common intuition and the results presented here both attest to the robustness with which speaker identity is encoded in a speaker's voice. A practical need for a test of speaker recognizability may be confined to the case of moderate- to low-rate coders, but final resolution of this issue will depend

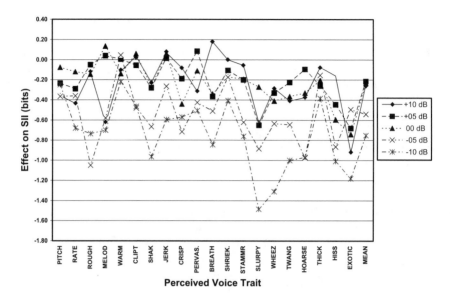

FIGURE 5.11
Effects of Gaussian noise masking on the speaker identity information content of 20 perceived voice traits (PVTs).

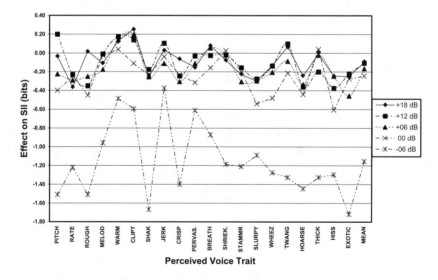

FIGURE 5.12
Effects of speech-modulated noise (MNRU) on the speaker identity information content of 20 perceived voice traits (PVTs).

on the results of continuing research and on future developments in the speech-coding field.

Conclusions

The results presented above attest strongly to the importance of knowledge regarding the effects of noise on the specific speech components on which intelligibility, acceptability, and speaker recognizability depend. Such knowledge should greatly facilitate the development of effective methods of noise reduction in speech communication systems.

References

1. Egan, J.P., Articulation testing methods, *Laryngoscope*, 58, 955–991, 1948.
2. American National Standards Institute, Method for Measuring the Intelligibility of Speech over Communication Systems (ANSI S3.2-1989) — A Revision of ANSI S3.2-1960, American Standards Association, New York.
3. Fairbanks, G., Test of phonemic differentiation: the Rhyme Test, *J. Acoust. Soc. Am.*, 30, 596–600, 1958.
4. House, A.S. et al., Articulation testing methods: consonantal differentiation with a closed response set, *J. Acoust. Soc. Am.*, 37, 158, 1965.
5. Voiers, W.D., Sharpley, A.D., and Panzer, I.L., A modified rhyme test with controlled item difficulty and foil attractiveness, in preparation.
6. Steeneken, H.J.M., *On Measuring and Predicting Speech Intelligibility*, H.J.M. Steeneken, Soesterberg, the Netherlands, 1992.
7. Miller, G.A. and Nicely, P.E., An analysis of perceptual confusions among some English consonants. *J. Acoust. Soc. Am.*, 30, 338–352, 1955.
8. Jackobson, R., Fant, C.G.M., and Halle, M., Preliminaries to speech analysis: the distinctive features and their correlates, Tech. Rep. No. 13, Acoustics Laboratory, MIT, Cambridge, MA, 1952.
9. Peters, R.W., Dimensions of perception for consonants, *J. Acoust. Soc. Am.*, 35, 1985–1989, 1963.
10. Wicklegren, W.A. Distinctive features and errors in short-term memory for English consonants, *J. Acoust. Soc. Am.*, 39, 388–398, 1965.
11. Singh, S., Woods, D.R., and Becker, G.M., Perceptual structure of 22 prevocalic English consonants, *J. Acoust. Soc. Am.*, 52, 1698–1713, 1972.
12. Voiers, W.D., Diagnostic evaluation of speech intelligibility, in *Speech Intelligibility and Speaker Recognition*, Hawley, M.E., ed., Dowden, Hutchinson, and Ross, Stroudsburg, PA, 1977, chap. 34.
13. Voiers, W.D., Evaluating processed speech using the Diagnostic Rhyme Test, *Speech Technol.*, January/February, 30–39, 1983.
14. Voiers, W.D., Measurement of intrinsic deficiency in transmitted speech: The Diagnostic Discrimination Test (DDT), *Proceedings of the 1982 IEEE International Conference on Acoustics, Speech, and Signal Processing*, Paris, France, 1982.

15. Voiers, W.D., Three new diagnostic intelligibility tests for communication systems, *J. Acoust. Soc. Am.*, 95, 3011(A), 1994.
16. Schmidt-Nielsen, A., Intelligibility and acceptability testing for speech technology, in *Applied Speech Technology*, Syrdal, A., Bennett, R., and Greenspan, S., eds. CRC Press, Boca Raton, FL, 1995, chap. 5.
17. Voiers, W.D., On the problems of reliability, sensitivity, and diagnostic value in tests of speech intelligibility, Conference preprints, *AFCRL-IEEE Conference on Speech Communication and Processing*, 1967.
18. Rothauser, E.H., Urbanek, G.E., and Pachl, W.P., Speech quality measurements. Final Scientific Report AF 61(052)–856, 1967.
19. Nakatani, L.H. and Dukes, K.D., A sensitive test of speech communication quality, *J. Acoust. Soc. Am.*, 53, 1083–1092, 1973.
20. Voiers, W.D., Diagnostic acceptability measure for speech communication systems, *Proc. IEEE ICASSP*, Hartford, CT, 1977.
21. Goodman, D.J. and Nash, R.D., Subjective quality of the same speech transmission conditions in seven different countries, *IEEE Trans. Commun. COM-30*, 642, 1982.
22. Combescure, P., Quality evaluation of speech coded at 32 kbit/sec by means of degradation category ratings, *Proc. ICASSP 82 (International Conference on Acoustics, Speech and Signal Processing)*, Vol. 2, Paris, May 1982.
23. Voiers, W.D., Subjective criteria of speech acceptability, in *An Analysis of Objective Measures for User Acceptance of Voice Communication Systems*, Barnwell, T.P. and Voiers, W.D., eds., Defense Communications Agency Report DCA100-78-C-0003, 1979, chaps. 2 and 5 (reproduced in Quackenbush, S.R., Barnwell, T.P., and Clements, M.A., *Objective Measures of Speech Quality*, Prentice-Hall, Englewood Cliffs, NJ, 1988).
24. International Telecommunication Union-Telecommunication Standardization Sector, Recommendation P.800, Methods for subjective determination of transmission quality, August 1996.
25. International Telecommunication Union-Telecommunication Standardization Sector, Recommendation P.830, Methods for objective and subjective assessment of quality, February 1996.
26. Panzer, I.L., Sharpley, A.D., and Voiers, W.D., A comparison of subjective methods for evaluating speech quality, in *Speech and Audio Coding for Wireless and Network Applications*, Atal, B.S., Cuperman, V., and Gersho, A., eds., Kluwer Academic Publishers, Boston, 1993, chap. 8.
27. McDermott, B.J., Multidimensional analysis of circuit quality judgments, *J. Acoust. Soc. Am.*, 45, 774–781, 1969.
28. Voiers, W. D., Effects of selected forms of degradation on speech acceptability and its perceptual correlates, in *An Analysis of Objective Measures for User Acceptance of Voice Communication Systems*, Barnwell, T. P. and Voiers, W. D., eds., Defense Communications Agency Report DCA100-78-C-0003, 1979, chap. 5 (reproduced in Quackenbush, S.R., Barnwell, T.P., and Clements, M.A., *Objective Measures of Speech Quality*, Prentice-Hall, Englewood Cliffs, NJ, 1988).
29. Quackenbush, S.R., Barnwell, T.P., and Clements, M.A., *Objective Measures of Speech Quality*, Prentice-Hall, Englewood Cliffs, NJ, 1988.
30. Kreiman, J. et al., Perceptual evaluation of voice quality: review, tutorial, and a framework for future research, *J. Speech Hear. Res.*, 36, 21–40, 1993.
31. Attneave, F., *Applications of Information Theory to Psychology: A Summary of Basic Concepts, Methods, and Results*, Henry Holt, New York, 1959.

Section III:

Digital Algorithms and Implementation

6

Single-Channel Speech Enhancement

Graham P. Eatwell

CONTENTS

0-8493-0949-2/02/$0.00+$1.50
© 2002 by CRC Press LLC

Characteristics of Speech and Noise

Single-channel speech enhancement relies upon differences between the characteristics of speech and noise. In some situations, these differences are clear. For example, telephone-quality speech lies in a frequency band 300 to 3300 Hz — any noise below 300 Hz or above 3300 Hz can be removed by passing the signal through a bandpass filter. Other noise, e.g., tonal noise, is contained in a number of very narrow frequency bands. This noise can be removed by filtering out these frequencies. If the bands are very narrow, the effect on speech will be small.

However, in many situations, the noise has a broadband, random nature with frequency components across the whole speech band. Some components of speech (such as vowel sounds in normal speech) are produced by a periodic vibration of the vocal chords. These components constitute voiced speech. Other components are produced by shaping a turbulent airflow. These are called unvoiced components. Whispers are unvoiced. The unvoiced components of speech are broadband, which makes the separation of speech from the noise a difficult task. To make any progress in this direction it is necessary to look at the statistical properties of the speech and noise signals.

The earliest statistical property used in speech enhancement is the autocorrelation function, or its equivalent, the power spectral density. An early review of such methods was done by Lim and Oppenheim.[1] Methods based on more sophisticated statistical models have been reviewed by Ephraim.[2]

Frames

Speech signals can be considered as stationary over periods of about 20 ms. This time scale is associated with changes in the shape of the vocal tract. It is common in speech processing to consider blocks or frames of speech of a similar duration. For some speech processing, such as pitch detection, longer frames are used to encompass multiple pitch periods. Since the transitions from one vocal tract shape to another are continuous, it is usual to process overlapping sections of speech.

We begin by considering a signal x, which is sampled at times nT, where n is an integer and T is the sampling period. We denote the sample value at time nT by $x(n)$. The dependence upon the sampling period is implicit. Each frame contains N samples, and the vector of samples for the frame ending at time sample n is defined as

$$\mathbf{x}_N(n) = \{x(n), x(n-1), x(n-2), ..., x(n-N+1)\}^T \tag{6.1}$$

In MATLAB notation,

```
x_sub_N  =  x(n:-1:n-N+1);
```

In general, where it does not cause confusion, the explicit dependence upon the frame length N will be dropped. The frame of N samples is called an N-vector and may be thought of as a vector in an N-dimensional vector space. Consecutive frames may be overlapped by N-M samples, so that a new frame is generated every M samples.

Models for Speech

Some speech enhancement techniques make no assumptions about the speech signal except that it is uncorrelated with the noise and that the noise statistics vary more slowly than the speech statistics. However, the speech signal can be reasonably well modeled as an autoregressive process. This provides some additional *a priori* knowledge of the speech signal, which can be used in an enhancement scheme. Before moving on to discuss speech enhancement, we shall look at how an autoregressive (AR) model for a signal may be formulated in terms of speech frames.

In an AR model, the current value of the signal is defined in terms of the previous value of the signal. In the AR model, the current sample $x(n)$ is written as

$$x(n) = \mathbf{x}_L^T(n-1)\mathbf{a}(n) + s(n) \tag{6.2}$$

where $\mathbf{a}(n)$ is an L-vector (a vector of length L) of linear prediction (LP) coefficients, $\mathbf{x}_L$ is an L-vector of previous samples and $s(n)$ is an excitation or source signal. $s(n)$ can also be interpreted as a prediction error signal (the difference between the signal and its prediction). It is also called the *innovation*, since it is the part of the signal that cannot be predicted from past measurements. The signal is thus modeled as passing the excitation signal $s(n)$ through a recursive (all-pole) filter with coefficients $\mathbf{a}(n)$. The excitation signal is often modeled as either a Gaussian random noise signal (for unvoiced speech or noise) or an impulse train (for voiced speech).

This model gives the current *sample* in terms of the previous samples. For speech enhancement, we would like to find expressions for the current *frame*. Several different formulations are described below.

Data Matrix Form

A frame of the signal can be written as

$$\mathbf{x}_N(n) = X(n-1)\mathbf{a}(n) + \mathbf{s}_N(n) \tag{6.3}$$

where $X(n-1)$ is an $N \times L$ *data matrix* of previous samples of the signal, defined as

$$X(n-1) = \begin{bmatrix} \mathbf{x}_L^T(n-1) \\ \mathbf{x}_L^T(n-2) \\ \vdots \\ \mathbf{x}_L^T(n-N) \end{bmatrix} \tag{6.4}$$

Frame-Recursive Form

The current frame can also be written in terms of previous frames, rather than in terms of a data matrix

$$\mathbf{x}_N(n) = A_F \mathbf{x}_N(n-M) + B_F \mathbf{s}_N(n) \tag{6.5}$$

where A_F and B_F are matrices that depend upon the time-varying LP coefficients.

Circular Convolution Approximation

An alternative representation is

$$A_C \mathbf{x}_N(n) = \mathbf{s}_N(n) + \mathbf{e}(n) \tag{6.6}$$

where A_C is an $N \times N$ circular matrix with first-row $c^T(n) = \{1, -a_1, -a_2, \ldots, 0, 0, \ldots 0\}$ and $\mathbf{e}(n)$ is an error vector. The first $N - L$ terms of the error vector are zero. If $N \gg L$, the error is small and can be neglected. An equivalent representation is

$$\mathbf{x}_N(n) \cong B_C \mathbf{s}_N(n) \tag{6.7}$$

where $B_C = A_C^{-1}$. This approximation is useful because of the special properties of circular matrices, which will be discussed later in the sections "Discrete Fourier Transform" and "Estimating the Characteristics of the Speech and Noise." It also allows us to develop some simple relationships between the autocorrelation matrices of the speech signal and the excitation signal. For example, we can construct the $N \times N$ autocorrelation matrix R of the signal from knowledge of the vector $\mathbf{a}$ and the power σ^2 of the excitation signal. Equivalently, we can think of $\mathbf{a}$ and σ^2 as a parameterization of the autocorrelation matrix. For zero mean Gaussian processes, the probability density function is determined from R, so $\mathbf{a}$ and σ^2 also provide a parameterization of the probability density function. The representation is valid for the period during which the signal is stationary and describes the intraframe relationship between samples.

For speech, the statistics can be considered stationary within each frame. However, they will vary from frame to frame. To obtain a better representation of the statistics of speech, we must also consider the relationships between frames. This will be considered later in "Vector Quantization."

Hidden Markov Model

Referring to the frame autoregressive model, we can write $n = mM$, where m is the frame index and M is the frame advance. This gives

$$\mathbf{x}_N(m) \cong B_M(\mathbf{a}_m, \mathbf{a}_{m-1})\mathbf{s}_N(m) \tag{6.8}$$

where the coefficient matrix B_M depends upon the LP coefficients in the current and previous frames. We refer to the LP coefficient vectors as the *state vectors* of the system. In this model, the state vectors are quantized. Rather than a continuum of states, it is assumed that a finite discrete set of state vectors is sufficient to describe the speech process. This is clearly an approximation. The probability density functions (pdf) of the states are assumed to form a Markov process, in which the pdf of the current state can be predicted from the pdf of the previous state alone. This form is called a *hidden* Markov model (HMM) for $\mathbf{x}_N$, since the state vectors cannot be observed directly. Instead, we observe the result of a random process whose statistics are determined by the state vector. The use of a HMM for speech enhancement is described in, e.g., Ephraim,[2] Ephraim, Malah, and Juang,[3] and Sameti, Sheikhzadeh, and Deng.[4]

The relationship between the state in one frame and the state in the next is described by a state transition matrix. The states contain information about how samples within each frame and the previous frame are related, while the state transition matrix contains information about how the state vector changes between frames.

General Speech Enhancement

Time-Domain Filtering

In many environments, the noise is *additive*. The noisy signal vector is written as

$$\mathbf{y}(m) = \mathbf{x}(m) + \mathbf{d}(m) \tag{6.9}$$

where $\mathbf{x}(m)$ is the vector of clean speech signals and $\mathbf{d}(m)$ is the vector of noise samples. Other types of noise exist. For example, distortion produces a noise that is dependent upon the speech signal. In this chapter we will be concerned with additive noise for which the speech and noise are assumed to be independent (i.e., the cross correlation of the speech and noise is equal to zero). The *autocorrelation* matrix R_y of the input for frame m satisfies

$$R_y(m) \equiv \left\langle \mathbf{y}(m)\mathbf{y}^T(m) \right\rangle = R_x(m) + R_d(m) \tag{6.10}$$

where R_x is the autocorrelation matrix of the speech and R_d is the autocorrelation matrix of the noise. The angled brackets denote the expected value.

The elements of the autocorrelation matrix can be estimated by a sample average over each frame. The mean square value (power) of an example speech signal is shown in Figure 6.1. The uppermost plot shows the power of the original speech signal in decibels (relative to full-scale input for this digitized signal). The vertical lines denote the beginning and end of the speech signal. Notice that there is some background noise at a level of –40 dB even in this relatively clean signal. The middle plot shows the power of a white noise signal. The noise is stationary, so the level is fairly constant. Even though the noise is stationary, the mean square value is only an approximation of the statistics and so is not exactly constant. The lowest plot shows the power of the combined speech and noise. This closely approximates the sum of the individual powers. Notice how the last section of speech is lost in the noise. This illustrates one of the difficulties in detecting speech in the presence of noise.

If the autocorrelation matrix of the noise is known, the autocorrelation matrix of the speech can be estimated as

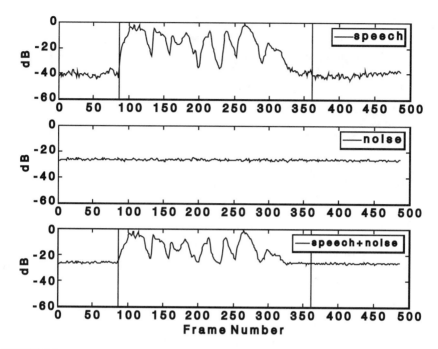

FIGURE 6.1
Signal powers.

$$R_x(m) = R_y(m) - R_d(m) \tag{6.11}$$

If the noise is stationary or almost stationary, the noise autocorrelation matrix R_d does not change much from frame to frame. R_d can be estimated by averaging during pauses in the speech. The speech is not stationary, and we only have one instance of the noisy speech. Estimation of $R_y(m)$ is therefore more difficult. However, this does provide the motivation for a variety of speech enhancement algorithms. The main problems are:

1. Estimation of the speech autocorrelation matrix
2. Estimation of the noise autocorrelation matrix
3. Use of the autocorrelation matrices to obtain enhanced speech

Spectral Filtering

We begin by considering the last of the problems described above. One way of using the autocorrelation matrices to obtain enhanced speech is to use them to design a filter. Filtering may be performed in the time domain or a spectral domain, such as the frequency domain. In spectral filtering, the estimate $\hat{x}(m)$ of the clean speech is

$$\hat{x}(m) = U(m)F(m)U^*(m)y(m) \qquad (6.12)$$

where U is a unitary transform matrix that satisfies $UU^* = I$, and $F(m)$ is a *gain* or *filter* matrix. The superposed star denotes the conjugate transpose of the matrix. This process is depicted in Figure 6.2.

The computation of $\hat{x}(m)$ is performed in three steps: analysis, filtering, and synthesis.

Analysis

The input spectrum is obtained by transforming the input vector

$$\mathbf{Y}(m) = U^*(m)\mathbf{y}(m) \qquad (6.13)$$

Although written as a matrix–vector multiplication, some transforms can be calculated using fast algorithms. Examples include wavelet, Fourier, and discrete cosine transforms.

Filtering

The output spectrum is calculated from the input spectrum

$$\hat{\mathbf{X}}(m) = F(m)\mathbf{Y}(m) \qquad (6.14)$$

Synthesis

The output vector is calculated by applying the inverse transform to the output spectrum

$$\hat{\mathbf{x}}(m) = U(m)\hat{\mathbf{X}}(m) \qquad (6.15)$$

The output time series is obtained from the output vectors. There are a number of different strategies that can be used to design a speech enhancement filter:

1. The unitary transform matrix U can be fixed and the filter matrix $F(m)$ adjusted for each frame (e.g., frequency domain spectral

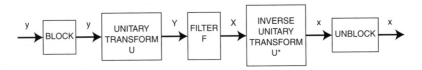

FIGURE 6.2
Spectral filtering.

filtering, wavelet denoising). In general, the matrix F can be a function of current and past inputs, $Y(m), Y(m-1), \ldots$, and past outputs, $\hat{X}(m), X(m-1), \ldots$. Often, the matrix F is diagonal. Indeed, the unitary transform should be chosen so that F is diagonal or almost diagonal.

2. The unitary transform matrix $U(m)$ can be adjusted for each frame and the weighting matrix F held fixed (e.g., subspace projection).

3. Both the unitary transform matrix $U(m)$ and the weighting matrix $F(m)$ can be adjusted for each frame (e.g., subspace filtering).

The underlying assumption is that the application of the unitary transform to the noisy vector helps to *separate* the speech and noise vectors in some sense.

We now consider some specific unitary transforms used in speech enhancement.

Spectral Filtering Examples

Discrete Fourier Transform

The discrete Fourier transform is widely used for several reasons. First, the fast Fourier transform (FFT) provides a computationally efficient way of computing the transform and its inverse and, second, the human ear performs a kind of Fourier analysis.

The kth column of U is the kth column of the Fourier transform matrix, namely,

$$\mathbf{u}_k(m) = \mathbf{u}_k = \frac{W_N^{-mM}}{\sqrt{N}} \{1, W_N^k, W_N^{2k}, \ldots, W_N^{(N-1)k}\}^T \qquad (6.16)$$

where $W_N = e^{-i2\pi/N}$. The factor $1/\sqrt{N}$ is introduced so that $\|\mathbf{u}_k\| = 1$. The factor W_N^{-mM} is included only if it is important to keep track of the phase between frames. If the factor is included, the transform is sometimes called the short-term discrete Fourier transform.

In MATLAB notation,

```
U = exp(i*(0:N-1)'*(0:N-1)*2*pi/N)/sqrt(N);
```

or

```
U = fft(eye(N,N))/sqrt(N);
```

In the *circular convolution approximation*, the Fourier transform diagonalizes the autocorrelation matrix. Furthermore, if the excitation is Gaussian, the statistics are completely determined by the autocorrelation matrix and the spectral components are statistically independent. For the approximation to be useful, the frame length (transform length) must be much greater than the maximum correlation time of the signal. This means that the frequency components can be treated independent of one another, which is a major simplification.

Karhunen-Loeve Transform

In the approximation above, the Fourier transform was found to diagonalize the autocorrelation matrix. The approximation may not be good for voiced speech, which has a long correlation time and where a long filter would be required to whiten the signal.

In the Karhunen-Loeve transform, the kth vector $\mathbf{u}_k(m)$ of the unitary transform matrix is the kth eigenvector of the $N \times N$ autocorrelation matrix R. Consequently, the unitary transform will diagonalize the autocorrelation matrix whatever its form. The autocorrelation matrix, and hence the transform, can be estimated from $N_1 > N$ samples, including the current frame.

In MATLAB notation,

$$[\text{U},\text{E}] \; = \; \text{eig}(\text{R}); \tag{6.17}$$

where R is the autocorrelation matrix and $E = D^2$ is the diagonal matrix of eigenvalues.

Subband Filtering/Wavelet Transform

In subband filtering, $\mathbf{u}_k(m)$ is the vector of coefficients of the kth synthesis filter. Although the filter is implemented in fast-form, as a tree of filters, with decimation at each stage (see, e.g., Reference 5), the result is mathematically equivalent to multiplication by a transform matrix.

Discrete Cosine Transform

In MATLAB notation,

```
U = diag([1 sqrt(2)*ones(1,N-1)])* ...
   cos((0:N-1)'*(0.5:N)*pi/N)/sqrt(N);
```

Estimating the Characteristics of the Speech and Noise

There are two approaches to the design of the spectral filter $F(m)$. The first approach is a direct approach, where the parameters of $F(m)$ are found directly from the data. The second approach is an indirect approach in which some characteristics of the speech and noise are estimated and then the filter is calculated from these characteristics. We shall consider the indirect approach first, and look at some methods for estimating the signal and noise characteristics.

Noise Estimators

Many noises can be modeled as filtered white noise. The autocorrelation function of the noise is constant, or at least slowly changing relative to the time scales of the speech. The autocorrelation matrix for the noise spectrum is

$$R_D(m) \equiv U(m)R_d U^*(m) \tag{6.18}$$

If U is a constant transform matrix, we can estimate $R_D(m)$ directly. Alternatively, $R_d(m)$ can be estimated and then $R_D(m)$ can be found using Equation (6.18). A special case is when the noise is white. In this case, $R_d(m) = R_D(m) = \sigma^2 I$. If the noise can be modeled as an autoregressive process, a pre-whitening filter can be used to ensure that the noise in the input to the speech enhancer is white. An inverse whitening filter is then used on the output to remove the distortion of the speech. This approach should be used with caution if the algorithm uses sophisticated speech models, since the speech will be distorted by the whitening filter.

For constant transforms, the autocorrelation matrix for the noise spectrum can be estimated recursively as

$$\hat{R}_D(m) = \hat{R}_D(m-1) + \varepsilon_m \mu[\mathbf{D}(m)\mathbf{D}^*(m) - \hat{R}_D(m-1)] \tag{6.19}$$

where

$$\varepsilon_m = \begin{cases} 0 & \text{if speech is present in the frame} \\ 1 & \text{otherwise} \end{cases} \tag{6.20}$$

$\mathbf{D}(m)$ is the transform of the noise vector, $\mathbf{d}(m)$, and μ is a small positive constant. The initial value can be set from the first nonspeech frame, for example.

For the Fourier transform, $R_D(m)$ is diagonal. We write the kth diagonal element as $R_D(m,k)$, which is updated according to

$$\hat{R}_D(m,k) = \hat{R}_D(m-1,k) + \varepsilon_m \mu[|D_k(m)|^2 - \hat{R}_D(m-1,k)] \tag{6.21}$$

These approaches require that the pauses in the speech be detected, so the periods of speech activity must be detected using a *voice activity detector*. A voice activity detector is a device (or algorithm) that determines if speech is present in a signal at a given time. A simple voice activity detector might monitor the power of the signal and compare it to a threshold level related to the ambient noise level.

An alternative approach[6] is to track minima in the spectral powers. This avoids the need for a voice activity detector but requires knowledge of the relationship between the sequence of minimum values and the expected value.

Signal Estimators

Spectral Subtraction

In the frequency domain, the autocorrelation matrices are assumed to be diagonal and the components satisfy

$$R_Y(m,k) = R_X(m,k) + R_D(m,k) \tag{6.22}$$

We can estimate $R_Y(m,k)$ from the current frame as simply

$$\hat{R}_Y(m,k) = \left| Y_k(m) \right|^2 \tag{6.23}$$

and we can estimate the speech power in the current frame as

$$\left| \hat{X}_k(m) \right|^2 = R_X(m,k) \cong \hat{R}_Y(m,k) - \hat{R}_D(m,k) \tag{6.24}$$

That is, the estimate of the power of the noise spectrum is subtracted from the spectrum of the power of the noisy input signal to yield an estimate of the power of the speech spectrum. Accordingly, this approach is called spectral subtraction.[7-10] One problem associated with this estimator is that the speech amplitude $\hat{R}_Y(m,k) - \hat{R}_D(m,k)$ can be negative. This is not a valid estimate, so it must be fixed in some way. The usual approach is to set the negative portions to zero. During pauses in the speech, this results in spectral components that appear and disappear, creating a "musical" artifact in the residual noise.

The spectral amplitude in the current frame can be estimated by simply taking the square root.

Interframe Smoothing

Ephraim and Malah[11] use a recursive estimate of the signal to noise ratio. If the noise is assumed constant, this is equivalent[12] to a recursive estimate of the signal power estimate $\hat{R}_X(m,k)$, namely,

$$\hat{R}_X(m,k) = (1-\beta)\max(\left| Y_k(m) \right|^2 - \hat{R}_D(m,k), 0) + \beta \left| \hat{X}(m-1,k) \right|^2 \tag{6.25}$$

where $\beta < 1$ is a parameter to be chosen. This estimate reduces to the simple spectral estimate when $\beta = 0$.

Speech Model–Based Signal Estimators

The algorithms described above treat each spectral component as being independent from the other. However, if the speech has been generated by an autoregressive process the autocorrelation function and hence the power spectrum have a limited number of free parameters. The spectral estimates above are not constrained to fit this parametric model.

Intraframe Smoothing

The idea here is to fit a parametric model to the power spectrum of the speech. The most common model is the LP model. The LP model has fewer free parameters than the full-power spectrum, and the net effect is a smoothing of the spectral components. This approach utilizes the relationship between spectral components in the same frame, so it is referred to as *intraframe smoothing*. There are, of course, many other types of intraframe smoothing that can be used.

Iterative Weiner Filter

The iterative Weiner filter[13] is an example of an enhancement algorithm that uses a parametric model for the power spectrum of the speech. The idea is to fit a parametric model to the output of a spectral filter (in "Intraframe Smoothing," the model was applied to the estimate of the speech power spectrum, which is used to design the filter). This parametrically constrained spectrum is then used as an improved estimate of the signal power spectrum. Based on this improved estimate, a new filter is defined, producing a new output. The process is then iterated until some stop criterion is met. In general, the iterative process does not converge, so various approaches have been used to determine the stop criterion or to constrain the parameters themselves. One method is to insist that the parameters remain close (in some sense) to the parameters on neighboring blocks. Tracking line spectral pairs is one method that has been used. Line spectral pairs are a function of the linear prediction coefficients of the frame. It has been shown that line spectral pairs change slowly between adjacent frames. Constraining the change between frames may be used to determine when the iteration should be halted.

Vector Quantization

The parametric modeling described above provides one method for constraining the estimate of the speech autospectrum. An alternative approach is to look at all of the possible speech spectra that can occur. Clearly, this is a continuum of possibilities, but the technique of vector quantization may be used to define a finite set of representative vectors. The vectors can be the spectral amplitudes themselves, but usually the dimension is reduced by considering some function of the spectral amplitude (often called a *feature vector*). The representation feature vectors (called *code words*) may be chosen by a clustering algorithm, such as the *K*-means algorithm. The feature vector is often associated with the LP parameters **a** and σ^2, or some other smoothed version of the spectrum. In the simplest algorithm, the feature vector of the input is calculated and the closest code word is selected and used as the estimate of the clean speech. This approach can be combined with the algorithms described above. Quatieri and McAulay[14] give an example of vector quantization.

Hidden Markov Model and State-Dependent Dynamical System Model

An enhancement of vector quantization is the state-dependent dynamical model. Here a vector quantization based on the LP parameters **a** is used, but in addition, a simple statistical model (a Markov model) of how the parameters are likely to change between frames is also used. The net result is a HMM or state-dependent dynamical system model. In the model, the states are related to the (quantized) parameters **a** (or the parameters of the particular feature vectors being used). However, the states themselves cannot be observed (they are hidden), instead we can only measure the input signal, which is the result of passing a random signal through the model and then adding noise.

In either form, if we know the states $\mathbf{a}_m$ and $\mathbf{a}_{m-1}$, and the statistics of the excitation signal vector $\mathbf{s}_N$, we can estimate the statistics of signal vector $\mathbf{x}_N$. However, we can only observe the noisy input signal. The probability density functions of the states are related by

$$p(\mathbf{a}_m) = p(\mathbf{a}_m \mid \mathbf{a}_{m-1})p(\mathbf{a}_{m-1}) = T(m, m-1)p(\mathbf{a}_{m-1}) \qquad (6.26)$$

where $p(\mathbf{a}_m)$ is the probability density for state $\mathbf{a}_m$. $T(m, m-1) = p(\mathbf{a}_m \mid \mathbf{a}_{m-1})$ is an element of the state transition matrix, which is the probability density for state $\mathbf{a}_m$, given that the system was in state $\mathbf{a}_{m-1}$ in the previous frame.

The problems are:

1. How do we determine the set of states, $\{\mathbf{a}_i\}$?
2. How do we determine the probabilities in the matrix, T?
3. How do we determine the probabilities of the model being in a particular state, given the observation of the noisy signal?
4. Given these probabilities, how do we design a filter?

Fortunately, the solutions to problems 1 and 2 are well documented, since this model has been used extensively in automatic speech recognition.

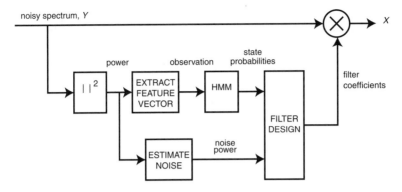

FIGURE 6.3
Spectral filter with hidden Markov model.

Problems 3 and 4 have been discussed by a number of investigators (see, e.g., Chapter 12 in Reference 15).

Several approaches have been used as solutions for problem 4. The first is simply to use the most likely state (i.e., the one with the highest probability) as the feature vector of the speech. This is the standard HMM approach used in speech recognition, where a single phoneme must be selected. A spectral filter or Kalman filter is then designed using this feature vector. An example is shown in Figure 6.3.

In speech enhancement, however, there is no need to make a hard decision. Accordingly, a second approach is to calculate filters for all of the states and apply all of them. The filter outputs are then combined according to the state probabilities, so that the most likely state has the highest weighting.[16]

It is important to note that a training phase is required in order to identify the set of state vectors to be used.

Spectral Filter Design

The estimation procedures described above result in estimates for the auto-spectra of the signal and the noise. Once these have been obtained, a corresponding filter (or gain) must be designed.

In this section, we look at some spectral filter designs. The speech and noise spectra can also be used to estimate statistical models of the speech and the noise. These models can also be used in time domain filters, such as the Kalman filter.

All of the signal estimators described above estimate the spectral amplitudes of the signal and the noise. No attempt is made to estimate the phase of the noise (since it is usually unpredictable). The phase of the speech is simply estimated as the phase of the noisy input spectrum, since it has been shown that the human ear is relatively insensitive to phase (except for binaural hearing). This is not surprising, since the listener must be able to understand speech in reverberant environments, where multiple reflections have a big influence on the phase. However, large, *time-varying* phase errors will produce significant distortion.

Spectral Subtraction

In the simple spectral subtraction approach, the final estimate of the speech spectrum is

$$\hat{X}_k(m) = [\max(|Y_k(m)|^2 - \hat{R}_D(m,k), 0]^{1/2} \frac{Y_k(m)}{|Y_k(m)|} \tag{6.27}$$

This can be written in the standard form $\hat{\mathbf{X}}(m) = F(m)\mathbf{Y}(m)$, where $F(m)$ is a diagonal matrix with diagonal elements

$$F_k(m) = \frac{[\max(|Y_k(m)|^2 - \hat{R}_D(m,k), 0)]^{1/2}}{|Y_k(m)|} \tag{6.28}$$

Notice that the filter has zero phase. Only the amplitude of the noise input is changed at each frequency. Although the ear is less sensitive to phase errors than to amplitude errors, this is a fundamental limitation on the performance of this type of filter. When the clean version of the speech signal is available, it is easy to generate a synthetic signal having the phase of the noisy signal and the amplitude of the clean signal. This should be one of the first tests performed when considering this type of filter for a particular application, since it represents a fundamental limitation on performance. In addition, a filter can be designed based upon the exact $|X_k(m)|$ rather than the estimate $|\hat{X}_k(m)|$. This is a good test of the filter design process.

Using the same approach as above, an alternative form for $F(m)$, motivated by the Wiener filter, is

$$F_k(m) = \frac{\max(|Y_k(m)|^2 - \hat{R}_D(m,k), 0)}{|Y_k(m)|^2} \tag{6.29}$$

A generic spectral subtraction/spectral filtering scheme is shown in Figure 6.4.

Examples of other gain functions are given in the example code that accompanies this chapter (available at http://www.crcpress.com/e_ products/download.asp?cat_no=0949). These include gain functions derived by McAulay and Malpass[17] and by Ephraim and Malah.[11] Ephraim and Malah[11] observe that the probability density function for the spectral amplitude is

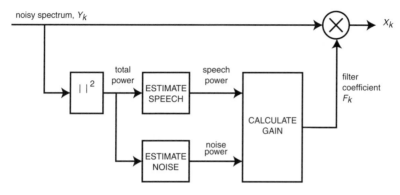

FIGURE 6.4
Spectral amplitude filtering.

not symmetric (it is bounded by zero on one side). Consequently, the Weiner filter is not an optimal estimator for the spectral amplitude. They derive an estimator for the spectral amplitude that yields the minimum mean square error (MMSE), assuming the complex spectral components are Gaussian.

In practical applications, the gain function may be computed via a lookup table.

There are many heuristic schemes for choosing the gain. These include both time (interframe) and frequency (intraframe) smoothing.

Threshold Filtering

In some stategies (e.g., see Reference 7 and wavelet denoising), only the largest spectral components are retained. The equivalent filter is

$$F_k(m) = \begin{cases} 1 & \text{if } |Y_k(m)| > T_k \\ 0 & \text{otherwise} \end{cases} \tag{6.30}$$

where the threshold T_k, is a constant or is dependent upon the expected noise.

Subspace Decomposition

Ephraim and Van Trees[18] and Asano et al.[19] used a Karhunen-Loeve transform instead of the Fourier transform. This transform is better suited to voiced speech than the Fourier transform, but it is seldom used for real-time speech processing because the transform requires a large amount of computation.

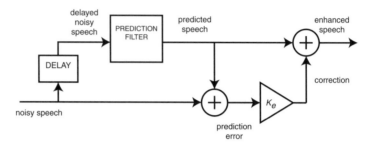

FIGURE 6.5
Feedforward prediction-correction filter.

Predictive Filtering

When the noise is white, the noisy input vector can be decomposed as

$$\mathbf{y}(m) = \mathbf{x}_p(m) + \mathbf{e}(m) \tag{6.31}$$

where

$$\mathbf{x}_p(m) = Y(Y^TY)^{-1}Y^T\mathbf{y}(m) \tag{6.32}$$

is the predictable part of the signal and $\mathbf{e}(m)$ is the error or the unpredictable part. This is known as Wold's decomposition and shows that the predictable part of $\mathbf{y}(m)$ is a projection of $\mathbf{y}(m)$ onto a subspace.

In the approach used by Eatwell,[20] for example, the speech estimate can be written in the general form

$$\hat{\mathbf{x}}(m) = K_p(m)\mathbf{x}_p(m) + K_e(m)\mathbf{e}(m) \tag{6.33}$$

The prewhitening filter is updated during pauses in the speech. One example for the choice of gains is

$$K_p = 1, \quad K_e(m) = \frac{\langle e^2(m)\rangle - \sigma^2}{\langle e^2(m)\rangle} \tag{6.34}$$

where σ^2 is an estimate of the noise power. The algorithm is implemented as shown in Figure 6.5.

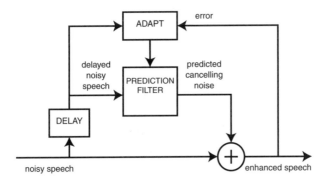

FIGURE 6.6
Adaptive interference canceller for predictable noise.

Predictable Noise

In the discussion above, we assumed that the speech was correlated while the noise was uncorrelated. A different situation occurs when the correlation time of the noise is much longer than that of the speech. This is the case for repetitive noise (such as hums or buzzes). In this case, the only part of the signal that can be predicted a long time ahead is the noise. The predictable part of the signal is then an estimate of the noise. The speech estimate is obtained by subtracting the noise estimate from the total input. This can be implemented as an adaptive interference canceller[21] as shown in Figure 6.6.

Recursive Algorithms

In the approach described above, the speech estimate is obtained by filtering the current noisy vector. An alternative approach is to use the previous estimate of the speech vector and the latest noisy sample. In this approach, the frame is advanced by a single sample at each iteration. The estimate is of the form

$$\hat{\mathbf{x}}(m) = G(m)\hat{\mathbf{x}}(m-1) + H(m)\mathbf{y}(m) \tag{6.35}$$

In state-space form, the AR model for the noisy input is

$$\mathbf{x}(m) = A\mathbf{x}(m-1) + \mathbf{g}s(m)$$
$$y(m) = \mathbf{h}^T\mathbf{x}(m) + d(m) \tag{6.36}$$

where

$$A = \begin{bmatrix} a_1 & a_2 & a_3 & \cdots & a_{L-1} & a_L \\ 1 & 0 & 0 & \cdots & 0 & 0 \\ 0 & 1 & 0 & \cdots & \vdots & \vdots \\ \vdots & \vdots & \vdots & \cdots & 0 & 0 \\ 0 & 0 & 0 & \cdots & 1 & 0 \end{bmatrix}, \quad \mathbf{g} = \mathbf{h} = \begin{bmatrix} 1 \\ 0 \\ \vdots \\ 0 \\ 0 \end{bmatrix} \qquad (6.37)$$

$s(m)$ is the excitation signal and $d(m)$ is the noise signal. (*Note:* The frame length is $N = L$ in this representation.) The prediction of $\mathbf{x}(m)$ is

$$\mathbf{x}_P(m) = A\hat{\mathbf{x}}(m-1) \qquad (6.38)$$

and the prediction error is

$$e(m) = y(m) - \mathbf{h}^T \mathbf{x}_P(m) \qquad (6.39)$$

The estimate takes the form

$$\hat{\mathbf{x}}(m) = \mathbf{x}_P(m) + \mathbf{k}(m)e(m) \qquad (6.40)$$

where $\mathbf{k}(m)$ is a gain vector to be determined. Comparing with the general form, we find that

$$G(m) = [I - \mathbf{k}(m)\mathbf{h}^T]A$$
$$H(m) = \mathbf{k}(m)\mathbf{h}^T \qquad (6.41)$$

Kalman Filter

The Kalman filter[22,23] provides a way for recursively calculating the optimal gain vector $\mathbf{k}(m)$ when the prediction filter coefficients $\mathbf{a}(m)$ for the clean speech and the statistics of $s(m)$ are known. For example, any of the spectral estimators described above could be used to estimate $\mathbf{a}(m)$. Alternatively, the coefficients can be found by iteration starting with the noisy input. The estimated speech is then used to recalculate the coefficients and the process is repeated.

Prediction Filter

In the prediction filter described above, the predicted signal was obtained by filtering the noisy input signal. An alternative[20] is to filter the output signal, so that

$$\mathbf{x}_P(m) = \hat{X}(\hat{X}^T \hat{X})^{-1} \hat{X}^T \hat{\mathbf{x}}(m) \qquad (6.42)$$

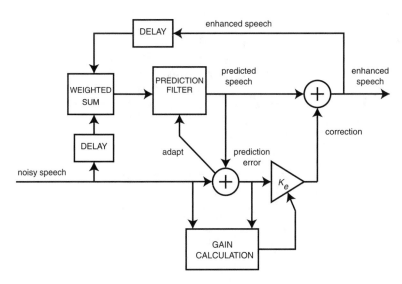

FIGURE 6.7
Recursive prediction-correction filter.

This approach, presented in Figure 6.7, shows the predicted speech is based on the previous outputs, and so the filter is recursive.

Artificial Neural Networks

An artificial neural network (ANN) provides a "model-free" approach for estimating a nonlinear function. The network is trained using noisy speech sequences for which the clean speech is available. The training vectors are the noisy input vectors $\mathbf{y}(m)$ and the associated error vectors

$$\mathbf{e}(m) = \mathbf{x}(m) - \hat{\mathbf{x}}(m) \tag{6.43}$$

The input vectors may be raw time samples[24] as shown in Figure 6.8.

Figure 6.8(a) shows a feedforward network in which the current and previous noisy input vectors are provided to the neural network, while Figure 6.8(b) shows a recurrent network where the current noisy input vector and past output vectors are fed to the neural network.

In addition, the vectors may be preprocessed before being input to the neural network. For example, the spectral amplitude vectors may be passed to the network rather than the vectors themselves, so that the network estimates the spectral amplitude of the speech.

Since the network is nonlinear, the input vectors are usually normalized in some way.

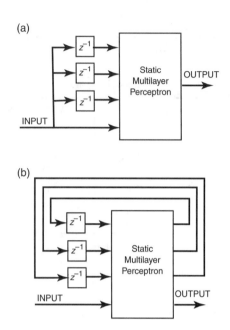

FIGURE 6.8
Artificial neural networks: (a) feedforward network, (b) recurrent network.

When spectral amplitudes are estimated, the speech spectrum itself can be estimated using the phase of the noisy input spectrum. In another example,[25] the LP coefficients are used as inputs to the neural network. The approach can be expected to work best when the characteristics of the speech and noise are similar to those in the training vectors. Haykin[26] provides a comprehensive foundation on ANNs.

Summary

Single-channel noise reduction relies upon differences between the properties of speech and noise. Since the characteristics of speech are well documented, the first step when confronted with a particular noise problem is to identify the characteristics of the noise. The types of noise that can be removed include:

- Noise outside of the speech frequency band, which can be removed by a bandpass filter
- Tonal or predictable noise, which can be removed by a notch filter or by subtracting an estimate of the predictable part of the noisy signal

- Stationary noise, which can be reduced by:
  1. Spectral filtering based on estimates of the power spectrum
  2. Time-domain filtering based on estimates of the autocorrelation of the noise and the speech or
  3. An ANN
- Impulse noise, which can be removed by an ANN or other pattern-recognition techniques.

Techniques for these types of noise are described in the literature, and software (for both real-time and postprocessing implementations) is available for many of the algorithms.

References

1. Lim, S.J. and Oppenheim, A.V., Enhancement and bandwidth compression of noisy speech, *Proc. IEEE*, 67(12), 1586–1604, 1979.
2. Ephraim, Y., Statistical-model-based speech enhancement systems, *Proc. IEEE*, 80(10), 1524–1555, 1992.
3. Ephraim, Y., Malah, D., and Juang, B.H., Speech enhancement based on hidden Markov modeling, in *Proc. IEEE Int. Conf. on Acoustics, Speech and Signal Processing*, Glasgow, Scotland, 1989, 353–356.
4. Sameti, H., Sheikhzadeh, H., and Deng, L., HMM-based strategies for enhancement of speech signals embedded in non-stationary noise, *IEEE Trans. on Speech and Audio Processing*, 6(5), 445–455, 1998.
5. Jarvinen, J.J., Noise attenuation system, U.S. Patent 5,406,635, April 1995.
6. Graupe, D. and Causey, G.D., Method and means for adaptively filtering near-stationary noise from an information bearing signal, U.S. Patent 4,185,168, January 1980.
7. Schroeder, M.R. Processing of communications signal to reduce effects of noise, U.S. Patent 3,403,224, September 1968.
8. Weiss, M.R., Aschkenasy, E., and Parsons, T.W., Processing speech signals to attenuate interference, in *Proc. IEEE Symposium on Speech Recognition*, Carnegie Mellon University, April 1974, 292–295.
9. Weiss, M.R., Aschkenasy, E., and Parsons, T.W., Study and development of the INTEL technique for improving speech intelligibility, Nicolet Scientific Corp., Final Report NSC-FR/4023, December 1974.
10. Boll, S., Suppression of acoustic noise in speech using spectral subtraction, *IEEE Trans. on Acoustics, Speech and Signal Processing*, ASSP-27(2), 113–120, April 1979.
11. Ephraim, Y. and Malah, D., Speech enhancement using a minimum mean-square error short-time spectral amplitude estimator, *IEEE Trans. on Acoustics, Speech and Signal Processing*, ASSP-32(6), 1109–1121, December 1984.
12. Eatwell, G.P. and Davis, K.P., Adaptive speech filter, U.S. Patent 5,768,473, June 1998.
13. Hansen, J.H.L. and Clements, M.A., Iterative speech enhancement with spectral constraints, in *Proc. IEEE Int. Conf. on Acoustics, Speech and Signal Processing*, April 1987, 189–192.

14. Quatieri, T.F. and McAulay, R.J., Noise reduction using a soft-decision sine-wave vector quantizer, in *Proc. IEEE Int. Conf. on Acoustics, Speech and Signal Processing*, Albuquerque, NM, 2, 821–824, 1990.

15. Deller, J.R. et al., *Discrete-Time Processing of Speech Signals*, Macmillan, New York, 1993, chap. 12.

16. Epraim, Y., Speech enhancement using state dependent dynamical system model, in *Proc. IEEE Int. Conf. on Acoustics, Speech and Signal Processing*, San Francisco, 1, 1992, 289–292.

17. McAulay, R.J. and Malpass, M.L., Speech enhancement using a soft-decision noise suppression filter, *IEEE Trans. on Acoustics, Speech and Signal Processing*, ASSP-28(2), April 1980.

18. Ephraim, Y. and Van Trees, H.L., A signal subspace approach for speech enhancement, *IEEE Trans. on Speech and Audio Processing*, 3(4), 251–266, 1995.

19. Asano, F. et al., Speech enhancement based on the subspace method, *IEEE Trans. on Speech and Audio Processing*, 8(5), 497–507, 2000.

20. Eatwell, G.P., Noise reduction filter, U.S. Patent 5,742,694, April 1998.

21. Widrow, B. and Stearns, S.D., *Adaptive Signal Processing*, Prentice-Hall, Englewood Cliffs, NJ, 1985.

22. Koo, B., Gibson, J.D., and Gray, S.D., Filtering of colored noise for speech enhancement and coding, *IEEE Trans. on Signal Processing*, 39(8), 1732–1742, 1991.

23. Chen, W.Y. and Haddad, R.A., Method and filter for enhancing a noisy speech signal, U.S. Patent 5,148,488, September 1992.

24. Tamura, S. and Nakamura, M., Improvements to the noise reduction neural network, in *Proc. IEEE Int. Conf. on Acoustics, Speech and Signal Processing*, 2, 825–828, 1990.

25. Aritsuka T. et al., Noise reduction system using neural network, U.S. Patent 5,185,848, February 1993.

26. Haykin, H., *Neural Networks*, IEEE Press, Macmillan College Publishing, New York, 1994.

7

Microphone Arrays*

Stephen J. Leese

CONTENTS

Introduction and Scope

This chapter is about noise reduction using two or more microphones. In actual applications, this will often be combined with additional functions such as stereophony or talker localization. Very large microphone systems have been designed for use in public spaces and auditoriums.[1,2] The emphasis in this chapter is, however, on relatively small enclosures and spaces, ranging from vehicle interiors up to a typical small conference room.

---

* The opinions expressed here are solely those of the author and are not necessarily those of NCT (Europe) Ltd. or its parent company NCT Group, Inc.

Microphones, by nature, have some directionality, and, in some cases, this can be maximized or minimized by the design of the hardware and its electronics. This directionality can be used to enhance the signal-to-noise ratio (SNR) of speech picked up by the microphone, but this enhancement may still not be sufficient for effective communication. Chapter 6 describes how the signal obtained from a single microphone may be enhanced by signal processing. If several microphones are available, it is possible to improve the quality of the signal yet further by using *digital signal processing* (DSP) to combine the inputs from each microphone. There is a cost penalty in doing this on account of the additional hardware required. However, the expected benefit is that greater noise reduction may be achieved without any distortion of the speech, so giving better intelligibility.

This chapter provides a brief introduction to what is a very large area of research and indicates what might reasonably be achieved in practice.

Beamforming

Basic Notions

The notion of beamforming was first developed in the field of radar and electronic warfare. First of all, it is necessary to define some technical terms. Figure 7.1 shows a typical "polar diagram" for an antenna. This is rather idealized: real polar diagrams can be much less regular, and there may be many more distinct sidelobes than are shown. The angle coordinate represents the direction of a reference source in the far field, and the radius coordinate is a measure of the power received from that source. The plot generally takes the form of a "main beam" and "sidelobes" that are separated by "nulls." Although the nulls may, in theory, be directions where no power is received, their actual "depth" in practice may vary considerably.

The behavior of a microphone in a sound field is in many ways analogous to that of an electromagnetic antenna, and its performance can be described by polar diagrams in the same way. Given an array of antennas, it is possible to combine their outputs so as to make a sharper beam than is achievable with a single antenna, and so increase the power of the received signal. This sharpening of the beam is achieved by modifying the amplitude and phase of the individual antenna feeds. Furthermore, if there is an interfering noise source or jammer, its effect can be minimized by steering a null so that it coincides with the direction of the jammer radiation. Newcomers to the subject may be confused by some of the terminology; for example, we often speak of a "radiation pattern" or a "main beam" for something that is actually "receiving" rather than "radiating." This use of words is permitted by the *principle of reciprocity*, which is defined in an acoustic context in Reference 3. The importance of this principle is that if we have experimental

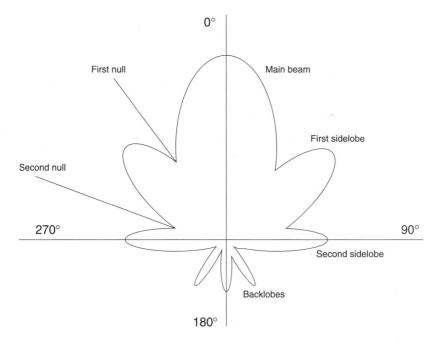

FIGURE 7.1
Features of antenna radiation pattern.

results or measurements about sound radiation (e.g., loudspeaker arrays), then we may (with caution) be able to make inferences about the corresponding microphone arrays.

Consider the case of a "broadside array" of M microphones, as shown in Figure 7.2. Suppose microphone i ($i = 0, ..., M - 1$) produces an electrical signal y_i. It is possible to produce a narrower main beam by applying scalar multipliers q_i to the microphone outputs before mixing them to obtain a combined signal e, thus,

$$e = \sum_{i=0}^{M-1} q_i y_i \qquad (7.1)$$

The classic paper by Dolph[4] calculates values for the coefficients q_i, which are now known as the *Dolph-Chebychev coefficients*. The motivations for this work were the needs of radar and radio communications, where the signals of interest were confined to a relatively small frequency range. However, it was already apparent that the art of compromise is crucial to the successful design of array sensors: there is a trade-off between the sharpness of the main beam and the level of the sidelobe peaks. For example, if our aim is to reduce the level of interfering noise in the transmitted signal e, it is pointless to make the beam sharper if we find that noise enters through the sidelobes.

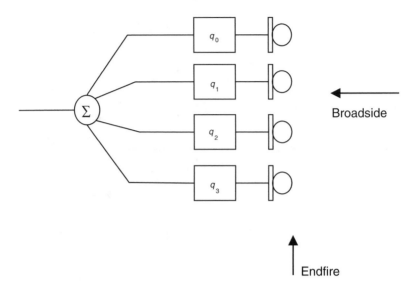

FIGURE 7.2
Beamforming basics.

Equation (7.1) can be implemented effectively using analog electronics. More complicated beamforming algorithms are possible, and these are usually implemented on DSP; this is the approach that is taken in the rest of this chapter. Equation (7.1) is a special case of the more general system where the q_i are vectors of length N and we calculate the *convolution* of each q_i with the corresponding y_i:

$$e_n = \sum_{i=0}^{M-1} \sum_{k=0}^{N-1} q_{i,k} y_{i,n-k} \tag{7.2}$$

where the subscript n denotes the sample point and $q_{i,k}$ is the kth element of the vector, q_i. The number of microphones is M throughout this chapter. We use uppercase letters to denote z-transformed variables, so that Equation (7.2) becomes

$$E(z^{-1}) = \sum_{i=0}^{M-1} Q_i(z^{-1}) Y_i(z^{-1}) \tag{7.3}$$

or more conveniently

$$E = \sum_{i=0}^{M-1} Q_i Y_i \tag{7.4}$$

Here, we are using the fact that a convolution in the time domain corresponds to a simple *multiplication* in the z-domain, as explained in Chapter 1. This is a great simplification, which helps us to undertake the theoretical analysis of various schemes and check on aspects of their feasibility.

The beam shape and hence the noise rejection properties of a microphone array will depend on frequency:[5] it follows, therefore, that if we want to achieve a consistent level of performance across a given frequency range some ingenuity may be needed. As an example, Mahieux et al.[6] describe how they use subarrays of a larger array in order to control array beamwidth over the speech band.

Delay-and-Sum Adaptive Beamforming

It has already been pointed out that Dolph's original work[4] postulated a narrow frequency band. It also assumed an anechoic operating environment of infinite extent, so that there were no interfering multipath reflections or reverberation. This "anechoic" assumption is approximately valid for certain situations, such as some instances of open-air use. We shall continue with it for the time being, as it is helpful in motivating the main approaches to the design of algorithms for microphone array processing.

First, suppose that the talker is not addressing the array from broadside on but from an angle; for example, the context could be teleconferencing (where perhaps only one participant can be broadside on), or it could be the casual use of a speech-recognition system where the user may not wish to sit in the same position all the time. Then, for best signal-to-noise ratio, it might be desirable to steer the beam so that it points directly to the talker. A simple way of doing this is delay-and-sum beamforming: the continuous-time output of the system at time t is calculated to be

$$e(t) = \tfrac{1}{M} \sum_{i=0}^{M-1} y_i(t - \delta_i) \tag{7.5}$$

where δ_i is the time-delay relative to microphone 0 (so that $\delta_0 = 0$), as illustrated in Figure 7.3, and M is the number of microphones. The angular response of the array to a signal at a given frequency depends on the distance between the microphones. A rough rule of thumb for the average noise reduction performance of the delay-sum beamformer described by Equation (7.5) is $10\log_{10}M$ dB, in the case of uncorrelated noise:[7] this estimate is independent of angle. The performance at specific frequencies or for particular localized noise sources may be better or worse than this.

In a practical implementation, it is necessary to estimate the time delays δ_i. These generally have to be calculated to the nearest whole number of samples. To understand this, recall that all the microphone signals have been sampled and digitized, as explained in Chapter 1 (if greater precision is

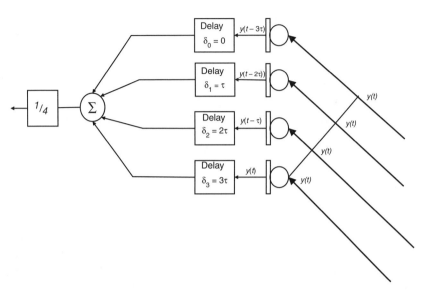

FIGURE 7.3
Principle of delay-sum beamforming.

required, it has to be obtained by interpolation, which adds to the DSP processing load). One method for estimating the time delays δ_i makes use of the *generalized cross correlation* (GCC) of the microphone signals. To do this, one of the microphone channels is arbitrarily selected to be the reference (say channel 0). Then the cross correlation between channels 0 and i at time t and time difference τ is given by

$$R_{0i}(t,\tau) = E(y_i(t)y_0(t-\tau)), \quad |\tau| \le \tau_{max} \qquad (7.6)$$

where E denotes statistical expectation, and τ_{max} is a practical bound on the variation of τ. In practice, the expectation E( ) has to be approximated by a time-domain filter. It may be found helpful to prefilter the microphone channels in order to remove out-of-band noise. Since at any one time we want to track the main talker, this cross-correlation estimate should be performed only during speech bursts. In situations where the interfering noise is not excessive, this could be decided by using a signal-to-noise estimate based on minimum statistics (i.e., tracking the recent minima in the input stream: one such scheme is described in Reference 8).

The above theory has been derived for conditions that are anechoic and *far field*, i.e., the talker is sufficiently far from the microphones for the wavefronts arriving at the array to be approximately planar. In practice, the sound received at the microphones will be affected by noise and multipath reflections, and various efforts have been made to design a more robust beamformer: one approach is to exploit the periodicity of the speech signal as explained elsewhere.[9] If there is a small number of microphones, the noise reduction provided by the delay-sum beamformer by itself is not expected to be dramatic. For example, if there are four microphones, the above formula predicts only 6-dB reduction of uncorrelated noise. It is most likely, therefore, that delay-sum beamforming will be used as a "front-end" for other noise reduction techniques, and some of these techniques are discussed in the following sections.

Superdirective Beamforming

The delay-sum beamformer is a special case of Equation (7.2) where each q_i consists of a series of zeros followed by a "1." *Superdirective beamforming* aims to produce a sharper beam from a wider choice of coefficients. In applying more general vectors q_i to the microphone outputs, we are in effect applying finite impulse response (FIR) filters to them (Chapter 1). It has been found that greatly improved directivity and hence noise reduction can be achieved by this means. For example, whereas a delay-sum beamformer will enhance the gain of a dual-mic array by 3 dB, superdirective beamforming will enhance it by up to 6 dB, for an endfire array[10] (*endfire* means that the microphones are positioned one in front of the other — see Figure 7.2).

There is, however, a price for this improved performance: the FIR filters may affect the spectral content of the speech. There is also a practical difficulty: the choice of filter coefficients for optimum performance depends on the response (i.e., gain and directivity) of the individual microphone elements. This microphone response can vary considerably (6 dB is not unknown), even within the same production batch, and so filter coefficients optimized for one particular set of microphones may not be appropriate for another. There is moreover a loss of performance when wide steering ranges are required (as, for example, in a conference phone, which talkers may address from any angle): the maximum SNR gain occurs for endfire arrays and steering away from the endfire direction entails substantial SNR loss.[10] This sensitivity of the optimal filter design suggests that an adaptive approach should be appropriate, and there are now a number of proprietary schemes for adjusting the filter coefficients, some of them built into commercially available products. Most of these products seek to enhance the SNR for a talker who is addressing the array from within a specified sector. There is also a considerable body of published work on adaptive array processing, and this is introduced below.

In fact, by using adaptive array processing, it is possible to do rather more than simply improve the directional gain of a microphone array. For example, it becomes possible to "null out" noise sources, as discussed in more detail below. This is not to imply that it is generally possible to reduce the interfering noise to silence: the depth of a null (noise reduction in dB) depends on a number of factors, such as the reverberance of the noise and the angular separation of the noise source and the talker,[11] but nevertheless useful reductions are possible in many instances.

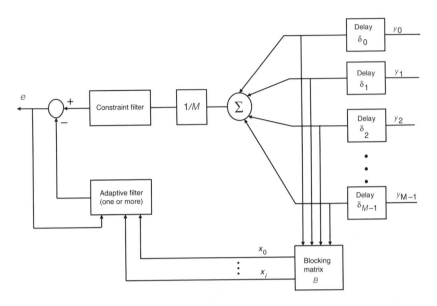

FIGURE 7.4
Griffiths-Jim beamformer.

Griffiths-Jim Beamformer

One scheme for adaptively adjusting the coefficients of Equation (7.4) is the Griffiths-Jim beamformer (GJB), as described, for example, in Reference 7. A simplified system diagram is shown in Figure 7.4. The input data is combined in two ways to give:

- Delay-sum combined signal, e
- One or more (up to $M - 1$) reference signals x_i

Noise reduction is then effected by means of an LMS-type algorithm, similar to that described in Chapter 8 for use in echo cancellation. The "reference signals" are linear combinations of the microphone inputs: the intention is that each reference signal should correspond to a different noise source and should also be relatively free of speech. This is the purpose of the "blocking matrix" B shown in Figure 7.4: clearly, we need some prior knowledge of the operating conditions in order to design B suitably. To take a very simple example, if the array consists of just two well-matched microphones, we could choose a blocking matrix, which simply calculates the reference signal to be the difference between the two microphone inputs. The "constraint filter," which is shown in Figure 7.4, is included for completeness; for example, it could be a simple delay (necessary to accommodate the adaptive parts of the system), or it could be used to filter out-of-band noise. If the noise source is situated roughly end-on to the two microphones, the present author has been able to achieve at least 10 dB of noise rejection without much difficulty with this scheme. This could be a practical method of combating noise from a source that is inaccessible to more direct ways of getting a reference signal (see Chapter 1 for the relevant theory). As with the echo cancellation algorithm described in Chapter 8, adaption will be adversely affected by speech. In order to complete the system, some form of speech detection is necessary so that adaption takes place only when no speech is present.

Griffiths-Jim Beamformer with Adaptive First Section

The difficulty with the GJB as described above is that, in a typical operating environment, various factors such as microphone mismatch and multipath reflections will ensure that there is leakage of speech into the reference signal. It is essential for successful operation of the GJB that the reference signal should be free of speech (in practice, a small amount may be tolerated, but more than this will introduce unwelcome distortion into the final output). One approach to achieving "clean" reference signals is to make the first section of the GJB adaptive: We use adaptive filters to cancel noise, so why not use an adaptive filter to cancel speech? This type of GJB is more fully explained in Reference 7, and Figure 7.5 shows a simplified schematic of such a system for two microphones (the system described in Reference 7 can have any number of microphones). Van Comperolle and Van Gerven[7] report, however, that they did not obtain significantly better results from this than with the simpler GJB discussed in the previous section. One possible reason for this is that adaption of the first section is affected by uncanceled noise mixed with the speech signal. This means that adaption can take place only when a high speech signal-to-noise ratio is detected; hence, convergence and tracking are very slow, and because of the noise only partial elimination of speech is achieved from the reference signal.

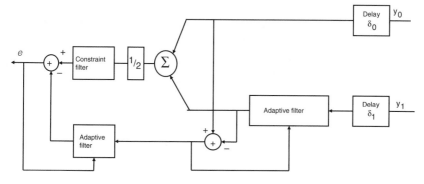

FIGURE 7.5
GJB with adaptive first stage.

Systems similar to GJB and GJB with adaptive first section, in that they contain an adaptive noise subtraction path, are frequently referred to in the literature as *generalized sidelobe cancellers*. The name comes from the general polar diagram shown in Figure 7.1: if the main beam of the array is pointing toward the main talker, and the main noise sources are "in the sidelobes," then the system can be said to reduce or cancel the sidelobe levels in the directions from which the noise is coming, hence, the name generalized sidelobe canceller.

Theoretical Background

This is a good point in the chapter at which to step back and look at the multimicrophone problem from a distance in order to gain a better understanding of potential noise cancellation performance and also of possible limitations. Suppose that we have several microphones (indexed by i), which provide data streams y_i. Assume also that there are several talkers producing speech streams s_j and noise sources producing noise streams n_k. Then the signal at microphone i can be represented by:

$$y_i = \sum_j G_{ij} s_j + \sum_k H_{ik} n_k \qquad (7.7)$$

where G_{ij} is the acoustic transfer function (however represented) from talker j to microphone i, and H_{ik} is the acoustic transfer function from noise source k to microphone i. The transfer functions G_{ij} and H_{ik}, as a matter of practicality, are assumed to be constant, or at least to vary very slowly relative to the speech data. Here, in order to keep the notation as simple as possible, I am using single symbols (y, s, n, etc.) to represent vectors or sequences, "multiplication" to represent linear operations (e.g., convolution), and subscripts as identifiers for microphones, talkers, and noise sources. As in the previous sections, we want to find a linear combination of the microphone data y_i so that the noise data n_k is as far as possible canceled and the speech data s_j enhanced. So, as before we seek vectors Q_i which optimize the linear combination

$$e = \sum_i Q_i y_i \qquad (7.8)$$

against certain criteria, which are discussed below. Substituting Equation (7.7) into Equation (7.8) gives

$$e = \sum_{i,j} Q_i G_{ij} s_j + \sum_{i,k} Q_i H_{ik} n_k \qquad (7.9)$$

It follows that for optimum quality (speech undistorted by room effects, noise "nulled") we must have:

$$\sum_i Q_i G_{ij} = 1, \quad \text{all } j$$

$$\sum_i Q_i H_{ik} = 0, \quad \text{all } k \tag{7.10}$$

If the first part of Equation (7.10) holds for any talker j, it means that the particular speech stream is transmitted with no reverberation effects. If the second part of Equation (7.10) holds for any particular noise source, k, then that source is effectively canceled. In practical applications, the transfer functions G_{ij} and H_{ik} are inaccessible: it is often difficult to measure them in laboratory conditions, let alone estimate them in real time.

Consider now the special case of two microphones with a single talker and a single noise source. Assume that one microphone gives an adequate response in the absence of noise, so that the corresponding transfer function may be assumed to be the identity. Then the microphone outputs can be described by

$$y_1 = s + H_1 n$$

$$y_2 = Gs + H_2 n \tag{7.11}$$

If we can get from this a data stream that consists of noise without speech, then we might be able to use it as the reference signal in an LMS-type adaption scheme. Therefore we define

$$r = y_2 - G y_1 \tag{7.12}$$

and attempt to find H_S such that

$$H_S r = H_1 n \tag{7.13}$$

and the output of the system is

$$e = y_1 - H_S r \tag{7.14}$$

This fits into our general scheme if we define

$$Q_1 = 1 + H_S G$$

$$Q_2 = -H_S G \tag{7.15}$$

If the two microphones are equidistant from the talker, then we can assume $G = 1$, and we have a special case of the GJB as described above. Although this algorithm is very similar to that described in Chapter 8, it requires a different interpretation. In particular, H_S does not represent a single physical quantity, such as an impulse response, and the optimum length for it may be much longer than the reverberation time of the room, enclosure, or space that contains the microphones and the talker. If the transfer functions H_i and G are represented in the z-domain, then manipulation of the above equations shows that H_S is a polynomial approximation to the rational function $H_1/(H_1 - GH_2)$. Consequently, it may be shorter or, more likely, longer than the length suggested by measurements of the room acoustics.

The above analysis makes a number of implicit assumptions. For example, if it were for a hands-free conference phone, then it would presuppose (apart from the ambient noise) a fairly calm and controlled working environment. For example, if there are two talkers, they should not talk at once (a different analysis will be needed if they do this habitually), but the conversation should flow naturally from one talker to the other with only a small overlap of speech. On the other hand, the analysis would allow intermittent or impulsive sounds, or other acoustic events (e.g., door slam, window open, workbench, serving hatch), as long as the *locations* of the noise source and of the talkers did not change.

The algorithm just described can be viewed as a special case of the generalized sidelobe canceller as described in "Griffiths-Jim Beamformer." As we noted in that section, the generalized sidelobe canceller can be very effective against discrete sources. In the case of a *diffuse* sound field, such as may be obtained in a very reverberant room, however, the theoretical noise reduction of this type of algorithm has been shown to be less than 1 dB.[11] To achieve good noise reduction performance in such an environment, a different algorithmic approach is required.

Microphone Placement

The theoretical analysis in the previous section holds for arbitrary arrays of microphones. The analysis does not, however, guarantee the existence of a useful solution to the basic signal processing problem: how to choose the coefficients for Equation (7.8), which give a substantial noise reduction while simultaneously preserving or enhancing speech quality. In this section, we explore how the choice of microphones, and how to arrange them, will improve our chances of success in solving this basic problem. We look at system designs that more closely integrate the microphone array design and the algorithm design in order to enhance the quality of the output, even with the most basic signal processing algorithm.

In general, a beamforming solution is frequency dependent, so that the output speech may have noticeable spectral coloring, or, even worse, suffer partial cancellation. Hoffman and Buckley[12] describe an adaptive scheme for optimizing the factors Q_i of Equation (7.8) in such a way that broadband noise reduction is achieved while avoiding the risk of canceling part of the desired signal. The particular example that they describe in detail is a seven-element head-worn array for hearing enhancement.

An alternative to adaptive beamforming is to start with a theoretical beam pattern, which defines the desired response, attainable only in ideal circumstances,[13] and then use a calibration procedure to calculate the filter coefficients, Q_i, which give a response close to the desired. A practical approach to calculating the coefficients for the ideal beamformer (before applying any procedure to compensate for the limitations of actual microphones) is given elsewhere.[14] These two papers are concerned with far-field effects, such as may be experienced outdoors or in a very large room.

The frequency–sensitivity of array geometry has already been mentioned: this is a function of the microphone separation. This suggests that an array with several different separation distances may be less sensitive to frequency. One such is the logarithmic array, where the spacing increases exponentially, as illustrated in Figure 7.6. Everything that has been discussed in previous sections applies to logarithmic arrays as much as any other kind. Besides flatter frequency response, another motivation for employing logarithmic arrays is that *spatial aliasing* is less likely to occur than with equally spaced microphones. Spatial aliasing is said to occur when an array, with its associated processing, has more than one main beam at a certain frequency. To put this another way, the polar diagram (as in Figure 7.1) has one or more sidelobes of size and width comparable to the main beam. The array is therefore very receptive to noise from these directions, and clearly this is a situation we would wish to avoid. The reasons why logarithmic arrays are less susceptible to spatial aliasing are treated in the literature on radar antenna theory and are beyond the scope of this chapter.

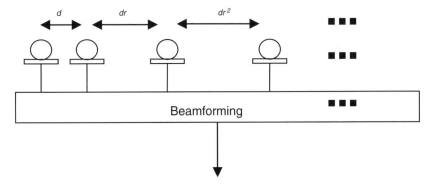

FIGURE 7.6
Logarithmic microphone array.

Combination with Other Noise Reduction Techniques

In many applications, the capabilities of multimicrophone processing can be extended by combining it with other noise reduction filters, such as single-channel speech filters as described in Chapter 6 or echo cancellers (see Chapter 8). The simplest approach is to apply these different processes in cascade. However, a number of investigators have explored the benefits that might come with closer integration. Consider for example the *coherence* between microphone channels. In probability theory, the coherence of two variables is a measure of the linearity of any relationship between them. For example, Meyer and Simmer[15] show that in the case of interior car noise the coherence between microphone channels generally decreases both as frequency increases and also as the distance between the microphones increases. Hence, it may be desirable to balance the system so that the lower frequencies, where coherence is higher and consequently adaptive filtering is more likely to be effective, are dealt with by beamforming or sidelobe canceling. Meanwhile, the upper frequencies, where coherence is lower and hence adaptive filtering is less effective, can be dealt with more appropriately by simple beamforming (e.g., delay-sum) and applying one of the single-channel methods described in Chapter 6.

For some applications, it will be desirable to combine microphone array processing with an acoustic echo canceller: one example would be a conference phone. Two distinct approaches to "cascading" come to mind. Conceptually, the simpler approach is to apply a separate echo canceller to each microphone input channel in turn. This approach can result in a heavy processing load, especially if the microphone array is a large one. An alternative approach is for the echo canceller to follow the microphone array processing. In order to get the optimum performance from this second approach, two issues need to be addressed:

1. The microphone array processing (e.g., adaptive beamforming) will extend the echo response that must be modeled within the echo canceller (see Chapter 8), and so it may be necessary to provide additional data memory.

2. Echo cancellers require this echo response to be *linear* (see Chapter 8), and so the microphone array processing must also be linear. The multimicrophone techniques discussed in this chapter generally meet the linearity condition, especially when they are "well-converged."

In addition, there may be some synergy that can be exploited between the microphone array processing and the echo cancellation in order to improve echo canceller performance further. For example, if the microphone array processing can steer a null in the direction of the loudspeaker, the acoustic echo will be reduced, greatly increasing the echo cancellation performance of the system during "double-talk" (see Chapter 8).

Inverse Filtering and Dereverberation

We have thus far discussed the second part of Equation (7.10), ignoring the first part. Consider now the first part of Equation (7.10), in the context of room reverberation, which can be considered to be a form of noise interference. If we can find a set of Q_i that solve this equation, then we shall have removed the effect of room reverberation completely. The effect of reverberation is familiar to anyone who has listened to a call made from an office speakerphone. The person who is speaking is often some distance from the microphone so that the ratio of direct to indirect transmission of sound to the microphone is reduced compared with a microphone held close to the mouth: the result is that the speech can become indistinct and hard to follow. With a single microphone, the resulting signal is

$$y = Gs \tag{7.16}$$

In order to remove the effect of reverberation, we need to find a transfer function H such that

$$Hy = s \tag{7.17}$$

This is not generally feasible, however, since the transfer function G is not always invertible. A detailed investigation of this point is given in Neely and Allen.[16]

Miyoshi and Kaneda[17] approached this problem of invertibility by using two microphones. Their approach is to find Q_i such that

$$Q_1 G_1 + Q_2 G_2 = 1 \tag{7.18}$$

where G_1 and G_2 are the transfer functions from the talker to each microphone. If the Q_i and G_i are interpreted as z-transforms, i.e., as polynomials in z^{-1}, then it is a well-known consequence of the Euclidean algorithm for polynomial division that a solution for Equation (7.18) can be found, provided that the G_i have no common factor. Moreover, Q_1 and Q_2 are of lower order than G_2 and G_1, respectively (this information about the orders of Q_1 and Q_2 is of practical value: it gives an indication of how much processor memory to allocate). Miyoshi and Kaneda[17] test their theory in an experiment using *one* microphone and *two* loudspeakers. It is interesting to note that their experimental approach uses the *principle of reciprocity* mentioned in

"Basic Notions,"[3] above. In a theoretical analysis of a system of microphones and loudspeakers, it may be more immediately practical to study the corresponding system in which microphones are replaced by loudspeakers and vice versa.

In order to solve Equation (7.18), we need to know the G_i: that is, the room needs to be calibrated. We also require that the G_i have no common factors, and that Equation (7.18) is "well conditioned" (i.e., the coefficients are such that a numerical solution is well defined with the precision of the DSP hardware available to us): any practical implementation designed for use in *any* room or enclosure must have some means to ensure this. The approach to dereverberation that I have outlined above has been taken further,[18,19] but it would appear that a practical implementation is some way off.

Signal Separation

Suppose the noise we wish to cancel or reduce is in fact not "noise" at all but another talker, who is either present in the room or transmitted via radio, television or public address system. The noise reduction solutions discussed thus far in this chapter depend crucially on the ability to distinguish between speech and nonspeech (in a practical system, it is usually necessary to include a "don't know" category wherein no adaption of any kind takes place: this is not usually an impediment to good performance). If, however, the interfering noise is speechlike, it becomes very difficult to control a generalized sidelobe canceller adequately: it may even enhance the interfering speech at the expense of the main talker. *Blind source separation* techniques seek to distinguish the principal sources of sound and to output each one as if the others were not there. Unlike echo cancellation, there may be no reference signal to identify one or other of the interfering sources, hence the term *blind* source separation.

The usual approaches to the source separation problem are based on the assumption that the sound from the various sources is uncorrelated. Microphones are arranged, not necessarily on beamforming principles, but so that each picks up a different mixture of the various sources.[7] The microphone inputs are processed to provide several outputs with minimal cross correlation. The user of the system can then select one or more of the separated outputs for listening, recording, transmission, or further processing. In many practical situations, it can be further assumed that the sources are non-stationary (in the statistical sense) and at least one investigator sees this as a promising line of attack.[20]

Blind source separation is a large research field in its own right: a detailed introduction can be found in Chapter 15 of this volume.

Small Arrays

This chapter has implicitly concentrated on small-scale applications of microphone array technology. However, I have avoided being too specific about how many microphones are being used. There is great interest in small arrays, typically of two to four microphones, for use in consumer products such as car phones, voice input to computers or other appliances, hearing aids, etc. The techniques that have been described so far can give good results in some of these situations: in other situations, as has been mentioned, there may be theoretical limits on what can be achieved.

Some actual acoustic measurements are reported in the paper by Hoffman et al.[21] They employ a four-microphone logarithmic array, 24.5 cm long, broadside on to the speech source. The outputs are processed by a generalized sidelobe canceller of their design. Their measurements show that, with a single interfering source, a SNR enhancement as much as 15 dB can be achieved. They found, however, that this performance level decreases as the talker direction and noise source direction get closer together. Their work illustrates that in order to get the best performance out of a given microphone array (as long as this is consistent with the manufacturer's instructions), the user should try to arrange things so that talker, noise source, and array are as far from collinear as possible; in other words, the talker and noise should be roughly 90° apart relative to the array. Greater noise reduction is expected from arrays with more microphones (see "Delay-and-Sum Beamforming," above).

The recordings that accompany this chapter (available at http://www.crc-press.com/e_products/download.asp?cat_no=0949) were made with commercially available multimicrophone systems and illustrate the kind of performance that is currently achievable with arrays of from four to eight microphones.

Future Developments

This chapter has been a very brief introduction to a vast subject, and it has not been possible to cover any of the points of detail that must be addressed in order to design a successful system. Many of these points are covered in the references at the end of this chapter. An extensive treatment of the subject of microphone arrays is given in Stefaan Van Gerven's thesis,[22] which at the time of writing is available as a download in PostScript format from http://www.esat.kuleuven.ac.be/~spch.

We have noted that multimicrophone systems depend for their success on the ability to distinguish between speech and noise: in simple systems, this is done spatially, for example, by using a fixed beam direction. This is excellent for situations where the talker/microphone geometry stays fairly constant (e.g., voice input to computer in a private office) but is not as good for situations where talkers and noise sources may overlap in direction (e.g., voice input in shared offices, conference phones, media interview, vehicle interiors). As already pointed out, many forms of array processing embody adaptive algorithms that require some form of speech/noise discrimination along the timeline. It would appear, therefore, that one route to improved performance is to increase the ability to distinguish speech from different kinds of noise: the development of more powerful processors and intelligent software is making this possible.

References

1. Silverman, H.F. et al., A digital processing system for source location and sound capture by large microphone arrays, in *ICASSP'97*, Vol. 1, IEEE, 1997, 251.
2. Rabinkin, D. et al., Optimal truncation time for matched filter array processing, in *ICASSP'98*, Vol. 6, IEEE, 1998, 3629.
3. Morse, P.M. and Ingard, K.U., *Theoretical Acoustics*, Princeton University Press, Princeton, NJ, 1986, 134.
4. Dolph, C.L., A current distribution for broadside arrays which optimizes the relationship between beam width and side-lobe level, *Proc. I.R.E and Waves and Electrons*, June 1946, 335.
5. Flanagan, J.L., Beamwidth and useable bandwidth of delay-steered microphone arrays, *AT&T Techn. J.*, 64, 983, 1985.
6. Mahieux, Y. et al., A microphone array for multimedia workstations, *J. Audio Eng. Soc.*, 44, 365, 1996.
7. Van Compernolle, D. and Van Gerven, S., eds., Beamforming with microphone arrays, in *Digital Signal Processing in Telecommunications: European Project COST#229 Technical Contributions*, Capellini, V. and Figueiras-Vidal, A.R., Ed., 107, European Union, 1995 (at time of writing downloadable from http://www.esat.kuleuven.ac.be/~spch).
8. Martin, R., An efficient algorithm to estimate the instantaneous SNR of speech signals, in *Eurospeech'93*, 1993.
9. Brandstein, M.S., Time-delay estimation of reverberated speech exploiting harmonic structure, *J. Acoust. Soc. Am.*, 105, 2914, 1999.
10. Dörbecker, M., Speech enhancement using small microphone arrays with optimized directivity, in *IWAENC'97*, IEEE, 1997, 100.
11. Bitzer, J. et al., Theoretical noise reduction limits of the generalized sidelobe canceller (GSC) for speech enhancement, in *ICASSP'99*, IEEE, 1999, Paper 1099.
12. Hoffman, M.W. and Buckley, K.M., Robust time-domain processing of broadband microphone array data, *IEEE Trans. Speech and Audio Processing*, 3, 193, 1995.
13. Sydow, C., Broadband beamforming for a microphone array, *J. Acoust. Soc. Am.*, 96, 845, 1994.
14. Ward, D.B. et al., Theory and design of broadband sensor arrays with frequency invariant far-field beam patterns, *J. Acoust. Soc. Am.*, 97, 1023, 1995.

15. Meyer, J. and Simmer, K., Multi-channel speech enhancement in a car environment using Wiener filtering and spectral subtraction, in *ICASSP'97*, Vol. 2, IEEE, 1995, 1167.

16. Neely, S.T. and Allen, J.B., Invertibility of a room impulse response, *J. Acoust. Soc. Am.*, 66, 165, 1979.

17. Miyoshi, M. and Kaneda, Y., Inverse filtering of room acoustics, *IEEE Trans. Acoustics, Speech, and Signal Processing*, 36, 145, 1988.

18. Wang, H., Multi-channel deconvolution using Padé approximation, in *ICASSP'95*, Vol. 5, IEEE, 1995, 3007.

19. Furuya, K. and Kaneda, Y., Two-channel blind deconvolution of nonminimum phase FIR systems, *IECE Trans. Fundamentals*, E80-A, 804, 1997.

20. Jones, D.L., A new method for blind source separation of nonstationary signals, in *ICASSP'99*, IEEE, 1999, Paper 2354.

21. Hoffman, M.W. et al., Comparison of microphone configurations for three and four microphone arrays, *Proc. SPIE*, 3162, 1997, 216.

22. Van Gerven, S., Adaptive Noise Cancellation and Signal Separation with Applications to Speech Enhancement, Ph.D. thesis, Katholieke Universiteit Leuven, Belgium, 1996.

8

Echo Cancellation

Stephen J. Leese

CONTENTS

0-8493-0949-2/02/$0.00+$1.50
© 2002 by CRC Press LLC

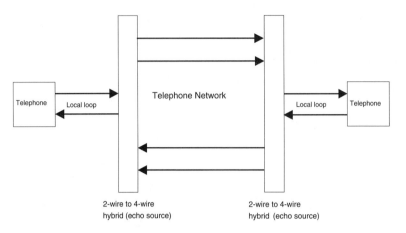

FIGURE 8.1
How echo arises in a telephone network.

Echo in Communications

The Problem of Echo

Historically, echo controllers were first developed for use in long-distance telecommunications in order to control *line echo*. Line echo arises when a signal is reflected from a junction point (such as a two- to four-wire hybrid; see Reference 1; Figure 8.1), so that talkers hear their own voices reflected back down the line. There can be a significant time delay from uttering a word to hearing the same word repeated back, which adds to the disruptive effect of the echo on the flow of conversation. This has been quantified, and "tolerance curves" of echo loudness vs. time delay[2] have been drawn up. In practical terms, the echo level must be reduced by at least 40 dB below normal speech level in order to restore conversational quality. The techniques that are used to control line echo have been successfully extended to control *acoustic echo*. A particular instance of acoustic echo is found in the hands-free telephone, which is being used increasingly in road vehicles. In this case, the microphone picks up sound from the loudspeaker and transmits it back to the far-end talker (Figure 8.2). Modern digital mobile networks have a significant time delay because of the need for speech compression and coding for transmission (see, for example, Reference 3), and so the requirement for acoustic echo cancellation is at least as great as for network or line echo cancellation. Other possible applications of acoustic echo cancellation include a voice-controlled audio system: in this case, the microphone will pick up sound from the audio loudspeakers which will tend to interfere with the recognition of voice commands. Although there is no "echo" here to be heard, the interfering sound still needs to be removed from the microphone input, and an echo canceller of the type described in this chapter is one way to do this.

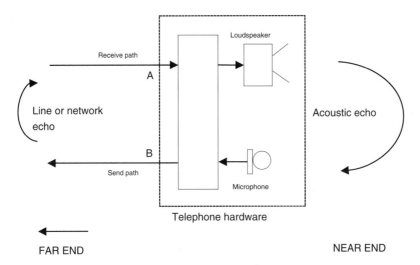

FIGURE 8.2
Definition of main terms relating to acoustic echo cancellation.

VOX, the Simple Solution

The simplest approach to echo suppression is to use a *voice-operated switch*, or VOX. This simply permits only the "send" or "receive" path to be open, not both as for a normal conversation. The switch is controlled by means of the speech level: the conversation goes to the end with the higher level. A well-designed VOX can be surprisingly unobtrusive in use, but there is always the possibility that a sudden loud noise can "snatch the link." Also, leading-edge "clipping" (not to be confused with "amplitude clipping") of the beginning of words may be heard, which means that initial consonants may be shortened or lost altogether. This tends to happen most often when the conversation passes from one talker to the other and may not even be noticed by the persons actually talking. If other persons are present, however, who are not speaking at that moment, as in a telephone conference, they generally will notice and may find the conversation more difficult to follow as a result of the clipping. True *double talk*, the overlap of both parties' speech as the conversation passes from one to the other, is not possible. We say that a system incorporating a VOX is *half duplex*. A system where both parties can speak at once and hear each other is *full duplex*. More formally, it is stated[4] that: "In [full] duplex operation, attenuation of the conversation partner either does not occur or is unnoticeable."

Echo Cancellation Definitions and Required Performance

An *echo canceller* is a device for canceling echo in speech communications which permits double talk. This implies that it employs some kind of filtering algorithm that blocks the echo while transmitting near-end speech. Any other sort of echo control device or VOX is generally called an *echo suppressor*. The most important performance measure of an echo canceller is its *echo return loss enhancement* (ERLE). This is defined as the attenuation of echo by the echo canceller, expressed in dB. The terminology comes from the origins of echo cancellation in networks: the *echo return loss* (ERL) is the natural attenuation of echo between two points, so that:

$$\text{ERLE} = \text{ERL}_{\textit{with echo canceller}} - \text{ERL}_{\textit{without echo canceller}} \qquad (8.1)$$

Performance measures for echo cancellers are generally frequency weighted: the weighting most often used is defined elsewhere.[5] For speech telephony, it consists of a $1/f$ weighting of frequencies in the range $f = 300$ to 3400 Hz. Frequencies outside this "telephone band" are ignored.

The performance that is required of echo cancellation algorithms is not always specified directly. In the case of a hands-free telephone, what is specified is the *terminal coupling loss* (TCL), which is the overall attenuation of the echo resulting from the acoustic coupling of the terminal combined with the effect of the echo canceller. In other words, TCL represents the echo attenuation across the hands-free equipment from receive input to send output (point A to point B in Figure 8.2), so that, in decibels,

$$\text{TCL}_{\textit{with echo canceller}} = \text{TCL}_{\textit{without echo canceller}} + \text{ERLE} \qquad (8.2)$$

The most commonly invoked standards for echo control are set out in ITU-T Recommendations G.167 and G.168.[6,7] These must be supplemented by the special requirements of particular types of networks, such as GSM.[3] A typical requirement is that, using a special test signal or "artificial voice,"[8] the TCL of a system using an acoustic echo canceller must reach 20 dB by 1 s after start-up and 40 dB by 2 s after start-up.

A DSP implementation of an echo canceller will introduce time delays: the microphone signal will be delayed as it passes through the processor, and in some implementations, the loudspeaker signal will also be delayed. Most network standards impose strict limits on these delays. For example, for a hands-free kit that is connected to a GSM network,[3] the two delays (microphone and loudspeaker) must add up to no more than 39 ms.

These published performance requirements are in effect a minimum. Echo at the prescribed minimum TCL levels is often still audible. This, together with the competition between suppliers of echo cancellation equipment and software, means that there should in practice be no audible echo at all. The artificial voice mentioned above may be generated using the P.50 Artificial Voice code, available at http://www.crcpress.com/e_products/download.asp?cat_no=0949. Typical wavefiles produced using this software are provided on this site as Audio 9 Male Artificial Voice.wav and Audio 10 Female Artificial Voice.wav.

Adaptive Echo Cancellation

Basic LMS Algorithm

The disadvantages of the VOX are becoming increasingly unacceptable to users. It is clearly desirable to somehow subtract the echo from the "near-end" signal so that conversation flows naturally. The existence of cheap digital signal processing (DSP) hardware makes this a practical reality. Suppose the far-end speech, in sampled form, is represented by the data stream x_n, $n = 0, 1, 2$, and so on. Then the sound received at the microphone can be represented by the data stream

$$y_n = s_n + \sum_{k=0}^{N-1} h_k x_{n-k} \tag{8.3}$$

where y_n denotes the data stream received by the microphone, h_k ($k = 1, ..., N - 1$) is a model, usually called the *impulse response*, of the acoustic path from loudspeaker to microphone, and s_n is the near-end speech. This can be represented more succinctly by the notation

$$y = s + h \otimes x \tag{8.4}$$

where $\otimes$ is used to denote the convolution operation written out more fully in Equation (8.3). It follows that if we have an estimate of the impulse response and access to a copy of the loudspeaker drive signal, then we can aim to cancel the echo.

Suppose that h_{est} is the estimate of h which is stored in our DSP and then we output to the "send" path the signal

$$e = y - h_{est} \otimes x \tag{8.5}$$

If there is no speech at the near end this is a measure of the error in the adaption routine that estimates the vector h and tracks changes. The adaption scheme most commonly used is a variant of the "least mean squares" (LMS) method described in detail elsewhere[1] (see also Chapter 1 of this volume). At each iteration, we update h_{est} to

$$h_{est\,k} = h_{est\,k} - \mu e_n x_{n-k}, \quad \text{for } k = 0, ..., N - 1 \tag{8.6}$$

where k is the index of elements of h_{est}, μ is the step-size, e_n is the error at sample point n, and x_n is the value of the speaker drive signal at sample point n. Most authors call x the *reference* signal. A simple echo canceller model, based on this principle and coded in ANSI C, accompanies this chapter as the Echo Canceller code, available at http://www.crcpress.com/ e_products/download.asp?cat_no=0949.

The basic LMS algorithm described above is simple and robust and is popular for these reasons. However, it needs some refinement in order to make a practical echo canceller. First of all there is the computational load. The path length N can be of the order of 512 sample points or more. With the data sampled at 8 kHz, this corresponds to a decay time of 64 ms, which is typical for a hands-free phone used in a car interior. Other applications, such as conference phones, may demand a greater path length due to the size of the room in which they are being used (see "Echo Path Length," below). Counting the number of "adds" and "multiplies" in Equations (8.5) and (8.6) and assuming a sample rate of 8 kHz, then we see that the processor has to perform something of the order of 24 million arithmetic operations per second. This processor is generally required to perform other functions simultaneously, such as speech coding, so this may well lead to an excessive computational load. This load can be reduced considerably by using a frequency domain algorithm. Frequency-domain methods for performing convolutions take advantage of the *fast Fourier transform* (FFT — see Chapter 1) and are described in the standard reference works on signal processing.[9] The disadvantage of frequency-domain methods is that they introduce an additional time delay, as the input data has to be accumulated into blocks of points before it can be used. An input block is typically 64 samples long, equivalent to a delay of 8 ms at 8 kHz sample rate. However, if the DSP is already running other programs that require a block input (a speech coder for instance), it may be possible to minimize or even eliminate the *additional* delay by designing the echo canceller to use a compatible block size. A frequency-domain echo canceller that does this has recently been patented.[10] A good general survey of the present state of adaptive algorithms for echo cancellation is given by Breining et al.[11]

Echo Path Length

The echo path length, N (as used in Equation [8.6]), is often quoted in manufacturers' data sheets. It can be said to correspond roughly to the reverberation time of the enclosure (e.g., room, office, vehicle interior) in which the echo canceller is to be used, and defined as the time for an impulsive sound to decay to 60 dB of its original level.[11] A typical impulse response is illustrated in Figure 8.3, which shows three main features: a pure delay (the propagation time from loudspeaker to microphone), reverberation from reflecting surfaces within the enclosure, with an exponentially decaying amplitude, and a long tail. The signals are generally digitized with 16-bit precision, so that the relative accuracy of the tail as estimated by h_{est} will be fairly low, especially as its amplitude approaches only a few bits. An adaptive echo canceller will, therefore, generally work best if the modeled impulse

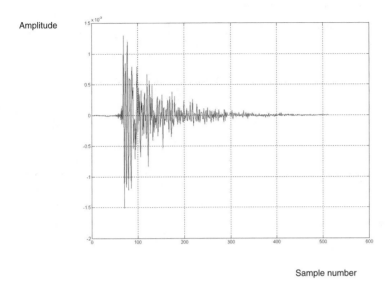

Sample number

FIGURE 8.3
Typical impulse response of car interior (Chevrolet Blazer), sampled at 8 kHz, showing main features.

response is somewhat shorter than the actual one: the data at the end of the echo tail is too inaccurate to be useful.

Stability

The standard texts and references on signal processing give detailed treatments of the stability of the LMS method. One point to be aware of concerning echo cancellers is that if the h_{est} is much longer than the actual echo path then there is the risk of numerical instability as errors build up in the redundant tail-end of h_{est}. The modeled echo $h_{est} \otimes x$ in Equation (8.5) then grows without limit: it no longer matches the echo, and the instability is observed as a constantly increasing output, unrelated to the microphone input level. This may appear unlikely to occur in practice, but it can often be induced by inputting a sine wave at the far end (modem and fax signals are modulated sine waves): the microphone receives the echo, which is a sine wave of the same frequency. It can easily be shown that for this particular case, in the absence of any other far-end data, the "h_{est}" needed for echo cancellation is much shorter than the actual one. This is especially so if the period of the sine wave corresponds to an exact number of sample points (e.g., a 1000-Hz sine wave sampled at 8 kHz has a period of eight sample points). The echo canceller must be designed so that any instability, which might result from this or any other cause, is "caught" before the output builds to an unacceptable level (see the section "Howling Detection," below).

Some Factors That Limit Performance

The type of echo cancellation being discussed here is a special case of adaptive filtering, in which the far-end talker is being used as the "calibration signal": we are using only a subspace of all the theoretical far-end signals. This means that the estimation of h_{est} is at best partial. The echo canceller may be achieving good performance for a particular talker, but it is often observed that if a second talker comes on the line at the far end, then echo may break through for a short time until the algorithm readapts. Another factor that limits performance is the presence of noise at the near end: this includes acoustic noise, which can be severe in road vehicles, and electrical noise in the circuitry, which precedes the DSP processor. The basic assumption of a linear impulse response in Equation (8.3) may be optimistic. It is quite common for users to turn up the volume on hands-free speakers to levels at which distortion occurs. In this situation, the adaptive algorithm will generally attempt a best least-squares fit, but inevitably some echo will leak through.

Double Talk

In normal conversation, there is frequent overlap of the speech from each party. A study[12] has found that, for the purposes of telephone engineering, conversational speech can be broken down into:

Talk-spurts (i.e., by one party): Average duration roughly 1 s (38.53% of the time)

Pauses (by one party): Average duration roughly 1.6 s (61.47% of the time)

Double-talk (both parties): Average duration roughly 0.23 s (6.59% of the time)

Mutual silence (both parties): Average duration roughly 0.5 s (22.48% of the time)

These figures, which are averaged from English, Italian, and Japanese studies, may seem surprising. In natural speech, individual words are not generally demarcated by silence as suggested by the written word, but they tend to follow one from another with very little break. A short sentence can easily last only 1 s. The degree of overlap, or double talk, may not seem very great, but experience shows that there is often enough double talk to undermine the adaption process of the echo canceller. To understand this, refer to Equations (8.4 through 8.6): the estimate h_{est} will converge to the "true" impulse response h only if the microphone input consists of loudspeaker echo and nothing else. The presence of near-end speech will increase the error in the h_{est}, which may result in breakthrough of echo. Furthermore, test engineers will use more aggressive "conversation,"[6,7] in which "speech" (both real and artificial) is played into both ends of the system simultaneously.

To prevent deconvergence, a robust echo canceller should suspend the adaption process while there is near-end speech. Meanwhile, the basic filtering of Equation (8.5) should carry on with the coefficients h_{est} temporarily frozen. In order to do this, the near-end speech has to be detected. However, if the filter is not well adapted — for example, at start-up, when the echo canceller is first switched on, or if the echo path has changed — then this speech has to be distinguished from far-end echo. The means by which an effective echo canceller can detect near-end speech and distinguish it from far-end echo is called a *double-talk detector.*

The simplest form of double-talk detector is designed to trigger when microphone and reference signal pass a preset threshold. This, however, is not always satisfactory for several reasons:

- Provision must be made for fast adaption at start-up and for changes in echo path when the echo level may be over the preset threshold.

- Near-end noise may be mistaken for speech. If this noise is persistent (e.g., car noise), the echo canceller may be more effective if the adaption is *not* suspended.

- Thresholds based on *actual* levels rather than on *ratios* of levels make product design and application engineering difficult; for example, microphone gains may vary considerably within a production batch.

The most successful approaches to double-talk detection look for correlation between the reference signal and the microphone signal.[13] It could be argued that this is exactly what the LMS adaption algorithm does. However, something is needed that will react much faster than LMS, not just to disable the adaption, but also to control the nonlinear features described in the following paragraphs. Moreover, the double-talk detector must not substantially add to the DSP processing load. A recent contribution to this subject has been given by Gänsler et al.[14] Also, as has been seen, an effective echo canceller needs some means, however elementary, of distinguishing near-end speech from noise and taking the appropriate action.

Impulse Response Change

The design of any adaptive echo canceling algorithm implicitly assumes that the impulse response, h, changes, if at all, much slower than the rate of adaption. In practice, rapid changes of echo path will occur: the microphone may get moved, the loudspeaker volume may be turned up, or the near-end talker may move. This will cause an increase in echo. This echo resembles speech, but an effective double-talk detector as discussed in the previous paragraph will not be deceived by this and adaption will continue. If, however, as often happens in the course of an impulse response change, the microphone is brushed, tapped, or struck, the noise this makes will almost certainly be treated as double talk (because it is clearly uncorrelated with any far-end speech). So, even with a well-designed double-talk detector, the recovery of full echo cancellation performance will be delayed.

More generally, if double talk and impulse response change occur together, there may be a prolonged breakthrough of echo. The solution to this problem, which is the subject of active research, will be a compromise allowing some adaption if the echo can be detected and separated from near-end speech: it is inevitable that there will either be breakthrough of echo in this case, or the risk of speech attenuation if nonlinear processing (NLP) is being used.

Practical Considerations for Hands-Free Telephony: Nonlinear Processing

The Need for NLP: The Center Clipper

The above discussion has shown that there are a number of factors that limit the performance of an adaptive echo canceller. Published standards typically require (by implication) an ERLE of at least 40 dB for far-end single talk, relaxing this to 30 dB during double talk. The convergence time of 2 s required by these standards is also difficult to achieve. In theory, it may be possible to reduce the convergence time to zero by storing the previously calculated h_{est} in nonvolatile memory, but there is no guarantee that the actual impulse response will stay constant between uses of the echo canceller. Typical operating conditions, with ordinary audio equipment, limit the achievable ERLE to no more than about 30 dB. Furthermore, as pointed out in "Echo Cancellation Definitions and Required Performance," above, echo at these levels is often clearly audible to the far-end talker, and competition between suppliers means that this is unacceptable. In practice, there must be *no* audible echo.

It follows that the echo canceller must be supplemented by additional processing in order to achieve the desired levels of performance. This is given the generic name of NLP. The simplest additional device of this kind is a *center clipper*. Figure 8.4 illustrates the principle of operation. Applied to the output signal e_n calculated by Equation (8.5), the new output becomes:

$$e_{nout} = \begin{cases} e_n - \delta, & e_n > \delta \\ 0 & |e_n| \le \delta \\ e_n + \delta, & e_n < -\delta \end{cases} \qquad (8.7)$$

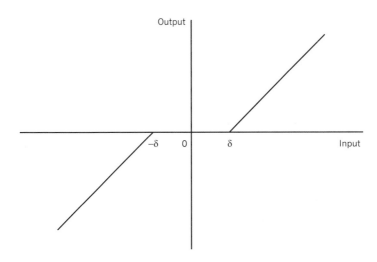

FIGURE 8.4
Transfer function of simple center clipper with threshold δ.

The threshold δ is adaptive, so that it is zero when there is near-end speech. Many refinements of this basic idea are possible, including frequency-domain versions. This simple model does, however, show the main features:

- Signals below the threshold are squelched completely.
- Large signals suffer a loss, which in terms of dB is relatively small, so the center clipper is forgiving of errors in the double-talk detection.
- The residual signal is whitened: this is a by-product of the process, which often makes it possible for a careful listener to determine whether a center clipper is being used.

These points are illustrated in the wavefiles (Audio 1–4) accompanying this chapter, available at http://www.crcpress.com/e_products/download.asp? cat_no=0949.

Additional Measures: Loss Insertion

Even more basic than center clipping is *loss insertion*, which is a loss inserted into the send path that is sufficient to achieve the desired ERLE. Note that the path should not be closed completely: users like to have some low-level sound to give the feeling that the line is "live." Another possibility is the addition of a single-channel speech filter as described in Chapter 6, to *precede* any center clipping or loss insertion. This has a squelching effect on residual echo, especially during double talk. Loss insertion and center clippers must come off during double talk (so that near-end speech is not attenuated), and so the speech filter is the main way of boosting ERLE at this time.

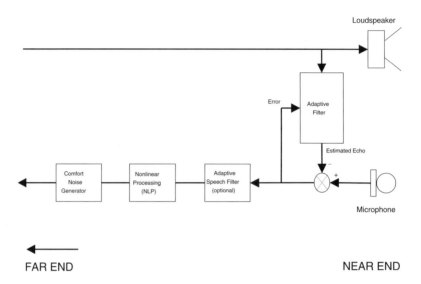

FIGURE 8.5
System diagram showing sequence of operations of echo canceller combined with speech filter, nonlinear processing and comfort noise.

These measures (center clipper, loss insertion, speech enhancement filter) can all be combined to ensure that the far-end talker hears no echo. Figure 8.5 shows how they fit together as a system. What distinguishes this approach from the old VOX is the level of control, so that the near-end talker can break in at will and enjoy conversation that is of full-duplex quality. In order to achieve good performance, there are some subtleties of design that can be employed:

- The clipper threshold and inserted loss are controlled by the presence of far-end speech. They should have a slight "hangover" so that when a far-end speech burst ends, the residual echo that follows is still caught.

- The NLP parameters must not be too aggressive or the near-end speech will be subject to "leading edge clipping." This is illustrated in the Audio 5 wavefile accompanying this chapter, available at http://www.crcpress.com/e_products/download.asp?cat_no=0949. Loss of initial consonants will reduce the intelligibility of speech and so is not acceptable.

- Note that the single-channel speech enhancer must come before the other NLP processes because this kind of filter generally works best with stationary noise: the NLP features tend to make the residual signal less stationary.

These features are generally controlled by proprietary methods that cannot be disclosed here. It can be said, however, that effective and unobtrusive use of NLP depends on a good adaptive algorithm, without which echo control reverts to a VOX.

Comfort Noise

Adding the above features does not, however, solve all the problems of acoustic echo cancellation. If there is background noise at the near end, applying NLP results in far-end talkers hearing this noise with breaks that match their speech pattern. Hearing breaks in the noise that are synchronized with one's own speech can be more unpleasant and annoying than hearing the noise continuously. This disjointed background noise is audible even when noise cancellation techniques are being applied (see Chapter 6), as these techniques will always leave some residual noise. One way to combat this effect is to generate pseudorandom noise which matches the spectral characteristics of the background noise and use it to fill the holes left by the removed echo. This artificial noise is called *comfort noise*. Figure 8.5 shows how comfort noise generation fits in with the rest of the system. The wavefiles Audio 6, 7, and 8 accompanying this chapter (available at http://www.crcpress.com/e_products/download.asp?cat_no=0949) illustrate the difference that comfort noise makes to the perceived quality of an echo canceller.

Voice over Internet Protocol Applications

Voice over Internet protocol (VoIP) is becoming increasingly important as a means of communication (see Chapter 11 and Reference 15). With VoIP goes the use of personal computers (PCs), with their sound cards, as communications terminals, often in hands-free mode. Acoustic echo cancellers of the type discussed in this chapter are equally applicable for this kind of telephony, but they need some modification to be effective in this context. This is basically because the processor chips used in PCs have a different architecture and a different function from the DSP chips used in communications applications. One result of this is that the microphone and loudspeaker signals will be subject to variable delays within the processor, which contrasts with a DSP in which signal delays are effectively constant, fixed by the software design. This means that, in the case of the PC processor, the impulse response h that must be estimated within the echo canceller algorithms (Equations [8.4 through 8.6]) is varying. The variation can, however, be compensated for: the precise details depend on the processor type, the operating system, and the VoIP software package that is being implemented.

Howling Detection

A modern acoustic echo canceller is designed to minimize the risk of instability, however, any particular design that is in service will accumulate millions of hours of continuous use. It is possible, even likely, that instability will occur at some time during the life of a particular design. Since it is not possible to predict every possible combination of circumstances, it is desirable to incorporate a feature which will detect instability and take action promptly. For example, the echo canceller could detect the situation where the output signal power exceeds the input power by a specified level over a number of iterations; this should catch the instability before amplitude clipping is reached. The appropriate action is to reset the h vector to zero (or recover the start-up h vector stored in nonvolatile memory) and begin the adaption afresh.

Performance Assessment

Standard tests of performance are given in ITU-T Recommendations G.167 and G.168 for acoustic echo cancellation and line echo cancellation, respectively;[6,7] these mostly measure ERLE in very specific circumstances. In addition to these tests, the time delay in the microphone path, which is due to the echo canceller, can be measured by seeking the maximum cross correlation between microphone input and echo canceller output. These "objective" tests, as they are called, only give snapshots of the system performance.

In the development of an actual echo cancellation product there will be extensive testing by both expert listeners and members of the public and their opinions will be used to inform the design process. Much use is made of recorded conversations taken from a hands-free terminal. This introduces the notion of *third-party listening*[16]: the participants in a conversation may be perfectly happy with the audio quality that they are getting because some of the artefacts may be inaudible to them. For example, as already mentioned in "VOX, the Simple Solution," above, a talker may be unaware of speech clipping if both parties are speaking simultaneously. However, this clipping may be very apparent to someone who is "listening in" to the conversation. The implication for future product development is that the quality required of an echo canceller for a conference phone application, where there may be several users at each end, is much greater than for a car phone with only a single user at each end.

Line Echo Cancellation

As already mentioned, echo can arise within communication networks, and it can be dealt with by essentially the same methods as for acoustic echo. Network operators place echo cancellers at switching centers and users are generally unaware of their operation. The echo typically consists of a relatively small number of single reflections. The precise timing of these reflections will not be known in advance because they depend on the routing of calls, so the algorithm still represents the impulse response as a vector h. The small number of reflection points means that the h vector will consist largely of zeros (in contrast, an acoustic echo path is characterized by a very large number of multiple reflections, so that most elements of the h vector have some significance — see Figure 8.3). Algorithm designers can exploit this sparseness of h in order to achieve faster adaption than for an acoustic echo canceller.[13]

The performance requirements for line echo cancellers[7] are very similar to those for acoustic echo cancellers. There are, however, important differences: the permitted group delay across a line echo canceller is less than 1 ms.[7] This means that algorithms that take block input are effectively precluded from use in line echo cancellation except when the echo canceller is used in conjunction with other processing that handles blocks of data. For example, at the interface between a cellular mobile network and a fixed network there may be equipment to decode speech signals from compressed form into pulse code modulated (PCM) form. The data will be in block form prior to transmission and there may be opportunity here to use block time-domain or frequency-domain algorithms for line echo cancellation.

Another important difference between line (or "network") and acoustic echo cancellers is that not all signals will consist of speech. There will be fax and data signals, which consist of modulated tones. For fax and low-rate data the echo canceller is required to perform in the same way as it does for speech[7] in order to prevent corruption of data interchanges. For higher-rate data signals, detected by the tone frequency and phase-switching pattern, the echo canceller is required to disable itself (full details are given in Reference 7). In these cases, there is special-purpose echo control built into the modems that originate these data signals.

In some mobile (cellular) networks there is a requirement for the mobile terminal to have a line echo canceller to cancel any speech echo that is reflected back via the radio link. This is not a universal requirement and is generally found only with the older analog networks. Modern digital networks (such as GSM and CDMA) use different frequencies and/or coding for uplink and downlink so that echo is not picked up at all. Nevertheless, the older network standards are still in operation in many places, and equipment manufacturers may require their DSP suppliers to provide both line and acoustic echo cancellation for this reason.

Stereophonic Echo Cancellation

Suppose that the voice input system that we are working with has several loudspeakers with different inputs (the systems already discussed will work for the case where a number of loudspeakers are fed with an *identical* monaural signal). In such a case, referring to Equation (8.5), the echo can be canceled by storing a separate impulse response vector h_i for each loudspeaker channel. The echo is canceled by performing the calculation

$$e = y - \sum_i h_i \otimes x_i \tag{8.8}$$

where x_i is the reference signal vector corresponding to loudspeaker channel i, y the microphone signal, and e the output signal before any NLP or other processing. Analogous to the simple case discussed earlier, the equation for coefficient update is:

$$h_{ik} = h_{ik} - \mu e_n x_{in-k} \quad \text{for} \quad k = 0, \dots, N-1 \quad \text{and all } i \tag{8.9}$$

The other features (e.g., NLP, speech filter, comfort noise) may be added to the output channel if desired. Thus far, this appears to be a straightforward extension of single-channel echo cancellation. Now suppose that the x_i are related stereophonically, that is, each loudspeaker channel corresponds to a different far-end microphone. The case of a single far-end talker can be represented as a single source s with microphone transfer functions g_i so that

$$x_i = g_i \otimes s, \quad \text{all } i \tag{8.10}$$

This means that the adaption algorithm is trying to solve the equation

$$\sum_i (h_i - \tilde{h}_i) \otimes g_i \otimes s = 0 \tag{8.11}$$

where $\tilde{h}_i$ is the actual impulse response of the ith loudspeaker–microphone path. This does not in general have a unique solution in the h_i (one equation, several unknowns). Since this is a linear system, the calculated h_i may (and probably will) grow without limit, leading to problems with adaption, including stability. A number of methods have been proposed to deal with this,[17] not all of them very satisfactory, as they entail modifications to the x_i to make them less correlated. The use of a *leakage* term in the adaption equation has been proposed (a standard DSP procedure[18]) so that Equation (8.9) becomes

$$h_{ik} = (1-\lambda)h_{ik} - \mu e_n x_{in-k}, \quad \text{for} \quad k = 0, \dots, N-1 \quad \text{and all } i \tag{8.12}$$

where the leakage factor λ is a small positive number. This will ensure a unique value of h, which will still give reasonable performance. It has been pointed out that Equation (8.10) may not hold in many instances on account of other factors, such as the use of nonlinear coding for transmission.[19] In this case, the x_i are *not* in a linear relationship, and stereophonic echo cancellation as described above may work well without the need for special features.

Current and Future Developments

Research and development of echo cancellers continues, as higher levels of performance are demanded and communications networks continue to grow. The ultimate goal is for speech communication and voice input (e.g., to a recording device or speech recognition engine) to become as easy and natural as a face-to-face conversation.

New developments are continually being announced by researchers and manufacturers. Two conference series are of particular relevance to this field: the annual IEEE-sponsored International Conference on Acoustics, Speech and Signal Processing (ICASSP) and the biennial International Workshop on Acoustic Echo and Noise Cancellation (IWAENC). The proceedings of IWAENC '99 have been published in book form.[20]

References

1. Widrow, B. and Stearns, S.D., *Adaptive Signal Processing*, Prentice-Hall, Englewood Cliffs, NJ, 1985, chap. 12.
2. ITU-T Recommendation G.131, Control of talker echo, International Telecommunication Union, 1996.
3. European Telecommunication Standard ETS 300 903, Digital cellular telecommunications system (Phase 2+); Transmission planning aspects of the speech service in the GSM Public Land Mobile Network (PLMN) system (GSM 03.50 version 5.0.2), ETSI, Sophia Antipolis, France, 1997.
4. ITU-T recommendation P.340, Transmission characteristics of hands-free telephones, International Telecommunication Union, 2000.
5. ITU-T Recommendation G.122, Influence of national systems on stability and talker echo in international connections, International Telecommunication Union, 1993.
6. ITU-T Recommendation G.167, Acoustic echo controllers, International Telecommunication Union, 1993.
7. ITU-T Recommendation G.168 Digital network echo cancellers, International Telecommunication Union, 1997.
8. ITU-T Recommendation P.50, Artificial Voices, International Telecommunication Union, 1999.
9. Oppenheim, A.V. and Schafer, R.W., *Discrete-Time Signal Processing*, Prentice-Hall, Englewood Cliffs, NJ, 1989, chap. 9.

10. Harley, T. and Leese, S., Acoustic Echo Canceller, U.S. Patent 6,091,813, July 2000.
11. Breining, S. et al., Acoustic echo control: an application of very-high-order adaptive filters, *IEEE Signal Processing Magazine*, July 1999, 42.
12. ITU-T Recommendation P.59, Artificial Conversational Speech, International Telecommunication Union, 1993.
13. Benesty, J. et al., A new class of doubletalk detectors based on cross-correlation, *IEEE Trans. Speech Audio Processing*, 8, 168, 2000.
14. Gänsler, T. et al., Double-talk robust fast converging algorithms for network echo cancellation, *IEEE Trans. Speech Audio Processing*, 8, 656, 2000.
15. Witowski, W.E., Understanding echo cancellation in voice over IP networks, in *ICSPAT'99 Proceedings*, International Conference of Signal Processing Applications and Technology, 1999, Paper W233.
16. ITU-T Recommendation P.831, Subjective performance evaluation of network echo cancellers, International Telecommunication Union, 1998.
17. Sondhi, M.M., Stereophonic acoustic echo cancellation — an overview of the fundamental problem, *IEE Signal Processing Letters*, 2, 148, 1995.
18. Loke, Y. and Chambers, J., Leakage factor: its application in stereophonic acoustic echo cancellation, in *IWAENC'97*, Vol. 1, IEEE, 1997, 112.
19. Gänsler, T. and Eneroth, P., Influence of audio coding on stereophonic acoustic echo cancellation, *ICASSP'98*, Vol. 6, IEEE, 3649.
20. Gay, S.L. and Benesty, J., *Acoustic Signal Processing for Telecommunication*, Kluwer Academic Publishers, Boston, 2000.

Section IV:

Special Applications

9

Signal and Feature Compensation Methods for Robust Speech Recognition

Rita Singh, Richard M. Stern, and Bhiksha Raj

CONTENTS

Introduction

As computing, communication, and other electronic devices become physically smaller and attempt to perform increasingly complicated functions, traditional interfaces such as buttons, keyboards, etc. become difficult to use. Speech is a much more natural and simpler interface for such devices, especially if they are to be remotely operated. Viable technology currently exists for the deployment of speech-enabled devices in controlled environmental conditions. However, as these devices are deployed in increasingly difficult operating conditions that are open to uncontrolled noises and acoustical disturbances, the performance of speech recognition systems degrades greatly. This chapter and Chapter 10 are concerned with the subject of development of techniques that reverse this degradation.

Broadly, techniques that enhance environmental robustness for speech recognition systems can be divided into two categories: techniques that operate on speech signals or the features derived from them prior to the recognition process, and techniques that modify the recognition system to perform optimally on incoming noisy speech signals. In this chapter, we review techniques that modify incoming signals or feature vectors. Techniques that modify the structure or parameters of the speech recognition system are discussed in Chapter 10.

For the benefit of readers with a limited background in speech recognition technologies, we begin by reviewing the formulation of automatic speech recognition as a statistical pattern classification process and by discussing how environmental disturbances adversely affect classifier performance. Later sections describe selected signal and feature compensation techniques in current usage, which were chosen on the basis of their efficiency and generality.

Speech Recognition as Statistical Pattern Classification

Automatic speech recognition systems are pattern classifiers designed to solve a rather specific statistical pattern classification problem. A simple example of a statistical pattern classification problem is that of determining the member of a set of N classes $C_1, C_2, \ldots, C_N$ to which a specific vector X_s belongs, knowing that it does belong to one of the classes.

Let $P(X|C_i)$ be the known distribution of all vectors belonging to class C_i. Let α_i be the fraction of all data points that belong to class C_i; α_i is also known as the *a priori* probability of C_i. It can be shown that if the data vector X_s is assigned to a class according to the following rule[1]:

$$X_s \rightarrow C_i \quad \text{if} \quad \alpha_i P(X_s|C_i) > \alpha_j P(X_s|C_j) \quad \text{for all } j \neq i \qquad (9.1)$$

the expected classification error is minimum. In other words, given an infinitely large set of data points to classify, the total number of misclassified points will be minimum if the classification rule above is followed. Pattern classifiers based on the above rule are known as Bayesian classifiers. If the criterion for classification is other than that of minimum expected classification error, e.g., that of minimizing the expected *cost* of classification (known as minimum risk classification), where the cost may be any function of the output of the classifier, the actual classification rule can vary from the one given above, but its *form* will still be very similar.

Speech recognition is the problem of determining the sequence of words that were spoken in an utterance, given the recorded signal for that utterance. We can consider the set of all signals that are instances of a particular word sequence to form the class of signals representing that word sequence. Hence, there is a class of signals associated with every possible sequence of words in a language. Statistical speech recognition can be stated as the problem of determining to which of these classes a given signal belongs. This problem can now be treated as an instance of the Bayesian classification problem.

Casting this problem in more formal terms, we let W represent an arbitrary sequence of words. Let $P(X|W)$ represent the distribution of the class of signals belonging to the word sequence W, where X now represents an arbitrary signal. The speech recognition problem can then be stated as:

$$X_s \to A \text{ if } P(A)P(X_s|A) > P(B)P(X_s|B) \text{ for all } A \neq B \qquad (9.2)$$

where A and B represent different instances of W. This can be rewritten as

$$X_s \to A : A = \arg\max_W \{P(W)P(X_s|W)\} \qquad (9.3)$$

where X_s is the signal for the utterance that must be recognized. $P(W)$ is the *a priori* probability of the word sequence W and is given by a *language model*. The language model and the *a priori* probabilities of word sequences are largely irrelevant to the subject matter of the rest of this chapter and will not be discussed here. The interested reader can find details of this topic elsewhere.[2]

Ordinarily, speech recognition systems do not store the value of $P(X_s|W)$ separately for every W. Instead, words are considered to be defined as sequences of elements from a small set of *subword units*, i.e., $W = W_a, W_b, W_c, \ldots$ and

$$P(X_s|W) = P(X_s|W_a, W_b, W_c, \ldots) \qquad (9.4)$$

The number of unique subword units is significantly less than the number of actual words that the system must recognize. The probability distributions for $W_a, W_b, W_c, \ldots$ are obtained by associating a statistical model with each of these units, and learning the parameters of those statistical models. Although any type of statistical model may be used for this purpose, the model that is currently the most popular for speech recognition is the hidden Markov model (HMM).[3]

Typically, recognition is performed on the basis of the observed vectors of a sequence of *feature vectors* derived from the digitized speech rather than the signal itself. Ideally, these feature vectors would be chosen to be representative of the characteristics of the speech signal that are most essential for recognition, while not representing other redundant information. The most commonly used features are *cepstral coefficients*. The cepstral representation of a speech signal is derived by segmenting it into a number of overlapping segments, called *frames*, and deriving a cepstral vector for each frame. Denoting a speech signal that is a function of time t, as s_t, the speech signal in the ith frame as s_t^i, and the feature vector for the ith frame as f_s^i, the cepstral representation f_s^i is

$$f_s^i = DCT(\log(DFT(s_t^i))) \qquad (9.5)$$

i.e., it is the discrete cosine transform of the log of the Fourier transform of the data. In practice, modified versions of cepstral coefficients such as Mel frequency cepstral coefficients (MFCC)[4] are used.

We use the notation $F(s_t^i)$ to represent the generalized transformation that is applied to s_t^i to derive the feature vector f_s^i. The entire sequence of feature vectors for s_t can be represented as

$$f_s = [f_s^1, f_s^2, f_s^3, \ldots] = [F(s_t^1), F(s_t^2), F(s_t^3), \ldots] \qquad (9.6)$$

Recognition is thus performed using $P(f_s|W)$ rather than $P(s_t|W)$ as

$$X_s \to A : A = \arg\max_W P(W)P(f_s|W) \qquad (9.7)$$

Effect of Noise on Speech Recognition Systems

The effect of corrupting noise is to transform the clean speech signal s_t into a noisy signal. We use the notation y_t to represent the noisy signal and $T_n(\ )$ to represent the corrupting transformation, so that $y_t = T_n(s_t)$. $T_n(\ )$ can range from the simple addition of a noise signal n_t to the speech signal s_t, to other linear or nonlinear transformations such as convolution, clipping, or compression. Features derived from the noisy signal are thus given by $F(y_t) = F(T_n(s_t))$.

Ideally, a recognition system used to recognize the noisy speech y_t would also be trained with noisy speech, since the distributions of word sequences for optimal recognition of y_t must be $P(F(y_t)|W)$, i.e., $P(F(T_n(s_t))|W)$ and not $P(F(s_t)|W)$. If a recognizer is trained with feature vectors of clean speech, the distribution learned for W, $P(F(s_t)|W)$ would differ from $P(F(y_t)|W)$, introducing a mismatch. This mismatch causes degradation of recognition performance. The greater the mismatch, the larger the degradation. Figure 9.1 shows some plots that are representative of this problem.

In a typical situation, however, the noise conditions and therefore the corrupting transformation $T_n(\ )$ affecting the speech signal can be diverse,

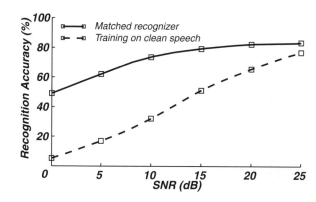

FIGURE 9.1

Typical plot of recognition accuracy as a function of the signal-to-noise ratio (SNR). The lower curve is for a "mismatched" recognizer, where the recognition system has been trained on clean speech, but the test speech is noisy. The upper curve is for a "matched" recognizer, where the recognition system has been trained with speech that had been subjected to the same level of noise as the test speech. (From Raj, B., Reconstruction of Incomplete Spectrograms for Robust Speech Recognition, Ph.D. thesis, Carnegie Mellon University, Pittsburgh, PA, 2000. With permission.)

and can vary from utterance to utterance. In many situations it may even be impossible to predict or determine the exact corrupting transformation affecting the speech signals, rendering it difficult, if not impossible, to train the recognizer optimally for every utterance to be recognized.

In the absence of reliable *a priori* information about the nature of the corrupting transformation $T_n(\ )$, it is common to train speech recognition systems using speech signals that are assumed to represent the entire range of operating conditions under which they must operate.

Even when $T_n(\ )$ is largely constant across the data on which the system is trained and which it recognizes, it still degrades recognition performance. This is because, as the level of noise in the signals increases, the different signals increasingly resemble the noise. As a result, the distinctions between the various sounds fade and they begin to resemble each other increasingly. As a result, even the optimal recognizer for noisy speech cannot perform as well on noisy speech as the optimal recognizer for clean speech performs on clean speech. The curve representing the "matched recognizer" in Figure 9.1 illustrates this point.

Different types of noises affect speech recognition systems differently. Figure 9.2 shows how a typical speech recognition system performs in a typical experiment when the speech to be recognized is affected by stationary (white) noise and by nonstationary noise (music). Music does not degrade recognition accuracy as severely as stationary noise. Nonstationary noises in general do not challenge recognition systems as much as stationary noises do because at a given average signal-to-noise ratio (SNR), the instantaneous SNR tends to be greater over large portions of the signal due to the

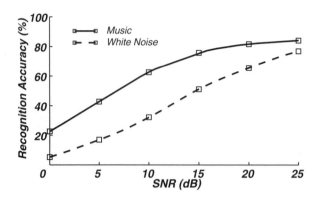

FIGURE 9.2
Recognition accuracy obtained for speech corrupted by white noise, and for speech corrupted by a segment of music, at various SNRs. SNRs have been measured as the average signal-to-noise ratio over entire utterances. (From Raj, B., Reconstruction of Incomplete Spectrograms for Robust Speech Recognition, Ph.D. thesis, Carnegie Mellon University, Pittsburgh, PA, 2000. With permission.)

time-varying nature of the noise. Nevertheless, it is generally more difficult to compensate for the effects of nonstationary noise.[5]

Compensating for the Effects of Noise

As stated above, the presence of noise generally causes the distributions characterizing the speech sounds to be recognized to differ from the corresponding distributions that were used to train the speech recognizer, which in turn causes recognition accuracy to degrade. This degradation in accuracy can only be reduced by reducing the difference between the actual distributions of the test data and those used by the recognizer. In terms of the notations used in the earlier section, to minimize the degradation in performance due to noise, the difference between $P(F(y_t)|W)$ and $P(F(s_t)|W)$ must be minimized for all W. This can be achieved in several ways:

1. **Signal compensation:** This is performed by applying a transformation $C(\ )$ to the degraded speech signal y_t such that $z_t = C(y_t) \approx s_t$. Recognition is then performed using $F(z_t)$, the features of the transformed signal z_t. The distribution of the test data for any word class W is now $P(F(z_t)|W) \approx P(F(s_t)|W)$. The more closely z_t approximates s_t, the closer the two corresponding distributions are, and the better the performance that is achieved on $F(z_t)$.

2. **Feature compensation:** This is done by transforming the features computed from the noisy speech, $F(y_t)$, with some transformation

$C(\ )$, such that $f_z = C(F(y_t)) \approx F(s_t)$. Recognition is then performed with the transformed features f_z instead of with $F(y_t)$. The distribution of word classes on the transformed features is now $P(f_z|W)$ $\approx P(F(s_t)|W)$. Once again, the better f_z approximates $F(s_t)$, the better the recognition performance that is achieved with it.

3. Model compensation: This is done by transforming the state distributions using some transformation $C(\ )$ such that $C(P(F(s_t))|W)$ $\approx P(F(y_t)|W)$. This is different from the previous two methods where the compensation was effected by modifying the incoming test data to reduce the mismatch between its distributions and those learned by the recognizer. Here, the distributions learned by the recognizer are transformed such that they better match those of the data to be recognized. Recognition is now performed using $C(P(F(s_t))|W)$. As before, the better this approximates $P(F(y_t)|W)$, the better the recognition performance that is achieved.

4. Matched condition techniques: In matched condition techniques we attempt to explicitly match the distributions used by the recognizer to those of the test data. In *matched training,* this is achieved by training the recognizer on data that have been corrupted in an identical manner to the test data. In *missing feature methods,* we attempt to use only those components of the data whose distributions explicitly match the distributions modeled by the recognizer, and treating the mismatched components as missing.

In the remaining sections of this chapter, we describe signal and feature compensation approaches in detail; we discuss the complementary model compensation and matched condition approaches in Chapter 10. We note that the transformations used in these methods are not necessarily simple and may not have a closed-form formulation, relying on iterative procedures for their solution.

Signal Compensation

In signal compensation, we attempt to modify the noisy signal y_t with a transformation $C(\ )$ such that the distributions of the modified noisy signal, $z_t = C(y_t)$, resemble those of s_t, the data used to train the recognizer. When the goal is to transform the noisy signal to resemble clean speech, these methods are also called signal *enhancement* methods. Many different transformation operators $C(\ )$ have been proposed over the years, motivated by, for example, differing assumptions about the type of noise or other sources of degradation, and the type of features to be extracted from the signal for recognition. In this section we discuss some of the most widely used signal compensation algorithms: linear spectral subtraction (LSS),[6] nonlinear spectral subtraction (NSS),[7] and Wiener filtering.[8] The complementary technique of enhancement through the use of multiple microphones is discussed in Chapter 7.

Linear Spectral Subtraction

Linear spectral subtraction, also referred to simply as *spectral subtraction*,[6] is a method of canceling additive uncorrelated noise from a noisy speech signal. In spectral subtraction, signals are separated into speech and nonspeech regions by a variety of techniques, and all regions deemed to be nonspeech are used to estimate the noise spectrum. Signal compensation is performed by subtracting the estimated noise spectrum from the spectrum of the noisy speech to obtain an estimate of the clean speech spectrum.

More specifically, let $Y(t)$ represent the Fourier transform of the tth frame of incoming noisy speech and let the kth component of $Y(t)$ be referred to as $Y_t(t, k)$. Since the Fourier spectrum is complex, it can be represented in polar coordinates as

$$Y(t, k) = |Y(t, k)|e^{i\angle Y(t, k)} = (|Y(t, k)|^\tau)^{\frac{1}{\tau}} e^{i\angle Y(t, k)} \tag{9.8}$$

where $|Y(t, k)|$ is the magnitude of $Y(t, k)$, $\angle Y(t, k)$ is its phase, and the parameter τ can take any value, since it does not modify the equation in any manner. Compensation is performed in two steps:

1. **Noise update step:** This step involves estimating the spectrum of the corrupting noise. The most common method used for this has the following rule for estimating the magnitude of the kth frequency component of the noise spectrum in the tth frame, $N(t, k)$:

$$|N(t, k)|^\tau = \begin{cases} (1 - \lambda)|N(t - 1, k)|^\tau + \lambda|Y(t, k)|^\tau & \text{if } |Y(t, k)| < \beta|N(t, k)| \\ |N(t - 1, k)|^\tau & \text{otherwise} \end{cases} \tag{9.9}$$

   The parameter λ, called the *noise update factor,* lies between 0 and 1 and controls how rapidly the noise estimate is permitted to change in response to changes in the time-varying noise. Large values of λ permit the noise estimate to change rapidly, but can result in poor estimates of the noise spectrum. Low values of λ, on the other hand, provide more robust estimates of the noise spectrum for stationary or slowly varying noise, but do not permit the algorithm to track noises that change quickly. β is a threshold factor used to identify the putative onset of speech.

2. **Noise cancellation step:** Once the estimate of the noise spectrum is known, the estimate of the magnitude of the clean speech spectrum $\hat{S}(t, k)$ is obtained from the noisy spectrum as

$$|\hat{S}(t, k)|^{\tau} = \begin{cases} |Y(t, k)|^{\tau} - \alpha|N(t, k)|^{\tau} & \text{if } |Y(t, k)|^{\tau} - \alpha|N(t, k)|^{\tau} > \gamma|Y(t, k)|^{\tau} \\ \gamma|Y(t, k)|^{\tau} & \text{otherwise} \end{cases} \quad (9.10)$$

where α is an *oversubtraction* factor that compensates for errors in the estimation of the noise spectrum and is typically set to be between 1 and 3. γ, the *spectral floor* factor, is a small positive constant that ensures that the spectral estimate never becomes negative. The entire spectrum is obtained by combining the compensated spectral magnitudes and the original phase as

$$\hat{S}(t, k) = |\hat{S}(t, k)| e^{i\angle Y(t, k)} \quad (9.11)$$

The spectrum can then be inverted to obtain the estimated clean speech signal for the tth frame or used directly to obtain cepstral coefficients or other features.

The noise spectrum must be initialized with nonzero values in some fashion. This is typically done by assuming that the initial portion of any utterance contains only noise, and using the average spectrum over the first few frames in the utterance to initialize the noise spectrum.

The value of τ determines the exact form of spectral subtraction being applied, with 1 and 2 being the most commonly used values. Equation (9.10) estimates the power spectrum of the noise (PSUB) when τ equals 2, and the magnitude spectrum of the noise (MSUB) when τ equals 1. MSUB is known to be more effective than PSUB for speech recognition.[9]

The noise estimate obtained for spectral subtraction can be very sensitive to variations in the noise spectrum. One common problem is that the estimated energy in the noise spectrum can be greater than the energy in the noisy speech itself, resulting in negative values for the estimated magnitude spectrum of clean speech. While the spectral floor factor γ eliminates negative magnitude spectra, the estimated noise spectrum may still not be representative of the true noise spectrum, rendering spectral subtraction ineffective. Since frequency bands are processed independently, the arbitrary occurrence of peaks and zeros in the spectrum can result in an unpleasant perceptual effect in the reconstructed speech commonly referred to as musical noise.[10]

Nonlinear Spectral Subtraction

Nonlinear spectral subtraction (NSS)[7] attempts to improve the noise cancellation of spectral subtraction by making the oversubtraction factor for any frequency band dependent on the local SNR. The estimate of the noise spectrum is made more robust by separating it into two components, an instantaneous component $N(t, k)$ and a long-term component $\overline{N}(t, k)$. Signal compensation is recast as a filtering operation. A *smoothed* spectral estimate of the noisy speech signal $\bar{Y}(t, k)$ is used along with the instantaneous spectrum $Y(t, k)$ in order to obtain robust estimates of the compensating filter. The instantaneous spectrum of the noisy signal $Y(t, k)$ is simply the Fourier spectrum of the tth frame of the noisy signal. The magnitude of the instantaneous spectrum of the noise is obtained as

$$|N(t, k)| = \begin{cases} |Y(t, k)| & \text{if } |Y(t, k)| < \beta|N(t, k)| \\ |N(t - 1, k)| & \text{otherwise} \end{cases} \tag{9.12}$$

where β is a threshold factor. $\bar{Y}(t, k)$, the smoothed spectrum for the noisy speech at frame t, is computed as a weighted average of the smoothed spectrum at $t - 1$ and the current instantaneous spectrum. Similarly, $\bar{N}(t, k)$, the long-term spectrum for the noise at t, is computed as a weighted average of the long-term spectrum at $t - 1$ and the instantaneous noise spectrum at t:

$$\bar{Y}(t, k) = \lambda_y Y(t, k) + (1 - \lambda_y)\bar{Y}(t - 1, k)$$
$$\bar{N}(t, k) = \lambda_n N(t, k) + (1 - \lambda_n)\bar{N}(t - 1, k) \tag{9.13}$$

where, typically, $0.1 \le \lambda_y \le 0.5$ and $0.5 \le \lambda_n \le 0.9$. To compensate the signal in the tth analysis frame, a filter $H(t, k)$ is constructed as

$$H(t, k) = \frac{|\bar{Y}(t, k)| - \Lambda(t, k)}{|\bar{Y}(t, k)|} \tag{9.14}$$

where $\Lambda(t, k)$ is a noise subtraction term computed using the following nonlinear function:

$$\Lambda(t, k) = \alpha(t, k)F(\rho(t, k), \alpha(t, k), \bar{Y}(t, k)) \tag{9.15}$$

where $\alpha(t, k)$, $\rho(t, k)$, and $F(\rho(t, k), \alpha(t, k), \bar{Y}(t, k))$ are defined as

$$\alpha(t, k) = \max_{t - M \le \eta \le t}\{N(\eta, k)\} \tag{9.16}$$

$$\rho(t, k) = \frac{|\bar{Y}(t, k)|}{|\bar{N}(t, k)|} \tag{9.17}$$

$$F(\rho(t, k), \alpha(t, k), \bar{Y}(t, k)) = 1 - g(t, k)\left(1 - \frac{|\bar{N}(t, k)|}{\alpha(t, k)}\right) \tag{9.18}$$

$g(t, k) = sigmoid(\rho(t, k))$. The denominator in Equation (9.18) represents the peak estimate of the noise. M represents the time window over which the peak noise spectra are estimated (typically about 40 frames). The parameters of the sigmoid control the degree of oversubtraction and are typically set such that $\bar{N}(t, k) \leq \Lambda(t, k) \leq 3\bar{N}(t, k)$. The spectrum of the clean speech signal can now be estimated as

$$
\begin{aligned}
|\hat{S}(t, k)| &= H(t, k)|Y(t, k)| \\
\hat{S}(t, k) &= |\hat{S}(t, k)|e^{i\angle Y(t, k)}
\end{aligned}
\tag{9.19}
$$

The estimated spectrum can now be inverted to obtain the estimated clean speech signal.

Although the above formulation reduces the incidence of many of the problems faced by regular spectral subtraction, negative estimates of the magnitude spectrum of clean speech are still possible unless they are explicitly excluded by forcing a positive minimum. In general, however, the reconstructed signal does not exhibit the musical tone artifacts that occur in linear spectral subtraction.

Wiener Filtering

The Wiener filter is a linear filter with an impulse response that is designed to minimize the expected squared error between the clean speech signal and the filtered noisy speech signal.[8] It can be shown that in the most generic case the frequency response of the optimal filter is given by

$$
H(\omega) = \frac{S_{xy}(\omega)}{S_y(\omega)}
\tag{9.20}
$$

where $S_{xy}(\omega)$ is the spectrum of the cross correlation between the clean speech signal s_t and the noisy signal y_t. $S_y(\omega)$ is the *power* spectrum of y_t. A Wiener filter can be implemented as a finite impulse response (FIR) filter with an impulse response that approximates the impulse response implied by Equation (9.20) most closely (and hence tends to minimize the squared error between the clean speech and the filtered noisy speech). However, it is more common to implement the Wiener filter directly in the spectral domain.

The design of the filter requires knowledge of $S_{xy}(\omega)$. If it is assumed that the noise corrupting the speech signal is additive and uncorrelated to the speech, it can be shown that

$$
S_{xy}(\omega) = S_y(\omega) - S_n(\omega)
\tag{9.21}
$$

where $S_n(\omega)$ is the power spectrum of the noise. The optimal Wiener filter in this case is given by

$$
H(\omega) = \frac{S_y(\omega) - S_n(\omega)}{S_y(\omega)}
\tag{9.22}
$$

In practice, the power spectrum of the signal and the noise are both unknown and must be estimated. The implementation is similar to that of spectral subtraction. The speech signal is segmented into overlapping frames and windowed. A separate filter is then built for every frame as

$$H(t, \omega) = \frac{|S_y(t, \omega)|^2 - |S_n(t, \omega)|^2}{|S_y(t, \omega)|^2} \qquad (9.23)$$

where $S_n(t, \omega)$ is the estimate of the spectrum of the noise in the tth frame and $S_y(t, \omega)$ is the corresponding spectrum of the noisy speech signal. The numerator in Equation (9.23) can sometimes become negative due to estimation errors. To prevent this, the equation for the filter is usually modified to

$$H(t, \omega) = \frac{\max(|S_y(t, \omega)|^2 - |S_n(t, \omega)|^2, \alpha|S_y(t, \omega)|^2)}{|S_y(t, \omega)|^2} \qquad (9.24)$$

where α is a small positive number. The spectrum of the clean speech signal is estimated simply as

$$\hat{S}_s(t, \omega) = H(t, \omega)S_y(t, \omega) \qquad (9.25)$$

The estimated clean speech spectrum can be inverted to obtain the clean speech signal. Alternately, the estimated power spectrum of the clean speech signal can be derived directly as

$$|\hat{S}_s(t, \omega)|^2 = |H(t, \omega)|^2|S_y(t, \omega)|^2 \qquad (9.26)$$

from which cepstra or other related features for recognition can be derived.

Wiener filtering can be viewed as a special case of LSS, since the estimated spectrum of the noise signal is effectively subtracted from the spectrum of the noisy speech. Estimates of the noise spectrum can be obtained using any of the heuristics that are used in spectral subtraction.

Feature Compensation

In feature compensation methods, the *features* $F(y_t)$ computed from the noisy speech are modified by a transformation $C(\ )$ such that the distributions of the transformed features $f_z = C(F(y_t))$ better match the distributions used by the recognizer. Feature compensation algorithms can be categorized as stereo based or parametric. Stereo-based algorithms use a small "adaptation" set of signals that have been simultaneously recorded in the noisy recording environment and over a clean channel (i.e., a head-mounted, close-talking, noise-canceling microphone). Such data are usually referred to as *stereo data*. The stereo adaptation data are used to learn statistical relationships between the feature vectors of the noisy utterances and their clean counterparts. These are then used to compensate for the noise in test utterances. In parametric algorithms, the effects of noise on the features of clean speech signals are analytically characterized by a function with a small set of statistical parameters. This function is used in conjunction with the noisy test utterance and the known distribution of the feature vectors of clean speech to estimate the statistical parameters that represent the noise which are then used to compensate the features of the noisy test utterance.

In the following sections, we describe a typical stereo-based algorithm, *multivariate Gaussian-based cepstral normalization* (RATZ),[11] and two parametric algorithms, *vector Taylor series* (VTS)[11] and *codeword-dependent cepstral normalization* (CDCN).[9] Several other feature compensation algorithms such as *probabilistic optimal filtering* (POF)[12] and Kalman filter-based compensation[13] have also been proposed in the literature. However, most of these algorithms are related to the algorithms described here in the general principles of their design.

Multivariate Gaussian-Based Cepstral Normalization

In multivariate Gaussian-based cepstral normalization (RATZ), an adaptation corpus of stereo recordings is used to learn the relationship between the feature vectors of clean and noisy speech. The feature vectors usually considered are cepstral vectors (hence the term "cepstral normalization").

RATZ is performed in two stages. In the first *learning* stage, the parameters of a Gaussian mixture distribution representing the distribution of the feature vectors of clean speech are learned. The likelihood of any feature vector f_s^i is represented by this distribution as

$$P(f_s^i) = \sum_{k=1}^{K} \frac{c_k}{\sqrt{(2\pi)^d |\Phi_k|}} e^{-0.5(f_s^i - \mu_k)^T \Phi_k^{-1}(f_s^i - \mu_k)} \tag{9.27}$$

where d is the dimensionality of f_s^i, K is the total number of Gaussians in the mixture, c_k is the *a priori* probability (or mixing weight) of the kth Gaussian, μ_k is the mean of the kth Gaussian, and Φ_k is the covariance matrix of the kth Gaussian in the Gaussian mixture distribution. Φ_k is usually assumed to be a diagonal matrix. The values of c_k, μ_k, and Φ_k are estimated from the feature vectors of a training corpus of clean speech (which could be the clean component of the stereo data) using the expectation maximization (EM) algorithm.[14] To simplify notation in the rest of this chapter, we will use the term $G(X;\mu, \Phi)$ to denote the value of a Gaussian with mean μ and variance Φ, at X. Using this notation, we can rewrite Equation (9.27) as

$$P(f_s^i) = \sum_k c_k G(f_s^i;\mu_k, \Phi_k) \tag{9.28}$$

The RATZ algorithm assumes that the corrupting noise modifies the distribution of clean speech by modifying the means and variances of the individual Gaussians in the mixture. The overall distribution itself remains a mixture of Gaussians. The modified mean of the kth Gaussian is computed from the stereo corpus as

$$\tilde{\mu}_k = \mu_k + r_k \tag{9.29}$$

where r_k is the correction term for the kth Gaussian given by

$$r_k = \frac{\sum_t P(k|f_c^t)(f_n^t - f_c^t)}{\sum_t P(k|f_c^t)} \tag{9.30}$$

f_c^t represents the tth feature vector of the clean component of the stereo corpus and f_n^t is the corresponding feature vector of the noisy component. The modified variance of the kth Gaussian is computed by:

$$\tilde{\Phi}_k = \frac{\sum_t P(k|f_c^t)(f_n^t - \tilde{\mu}_k)(f_n^t - \tilde{\mu}_k)^T}{\sum_t P(k|f_c^t)} \tag{9.31}$$

When Φ_k is assumed to be a diagonal matrix, $\tilde{\Phi}_k$ is also diagonal. $P(k|f_c^t)$ represents the *a posteriori* probability that f_c^t belongs to the kth Gaussian in the mixture and is given by

$$P(k|f_c^t) = \frac{c_k G(f_c^t;\mu_k, \Phi_k)}{\sum_j c_j G(f_c^t;\mu_j, \Phi_j)} \tag{9.32}$$

In the second *compensation* stage of RATZ, the parameters of the distributions of clean speech, and those of the noisy speech are used to compensate the feature vectors of incoming noisy test data for the noise. This is done by estimating the clean counterpart of every feature vector in the noisy utterance using a minimum mean squared error (MMSE) estimator. Given a noisy

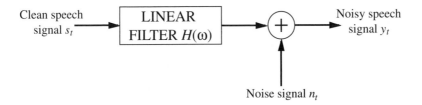

FIGURE 9.3
Block diagram representation of the linear filter-additive noise model of noise corruption.

feature vector f_n^t, the MMSE estimate of the clean counterpart $\hat{f}_c^t$ computed by RATZ is given by

$$\hat{f}_c^t = f_n^t - \sum_k P(k|f_n^t) r_k \tag{9.33}$$

where $P(k|f_n^t)$ is the *a posteriori* probability of the kth Gaussian, given the noisy feature vector f_n^t, and is given by

$$P(k|f_n^t) = \frac{c_k G(f_n^t; \tilde{\mu}_k, \tilde{\Phi}_k)}{\sum_j c_j G(f_n^t; \tilde{\mu}_j, \tilde{\Phi}_j)} \tag{9.33}$$

All the feature vectors of the noisy utterance are replaced by their estimated clean counterparts computed using Equation (9.33). The sequence of estimated clean vectors is used for recognition.

Vector Taylor Series Compensation

The vector Taylor series (VTS) algorithm[11] assumes that the corrupting influences on the clean speech signal are an unknown linear time-invariant filter and an unknown additive stationary noise signal that is uncorrelated with the speech signal (Figure 9.3).

Under these assumptions, the effect of the noise and the filter on the power spectrum of the utterance is given by

$$S_y(\omega) = |H(\omega)|^2 S_s(\omega) + S_n(\omega) \tag{9.35}$$

where $S_y(\omega)$ is the power spectrum of the noisy speech, $S_s(\omega)$ is the power spectrum of the clean speech, $H(\omega)$ is the frequency response of the linear filter and $S_n(\omega)$ is the power spectrum of the additive noise. It can easily be shown that the relationship between the log spectrum (i.e., the logarithm of the power spectrum) of the noisy speech and that of the clean speech is given by

$$Y = S + H + \log(1 + e^{N - H - S}) \tag{9.36}$$

where $S = \log(S_s(\omega))$, $H = \log(|H(\omega)|^2)$ and $N = \log(S_n(\omega))$. Note the slight change in notation here. Y, S, and H now represent *log* spectra, and ω has been dropped for brevity. For notational simplicity, we will also denote the term to the right in Equation (9.36) as $F(S, H, N)$ in the rest of this section. If the channel parameter H and noise parameter N were precisely known, Equation (9.36) could be inverted to obtain S from Y. However, they are not usually known beforehand and must be estimated.

For the purposes of estimation, it is assumed that the linear filter affecting the speech signal is constant across all analysis frames and can be characterized by a single channel parameter H. On the other hand, the noise log-spectrum in Equation (9.36) is assumed to vary from frame to frame. This accounts for the fact that the precise noise sample affecting the speech signal varies from analysis frame to analysis frame even when the noise is stationary. The distribution of the log-spectra of the noise in the various analysis frames is assumed to be Gaussian with mean n_μ and variance n_Φ. Thus, the parameters that must actually be estimated are H, n_μ, and n_Φ.

In order to estimate these parameters, it is assumed that the probability distribution function (PDF) of the log-spectral vectors of clean speech can be characterized by a mixture of Gaussians.

$$P(S) = \sum_k c_k G(S; \mu_k, \Phi_k) \tag{9.37}$$

The covariance matrices, Φ_k, are assumed to be diagonal. The parameters of this distribution can be learned from the log-spectral vectors of a corpus of clean speech using the EM algorithm.[14]

The VTS algorithm assumes that the effect of the noise and the linear filter is to change the means and variances of the Gaussians in Equation (9.37) to $\tilde{\mu}_k(H, n_\mu)$ and $\tilde{\Phi}_k(H, n_\mu, n_\Phi)$, which are given by

$$\tilde{\mu}_k(H, n_\mu) = F(\mu_k, H, n_\mu) \tag{9.38}$$

$$\tilde{\Phi}_k(H, n_\mu, n_\Phi) = \nabla_S F(\mu_k, H, n_\mu)^T \Phi_k \nabla_S F(\mu_k, H, n_\mu) +$$
$$\nabla_N F(\mu_k, H, n_\mu)^T n_\Phi \nabla_N F(\mu_k, H, n_\mu) \tag{9.39}$$

where $\nabla_S F(\mu_k, H, n_\mu)$ and $\nabla_N F(\mu_k, H, n_\mu)$ are the derivatives of $F(S, H, N)$ with respect to S and N, respectively, at (μ_k, H, n_μ). The likelihood of any noisy vector Y is therefore assumed to be

$$P(Y) = \sum_k c_k G(S; \tilde{\mu}_k(H, n_\mu), \tilde{\Phi}_k(H, n_\mu, n_\Phi)) \tag{9.40}$$

H, n_μ, and n_Φ are estimated such that the total likelihood of the log-spectral vectors in the noisy utterance, as computed using Equation (9.40), is maximized. Direct solutions to this estimation problem are not possible. Estimation is performed, instead, by approximating $F(H, N, S)$ by a truncated Taylor series expansion. This results in an iterative algorithm, the derivation of which is beyond the scope of this chapter and can be found elsewhere.[11] Here, we only describe the key steps of the algorithm, which are as follows.

We represent the ith estimate of the parameters as H^i, n_μ^i, and n_Φ^i, respectively. To obtain the $(i + 1)$th estimates, we define the variables A_k, B_k, and C_k as

$$A_k = F(\mu_k, H^i, n_\mu^i) - \nabla_H F(\mu_k, H^i, n_\mu^i) H^i - \nabla_N F(\mu_k, H^i, n_\mu^i) n_\mu^i$$

$$B_k = \nabla_H F(\mu_k, H^i, n_\mu^i) \tag{9.41}$$

$$C_k = \nabla_N F(\mu_k, H^i, n_\mu^i)$$

We denote the *a posteriori* probability of the kth Gaussian, given any noisy vector Y, as $P^i(k|Y)$, where

$$P^i(k|Y) = \frac{c_k G(S; \tilde{\mu}_k(H^i, n_\mu^i), \tilde{\Phi}_k(H^i, n_\mu^i, n_\Phi^i))}{\sum_j c_j G(S; \tilde{\mu}_j(H^i, n_\mu^i), \tilde{\Phi}_j(H^i, n_\mu^i, n_\Phi^i))} \tag{9.42}$$

We denote the tth vector in the sequence of noisy log-spectral vectors as $Y(t)$. To simplify notation, we denote $\tilde{\Phi}_j(H^i, n_\mu^i, n_\Phi^i)$ as $\tilde{\Phi}_j^i$ and define the following set of variables:

$$D = \sum_t \sum_k P^i(k|Y(t)) C_k^T (\tilde{\Phi}_k^i)^{-1} B_k \qquad E = \sum_t \sum_k P^i(k|Y(t)) C_k^T (\tilde{\Phi}_k^i)^{-1} (Y(t) - A_k)$$

$$F = \sum_t \sum_k P^i(k|Y(t)) C_k^T (\tilde{\Phi}_k^i)^{-1} C_k \qquad G = \sum_t \sum_k P^i(k|Y(t)) B_k^T (\tilde{\Phi}_k^i)^{-1} (Y(t) - A_k)$$

$$J = \sum_t \sum_k P^i(k|Y(t)) B_k^T (\tilde{\Phi}_k^i)^{-1} B_k \qquad O = \sum_t \sum_k P^i(k|Y(t)) B_k^T (\tilde{\Phi}_k^i)^{-1} C_k$$

$$\tag{9.43}$$

The $(i + 1)$th estimate of H and n_μ are now obtained as

$$H^{i+1} = (J - OF^{-1}D)^{-1}(G - OF^{-1}E) \tag{9.44}$$

$$n_\mu^{i+1} = (O - JF^{-1}D)^{-1}(G - JF^{-1}E) \tag{9.45}$$

To obtain the $(i + 1)$th estimate of n_Φ, we first define the following variables:

$$U = \sum_t \sum_k P^i(k|Y(t))(\tilde{\Phi}_k^i)^{-2}((Y(t) - \tilde{\mu}_k(H^i, n_\mu^i))C_k(Y(t) - \tilde{\mu}_k(H^i, n_\mu^i))^T - C_k\Phi_kC_k^T)$$

$$V = \sum_t \sum_k P^i(k|Y(t))(\tilde{\Phi}_k^i)^{-2}C_kC_k^T$$

(9.46)

The estimate is then defined simply as

$$n_\Phi^{i+1} = V^{-1}U \qquad (9.47)$$

The channel parameter is usually initialized as the difference between the average value of the log spectral vectors of the noisy speech and the global mean of the PDF of the log spectra of clean speech. The mean and variance of the noise are initialized as the mean and variance of all noisy speech vectors whose total energy lies below a threshold. Updated estimates of the noise and channel parameters are obtained by iterations of Equations (9.44), (9.45), and (9.47). Each iteration is expected to result in estimates of the parameters that increase the total likelihood of the noisy vectors as computed using Equation (9.40). Iterations are continued until the likelihood converges, i.e., it does not increase further with additional iterations. In practice, only two or three iterations of the algorithm need be run and likelihood convergence is not required. The estimation process in VTS is sometimes prone to instability. To account for this all estimates are constrained to lie within an empirically determined range of values.

Once the final estimates of the noise and channel parameters, H^{fin}, n_μ^{fin}, and n_Φ^{fin} have been obtained, every noisy log-spectral vector is replaced by an MMSE estimate of the corresponding log-spectral vector of the underlying clean speech signal, which is obtained as:

$$\hat{S}(t) = Y(t) - \sum_k P^{fin}(k|Y(t))(\tilde{\mu}_k^{fin}(H^{fin}, n_\mu^{fin}) - \mu_k) \qquad (9.48)$$

where $P^{fin}(k|Y(t))$ is the *a posteriori* probability of the kth Gaussian computed using the final estimates of the noise and channel parameters in Equation (9.42). Recognition is now performed with features derived from the estimated clean speech log-spectral vectors.

Codeword-Dependent Cepstral Normalization

Codeword-dependent cepstral normalization (CDCN)[9] is very similar to VTS in that the corrupting influences on the speech are modeled as a constant linear filter and stationary additive noise. The major difference between VTS and CDCN is that CDCN compensation is based on modeling the effect of noise on the mean values of speech features, while VTS incorporates an explicit characterization of the effects of noise on the variance of these features as well. In addition, CDCN compensation is performed directly on the cepstra of noisy speech rather than on log-spectra as is the case for VTS. The relationship between the cepstrum of a noisy segment of speech y, and that of the underlying clean speech s, is easily obtained by taking the discrete cosine transform (DCT) of both sides of Equation (9.36) as

$$y = s + h + DCT(\log(1 + e^{IDCT(n-s-h)})) \tag{9.49}$$

where $y = DCT(Y)$, $s = DCT(S)$, $h = DCT(H)$, $n = DCT(N)$, and $IDCT$ denotes inverse DCT. The same relation can also be expressed as

$$y = n + DCT(\log(1 + e^{IDCT(s+h-n)})) \tag{9.50}$$

Both forms of the relation are used in the CDCN algorithm. Equation (9.49) is used in the estimation of the channel parameters, and Equation (9.50) is used to estimate the noise. h is assumed to be relatively constant from frame to frame. The noise is assumed to be characterized entirely by the vector n and no explicit variance is associated with it, although it is acknowledged that the exact noise sequence can vary from analysis window to analysis window. Both h and n are unknown and must be estimated.

In order to estimate these parameters, it is assumed that the probability distribution of the cepstral vectors of clean speech can be characterized by a mixture Gaussian:

$$P(s) = \sum_k c_k G(s; \mu_k, \Phi_k) \tag{9.51}$$

where, as in VTS, all covariance matrices Φ_k, are assumed to be diagonal. CDCN assumes that the effect of the noise and linear filter is to change the means of the Gaussians in Equation (9.51) to $\tilde{\mu}_k(h, n)$, which is given by

$$\tilde{\mu}_k(h, n) = \mu_k + h + IDCT(\log(1 + e^{DCT(n-\mu_k-h)})) \tag{9.52}$$

The likelihood of any noisy vector y is therefore assumed to be given by

$$P(y) = \sum_k c_k G(y; \tilde{\mu}_k(h, n), \Phi_k) \tag{9.53}$$

Both h and n are estimated such that the total likelihood of all the cepstral vectors in the noisy utterance, as computed using Equation (9.53), is maximized. This leads to the following iterative solution.[9]

We represent the ith estimate of h and n as h^i and n^i, respectively. The *a posteriori* probability of the kth Gaussian for any noisy cepstral vector y, based on these values, is given by

$$P^i(k|y) = \frac{c_k G(y;\tilde{\mu}_k(h^i, n^i), \Phi_k)}{\sum_j c_j G(y;\tilde{\mu}_j(h^i, n^i), \Phi_j)} \tag{9.54}$$

In order to obtain the $(i + 1)$th estimate of the channel h and noise n, we represent the tth cepstral vector of the noisy utterance as $y(t)$, and define the variables R_k and W_k as

$$R_k = DCT(\log(1 + e^{IDCT(n^i - \mu_k - h^i)})) \tag{9.55}$$

$$W_k = DCT(\log(1 + e^{IDCT(\mu_k + h^i - n^i)})) \tag{9.56}$$

The $(i + 1)$th estimates h and n are now given by

$$h^{i+1} = \left(\sum_t \sum_k P^i(k|y(t))\Phi_k^{-1}\right)^{-1}\left(\sum_t \sum_k P^i(k|y(t))\Phi_k^{-1}(y(t) - \mu_k - R_k)\right) \tag{9.57}$$

$$n^{i+1} = \left(\sum_t \sum_k P^i(k|y(t))\Phi_k^{-1}\right)^{-1}\left(\sum_t \sum_k P^i(k|y(t))\Phi_k^{-1}(y(t) - W_k)\right) \tag{9.58}$$

In order to initialize the algorithm, the channel is usually initialized as the difference between the average value of the noisy cepstral vectors and the global mean of the distribution of the cepstral vectors of clean speech. The noise is initialized as the mean of the cepstral vectors of speech frames whose total energy lies below a threshold. Equations (9.57) and (9.58) are then iterated until the likelihood of the noisy utterance converges. As in VTS, in practice, it is sufficient to run two or three iterations of the algorithm and likelihood convergence is not required.

Once the final estimates of the noise and channel parameters, n^{fin} and h^{fin} have been obtained, every noisy cepstral vector is replaced by an MMSE estimate of the corresponding cepstral vector of the underlying clean speech signal:

$$\hat{s}(t) = y(t) - \sum_k P^{fin}(k|y(t))(\tilde{\mu}_k(h^{fin}, n^{fin}) - \mu_k) \tag{9.59}$$

where $P^{fin}(k|y(t))$ is the *a posteriori* probability of the kth Gaussian computed using the final estimates of the noise and channel parameters in Equation (9.54). Recognition is now performed with the estimated clean speech cepstra.

Cepstral and Spectral High-Pass Filtering

Cepstral high-pass filtering provides a remarkable amount of robustness at almost zero computational cost. Approaches based on cepstral high-pass filtering can compensate for the effects of both linear filtering and additive noise.

Cepstral mean normalization (CMN) is one of the most common ways of accomplishing cepstral high-pass filtering. It is a very simple procedure whereby the mean of the cepstral vectors in any utterance is subtracted out of all the cepstral vectors. If we represent the tth cepstral vector in an utterance with T cepstral vectors as x_t, CMN can be denoted as

$$\bar{x} = \frac{1}{T} \sum_{t=1}^{T} x_t$$

$$y_t = x_t - \bar{x}$$

(9.60)

The recognition system is now both trained and tested using y_t instead of x_t. CMN effectively subtracts the short-term average of cepstral vectors from the incoming cepstral coefficients, which cancels out any long-term average shift in the values of the vectors that has been introduced by environmental effects. Because the operation of convolution in the time domain maps into addition in the cepstral domain, cepstral high-pass filtering can, in principle, compensate completely for the effects of stationary unknown linear filtering, provided that the impulse response of the filter characterizing the degradation is not significantly greater in duration than the frame length.

The other popular high-pass filtering approach is the well-known *relative spectral processing* (RASTA),[15] which was originally motivated by a desire to emphasize the transient components of the speech signal, as is done by the peripheral auditory system. The difference equation

$$y_t = 0.98 y_{t-1} + 0.2 x_t + 0.1 x_{t-1} - 0.1 x_{t-3} - 0.2 x_{t-4}$$

(9.61)

implements a causal version of the original RASTA filter. Although this difference equation actually specifies a bandpass filter, its high-pass section cancels long-term shifts in cepstral coefficients in a fashion similar to that of CMN.

High-pass filtering can also be applied directly in the power spectral domain to compensate for the effects of additive noise in a similar fashion. Morgan and Hermansky have proposed the *J-RASTA* method,[16] which uses high-pass filtering approaches to compensate jointly for the effects of additive noise and linear filtering by passing incoming *spectral* coefficients through the transformation

$$w_t = \log(1 + J x_t)$$

(9.62)

before applying Equation (9.61) or Equation (9.60) to the output coefficients w_t. This transformation is linear for small values of $J x_t$, causing spectral coefficients to be high-pass filtered directly at low SNRs which tends to compensate for the effects of additive noise. At larger values of Jx_t the spectral coefficients are transformed into log-spectral coefficients, and the high-pass filtering operation now compensates for the effects of linear filtering.

Discussion of Relative Merits of the Methods

Although several methods have been described in this chapter, not all of them are equally applicable in all situations. For example, in situations where the speech recognizer is an externally provided module (as in the case of commercial recognition engines) and no access has been provided to the features that the recognizer computes from the speech signal, it is only possible to use signal compensation methods. Even in systems where one has access to the features, it may still not be possible to use the feature compensation methods described in this chapter if the features used are not based on log-spectra or cepstra. Once again, only signal compensation methods can be applied.

Among the signal compensation methods, linear spectral subtraction (LSS) is easily the simplest to implement and the least expensive in terms of computational resources. It is very effective where the noise corrupting the signal is actually additive and varies slowly with time, as in the case of automotive noise. Table 9.1 shows the recognition performance obtained on noisy speech with and without LSS compensation. LSS is observed to result in large improvements in accuracies over the baseline (where no compensation is performed), although a large component of the corruption is due to linear filtering, which it is not expected to compensate for.

TABLE 9.1

Word-Recognition Accuracies on Speech Recorded Using a Close-Talking Microphone (CLSTK) and Crown PZM (CRPZM) Desktop Microphone in a Noisy Environment

	CLSTK (%)	CRPZM (%)
BASELINE	85.3	18.6
LSS (MSUB)	N/A	63.6
CDCN	85.3	74.9

Note: The recognizer was trained using CLSTK speech. Recognition accuracies obtained without compensation (baseline) and with LSS and CDCN compensation of the test data are shown.

Source: From Acero, A., Acoustic and Environmental Robustness in Automatic Speech Recognition, Ph.D. thesis, Carnegie Mellon University, Pittsburgh, PA, 1990. With permission.

TABLE 9.2

Recognition Accuracy on Speech Recorded in Cars
Moving at Various Speeds

Database	NO COMP. (%)	LSS (%)	NSS (%)
MATRA 90	74	96.2	98.6
MATRA 130	54	93.3	96.8
ENST 110	64.5	90	94.9
CSELT 70	99.2	100	100
CSELT 130	83	98.4	99.7
Average	94.8	95.6	98
Std. Dev.	20.7	5.6	2.9

Note: The acronyms in the name of the database refer to
the organization that collected the data and the
numbers beside the acronyms indicate the speed
of the car in kilometers per hour. The recognizer
was trained on data collected in a parked car. NSS
is observed to halve the percentage of wrongly
recognized words as compared to LSS.

Source: From Lockwood, P. and Boudy, J., Experiments
with a nonlinear spectral subtractor (NSS), hidden Markov models and the projection, for robust speech recognition in cars, *Speech Communication*, 11, 215–228, 1992.
© 1992 by Elsevier Science. With permission.

NSS is essentially an enhanced version of LSS, and performs better. Table 9.2 compares the performance of LSS and NSS on speech corrupted by the noises of moving cars. NSS is, however, rather more expensive computationally than LSS and the implementation is somewhat more involved. Both LSS and NSS are ineffective when the characteristics of the corrupting noise change quickly with time, or when the noise is not strictly additive (e.g., if the signal has undergone any kind of nonlinearity).

Wiener filtering, in the form presented in this chapter, is essentially a form of spectral subtraction. However, several enhanced versions of the algorithm have been proposed in the literature (e.g., References 17 and 18). These methods have generally proved to be more effective than spectral subtraction, but are also significantly more expensive computationally. They also fail when the noise is not additive or if it is nonstationary.

Feature compensation methods are generally much more effective than signal compensation methods. For example, we see in Table 9.1 that CDCN, which is a feature compensation method, results in much better recognition than LSS, which is a signal compensation method. Feature compensation methods are useful when one has control over the features going into the recognizer, but not on the parameters of the statistical models used by the recognizer themselves. Figure 9.4 compares the performances of RATZ, VTS, and CDCN on speech corrupted by white noise when the recognizer has been trained with clean speech. Of the feature compensation methods, RATZ is computationally the cheapest, since there is no estimation of parameters performed during recognition. It is also at least as effective as the other

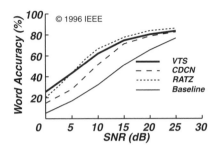

FIGURE 9.4

Recognition accuracy on speech corrupted by white noise as a function of the SNR when various feature compensation algorithms are applied. The recognition system was trained with clean speech. The baseline recognition accuracy obtained when no compensation is applied is also shown. (From Raj, B. et al., Cepstral compensation by polynomial approximation for environment-independent speech recognition, *Proc. ICSLP*, 1996. © IEEE. With permission.)

methods and has the added advantage that it does not explicitly assume any particular form for the corrupting transformation. However, the requirement of stereo adaptation data for learning the relationship between the feature vectors of clean and noisy speech may sometimes be unrealistic. VTS and CDCN do not require stereo data and estimate the noise and channel parameters based only on the noisy utterance. The trade-off is that this estimation is computationally expensive. Of the two, CDCN is computationally less expensive, and has the added advantage that it has been observed to be effective even when the recognizer has not been trained with clean speech. VTS, on the other hand, is more effective than CDCN, but only when the recognizer has been trained with clean speech. However, both VTS and CDCN become ineffective if the corrupting transformations on the signal cannot be properly modeled as the combination of a linear filter and additive noise. Even when the model *is* appropriate, they are ineffective if the noise is nonstationary, although enhancements to the VTS algorithm have been proposed that are effective for time-varying noises.[13]

Of all the methods presented in this chapter, CMN is perhaps the most ubiquitous. It has been shown to improve recognition performance across the board under all kinds of conditions even when other compensation methods are applied or when there is no noise. Most speech recognition systems apply CMN or RASTA by default. All results in Figure 9.4 were obtained, for example, on a system where CMN was also performed. Figure 9.5 shows a typical comparison of performances obtained with and without CMN.

It must be noted that there are several other important techniques (e.g., References 19 and 20) within the categories mentioned in this chapter. The treatment in this chapter is not exhaustive and is only meant to give the reader a first glimpse of the kinds of algorithms and methods used for noise robustness in speech recognition systems. Several review papers and books (e.g., References 21 through 23) provide more comprehensive accounts of other techniques used.

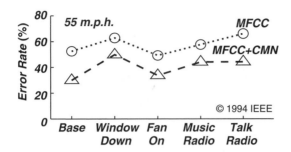

FIGURE 9.5
Recognition accuracies on speech recorded under various conditions in a car moving at 55 mph, with and without CMN. MFCC features were used in this case. (From Hanai, N. and Stern, R. M., Robust speech recognition in the automobile, *Proc. ICSLP*, 1994. © IEEE. With permission.)

Acknowledgments

We thank Sam-Joo Doh of Speechworks Inc. and Bent Schmidt-Nielsen of Mitsubishi Electric Research Labs for their insightful suggestions, which improved the correctness and quality of several sections of this chapter. We also thank Gillian Davis for inviting us to write this chapter.

Preparation of this report was sponsored by the Space and Naval Warfare Systems Center, San Diego, under Grant No. N66001-99-1-8905. The content of the information in this publication does not necessarily reflect the position or the policy of the U.S. Government, and no official endorsement should be inferred.

References

1. Duda, R.O., Hart, P.E., and Stork, D.G., *Pattern Classification*, 2nd ed., John Wiley & Sons, New York, 2000.
2. Jelinek, F., *Statistical Methods for Speech Recognition*, MIT Press, Cambridge, MA, 1998.
3. Juang, B.H., Levinson, S.E., and Sondhi, M.M., Maximum likelihood estimation for multivariate mixture observations of Markov chains, *IEEE Trans. Info. Theory*, IT-32(2), 1986, 307–309.
4. Davis, S.B. and Mermelstein, P., Comparison of parametric representations for monosyllabic word recognition in continuously spoken sentences, *IEEE Trans. Acoustics, Speech, and Signal Proc.*, ASSP-28, 4, 357–366, 1980.
5. Raj, B., Reconstruction of Incomplete Spectrograms for Robust Speech Recognition, Ph.D. thesis, Carnegie Mellon University, Pittsburgh, PA, 2000.
6. Boll, S.F., Suppression of acoustic noise in speech using spectral subtraction, *IEEE Trans. Acoustics, Speech, and Signal Proc.*, ASSP-27, 2, 113–120, 1979.

7. Lockwood, P. and Boudy, J., Experiments with a nonlinear spectral subtractor (NSS), hidden Markov models and the projection, for robust speech recognition in cars, *Speech Commun.*, 11, 215–228, 1992.
8. Lim, J.S. and Oppenheim, A.V., All-pole modeling of degraded speech, *IEEE Trans. Acoustics, Speech, and Signal Proc.*, ASSP-26, 3, 197–210, 1978.
9. Acero, A., Acoustic and Environmental Robustness in Automatic Speech Recognition, Ph.D. thesis, Carnegie Mellon University, Pittsburgh, PA, 1990.
10. Berouti, M., Schwartz, R., and Makhoul, J., Enhancement of speech corrupted by acoustic noise, *Proc. ICASSP79*, 1979, 215–228.
11. Moreno, P.J., Speech Recognition in Noisy Environments, Ph.D. thesis, Carnegie Mellon University, Pittsburgh, PA, 1996.
12. Neumeyer, L. and Weintraub, M., Probabilistic optimum filtering for robust speech recognition, *Proc. ICASSP94*, 1994, 417–420.
13. Kim, N.S., IMM-based estimation for slowly evolving environments, *IEEE Signal Processing Letters*, 5(6), 146–149, 1998.
14. Dempster, A.P., Laird, N.M., and Rubin, D.B., Maximum likelihood from incomplete data via the EM algorithm (with discussion), *J. Royal Stat. Soc., Series B*, 39, 1–38, 1977.
15. Hermansky, H. and Morgan, N., RASTA processing of speech, *IEEE Trans. on Speech and Audio Processing*, 2, 578–589, 1994.
16. Morgan, N. and Hermansky, H., RASTA extensions: robustness to additive and convolutional noise, *Proc. ESCA Workshop on Speech Processing in Adverse Conditions*, 1992, 115–118.
17. Ephraim, Y., A minimum mean square error approach for speech enhancement, *Proc. IEEE Conf. on Acoustics, Speech and Signal Processing*, 1990, 829–832.
18. Beattie, V.L., Hidden Markov Model State-Based Noise Compensation, Ph.D. thesis, Cambridge University, Cambridge, U.K., 1992.
19. Ephraim, Y. and Van Trees, H.L., A signal subspace approach for speech enhancement, *IEEE Trans. Speech, Audio Processing*, 3, 251–266, 1995.
20. Mansour, D. and Juang, B.H., A family of distortion measures based upon projection operation for robust speech recognition, *IEEE Trans. Acoustics, Speech, and Signal Proc.*, ASSP-37, 1989, 1659–1671.
21. Gong, Y., Speech recognition in noisy environments: a survey, *Speech Commun.*, 16, 261–291, 1995.
22. Huang, X., Acero, A., and Hon, H.W., *Spoken Language Processing: A Guide to Theory, Algorithm, and System Development*, Prentice-Hall, Englewood Cliffs, NJ, 2001.
23. Junqua, J., *Robustness in Automatic Speech Recognition*, Kluwer Academic Publishers, Boston, 1995.

10

Model Compensation and Matched Condition Methods for Robust Speech Recognition

Rita Singh, Bhiksha Raj, and Richard M. Stern

CONTENTS

Introduction

In this chapter we describe some important model compensation and matched condition techniques used in speech recognition systems. As mentioned in Chapter 9, model compensation methods modify the statistical models used to recognize noisy speech, in order to reduce the mismatch between the distributions of the noisy speech and those used by the recognizer. Matched condition techniques, on the other hand, attempt to perform recognition based only on those components of the noisy speech that are matched with the recognizer. Some key model compensation techniques that we describe in this chapter are model decomposition, parallel model combination (PMC), maximum likelihood linear regression (MLLR), maximum *a posteriori* (MAP) adaptation, and extended maximum *a posteriori* (EMAP) adaptation. Under matched condition techniques, we primarily describe missing feature methods.

Most of the compensation methods described in this chapter assume that the recognizer uses hidden Markov models (HMMs) to model speech. Since most model compensation and matched condition techniques take into consideration the specific model used by the recognizer, we begin with a brief description of HMMs.

Hidden Markov Models

In the HMM representation of speech, it is assumed that each sound unit has several states. Every instance of the unit is assumed to consist of subsegments, each belonging to one of these states. The distribution of the data vectors in subsegments belonging to any state is called the *state output distribution* of that state. The model assumes that the underlying process that generates the sounds follows a sequence of states, generating a sequence of stochastic observation vectors with characteristics that are specified by the state output distributions of these states. The model follows the Markovian assumption that the *a priori* probability that a vector belongs to a state j is dependent only on the identity of the state i that the previous vector belonged to. This is called the *state transition probability* of transiting from state i to state j. The pattern of allowed state transitions within an HMM is called its *topology*. Figure 10.1 shows the topology of a typical three-state HMM.

The statistical parameters of an HMM are its set of state transition probabilities and the set of state output distributions. For any state, the transition probabilities from that state to any other state must sum to 1. In speech recognition systems, the state output distributions are usually modeled as Gaussians or mixtures of Gaussians (e.g., see Reference 1). Typically, these

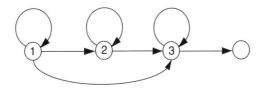

FIGURE 10.1
A three-state Bakis topology HMM with a nonemitting terminating state. In this topology, the system can skip from state 1 to state 3, bypassing state 2 completely. This implies that although the most general realization of the sound has three distinct stages, some realizations may not exhibit the second stage.

Gaussians are assumed to have diagonal covariance matrices for computational efficiency. If we represent the state output probability of the kth state of the HMM for unit U by $P_{U,k}(X)$, where X represents any feature vector that belongs to that state then

$$P_{U,k}(X) = M(X;\Lambda_{U,k}) \tag{10.1}$$

where $M(X;\Lambda_{U,k})$ denotes a Gaussian mixture distribution corresponding to the kth state with parameters $\Lambda_{U,k}$. The Gaussian mixture has the form

$$
\begin{aligned}
M(X;\Lambda_{U,k}) &= \sum_j c_{U,k,j} G(X;\mu_{U,k,j}, \Phi_{U,k,j}) \\
&= \sum_j \frac{c_{U,k,j}}{\sqrt{(2\pi)^d |\Phi_{U,k,j}|}} e^{-0.5(X-\mu_{U,k,j})^T \Phi_{U,k,j}^{-1}(X-\mu_{U,k,j})}
\end{aligned} \tag{10.2}
$$

where $c_{U,k,j}$ is the *mixture weight* of the jth Gaussian $G(X;\mu_{U,k,j}, \Phi_{U,k,j})$, with mean $\mu_{U,k,j}$ and variance $\Phi_{U,k,j}$.

In most speech recognition systems the basic sound units modeled by HMMs are *subword* units, i.e., sound units that are smaller than words, which can be used to compose words. HMMs for longer units, such as words or word sequences, are obtained by concatenating the HMMs for the constituent subword units. Figure 10.2 shows such a construction.

Although the topology of the HMM can be learned from data,[2] it is usually prespecified. The state transition probabilities and the parameters of the state output distributions for the HMM of any unit are learned from examples of that unit. The examples used are referred to as "model training data" or

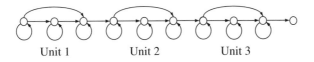

Unit 1   Unit 2   Unit 3

FIGURE 10.2
Concatenation of HMMs representing subword units to create a longer HMM.

simply "training data." Any speech signal, or *utterance*, is now recognized using the following rule:

$$\hat{W} = \arg\max_W \{ P(W)P(X|W) \} \tag{10.3}$$

where X represents the sequence of feature vectors for that utterance, W represents word sequences, and $\hat{W}$ represents the recognized word sequence. $P(X|W)$ represents the probability of X given by the HMM for W and is given by

$$P(X|W) = \sum_k P(X, state(t) = k|W) \tag{10.4}$$

where $P(X, state(t) = k|W)$ is the joint probability of observing vector sequence X, and the tth vector in the sequence belonging to state k. This can be derived using the Baum-Welch algorithm,[1] which can also be used to derive $\gamma_W(t, k)$, the *a posteriori* probability that the tth vector belongs to state k

$$\gamma_W(t, k) = P(state(t) = k|X, W) = \frac{P(X, state(t) = k|W)}{\sum_k P(X, state(t) = k|W)} \tag{10.5}$$

The *a posteriori* probability of the jth Gaussian in the state output distribution of state k is given by

$$\gamma_W(t, k, j) = \gamma_W(t, k) \frac{c_{W,k,j} G(X(t); \mu_{W,k,j}, \Phi_{W,k,j})}{M(X(t); \Lambda_{W,k})} \tag{10.6}$$

In practice, $P(X|W)$ in Equation (10.3) is approximated as:

$$\hat{P}(X|W) = \arg\max_s \{ P(X, s|W) \} \tag{10.7}$$

where s represents a state sequence, i.e., a sequence of states, one for every feature vector in X:

$$s = [state(1), state(2), state(3), ..., state(N)] \tag{10.8}$$

where N is the total number of vectors in X. Equation (10.7) can be computed using the Viterbi algorithm,[3] which is much less computationally expensive than the Baum-Welch algorithm.

Model Compensation

In model compensation methods, the distributions of speech vectors that are modeled by the recognizer are transformed to represent the distributions of the noisy test speech. The transformation may be such that it modifies the basic structure of the statistical model, i.e., the HMM, e.g., model decomposition,[4] or merely modifies its state distributions, without affecting the topology. The latter type of transformations can either be based on analytical characterizations of the effect of noise, e.g., PMC, or on empirical evidence obtained from noisy data. Methods based on empirical evidence can further be categorized as those that modify parameters based *only* on the empirical evidence obtained from noisy data, e.g., MLLR,[5] and those that use *a priori* information about the statistical distribution of parameters of state distributions, e.g., MAP[6] and EMAP[7,8] adaptation. In the following section we briefly describe the model decomposition, PMC, MLLR, MAP, and EMAP methods.

Model Decomposition

In model decomposition,[4] corrupting noises and clean speech are both modeled explicitly using HMMs. The HMMs for the speech and the noises are trained independently of each other. During recognition the noise and the speech are simultaneously recognized using the noisy speech samples by constructing "compound" HMMs from the HMMs for the noise and speech.

Consider a speech HMM with K states and a noise HMM with L states. At each instant of time, the noise could belong to any of the L noise states. The clean speech could belong to any of the K clean speech states. Since the observed noisy speech is a combination of clean speech and noise, it can belong to one of $K \times L$ states. Consequently, the compound HMM for the noisy speech has $K \times L$ states, each of which is a combination of a state from the speech HMM and one from the noise HMM. Transitions are allowed between any two states of the compound HMM if transitions are also permitted between the corresponding component states of the noise and clean speech HMMs. If we represent the state obtained by combining state i_n of the noise HMM and state j_s of the clean speech HMM as state (i_n, j_s), and the transition probability between any state i and any state j as $T(i, j)$, the transition probability between state (i_n, j_s) and state (k_n, l_s) is given by

$$T((i_n, j_s), (k_n, l_s)) = T(i_n, k_n)T(j_s, l_s) \qquad (10.9)$$

Figure 10.3 shows the topology of a typical compound HMM. The state output distribution of any state (i_n, j_s) must be computed from the distributions of states i_n and j_s, based on the relationship between the feature

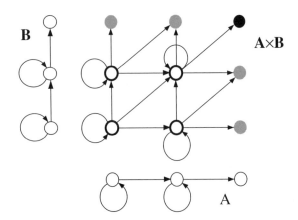

FIGURE 10.3
Combining two-component HMMs **A** and **B** to construct the compound HMM **A×B**. In the compound HMM, the empty circles represent states that are combinations of states with self-transitions, shaded circles are combinations of a state with self-transition and a state with no self-transition. The solid circle represents the combination of two states with no self-transitions.

vectors of the noisy speech and those of the clean speech and the noise. When the feature vectors used are log-spectral vectors, and the noise is assumed to be additive and uncorrelated with the speech, the relationship between any log-spectral vector Y of the noisy speech and the corresponding log-spectral vectors of the clean speech S and the noise N is given by

$$Y = \log(e^S + e^N) \tag{10.10}$$

For model decomposition this is usually approximated as

$$Y = \max(S, N) \tag{10.11}$$

If the state distribution of state i_n of the noise HMM is represented as $P_n(N|i_n)$ and that of state j_s of the clean speech HMM as $P_s(S|j_s)$, then the approximation in Equation (10.11) gives us the following form for the state probability of the compound state (i_n, j_s)

$$P(Y|i_n, j_s) = P_s(Y|j_s) \int_{-\infty}^{Y} P_n(N|i_n)dN + P_n(Y|i_n) \int_{-\infty}^{Y} P_s(S|j_s)dS \tag{10.12}$$

When the state distributions are Gaussians or Gaussian mixtures, Equation (10.12) is easy to compute. Recognition is now performed using the compound HMMs for the noisy data. In practice, compound HMMs are not explicitly constructed. Rather, compound states are composed dynamically during recognition. Since the speech recognizer would derive the best state sequence in the compound HMM along with the hypothesized word sequence, the best sequence of states through the noise HMM is also implicitly obtained. This can be used to obtain an estimate of the dynamics followed by the noise along with the hypothesized word sequence.

Parallel Model Combination

In PMC,[9] the distributions of the clean speech are modified to approximate the distributions that optimally represent the noisy test speech. These distributions are modeled by HMMs whose state distributions are Gaussian mixtures. Once again, it is assumed that the noise is additive and uncorrelated with the speech and that any filtering effects on the signal can be modeled by a linear filter. The HMMs for the various sound units are trained with the cepstral vectors of clean speech. The HMM for the noise signal is trained using the cepstra of separately recorded instances of the noise. In most practical implementations of PMC, however, the noise is assumed to be modeled by a Gaussian mixture, and not an HMM, which can be viewed as an HMM with only one state. The parameters of the noise and clean speech HMMs are used along with an analytical model of the effect of noise on the feature vectors of clean speech to determine the optimal distributions to represent the noisy speech.

In a speech recognition system, in addition to cepstra, higher-order features such as difference and double-difference cepstra are also used. The tth difference cepstral vector, $s'(t)$, is computed as $s'(t) = s(t + \tau) - s(t - \tau)$. The tth double-difference cepstral vector, $s''(t)$, is computed as $s''(t) = s'(t + \tau_d) - s'(t - \tau_d)$. Typically $\tau = 2$. For PMC, it is assumed that $\tau_d = \tau$. The entire tth feature vector actually used by the recognizer is given by

$$\bar{s}(t) = [s(t); s'(t); s''(t)] \tag{10.13}$$

It is assumed that any feature vector $\bar{y}(t)$ derived from the noisy speech has two components, a clean speech feature vector $\bar{s}(t)$ and a noise feature vector $\bar{n}(t)$. The cepstral components of these feature vectors are assumed to be related in the following manner:

$$y(t) = D(\ln(e^{D^{-1}s(t) + D^{-1}h} + e^{D^{-1}n(t)})) \tag{10.14}$$

where $\ln( )$ represents the natural logarithm, h represents the cepstral vector representing the time-invariant linear filter, and D and D^{-1} represent the transform matrices for the discrete cosine transform and the inverse discrete cosine transform, respectively. The following relations can now be derived:[9]

$$y'(t) =$$
$$D(\ln(e^{D^{-1}(s'(t) + s(t - \tau) + h)} + e^{D^{-1}(n'(t) + n(t - \tau))}) - \ln(e^{D^{-1}(s(t - \tau) + h)} + e^{D^{-1}n(t - \tau)})) \tag{10.15}$$

$$y''(t) =$$

$$D\ln(e^{D^{-1}(s''(t) + 2(s(t) - y(t)))} + e^{D^{-1}(n''(t) + n'(t - \tau) - s'(t - \tau) + s(t) + n(t) - 2y(t))})$$ (10.16)

$$+ (e^{D^{-1}(n''(t) + 2(n(t) - y(t)))} + e^{D^{-1}(s''(t) + s'(t - \tau) - n'(t - \tau) + s(t) + n(t) - 2y(t))})$$

The complete noisy feature vector is given by $\bar{y}(t) = [y(t); y'(t); y''(t)]$. The distribution of $\bar{y}(t)$ clearly depends on the distributions of $\bar{s}(t)$ and $\bar{n}(t)$. If the HMM for clean speech has K states and that for the noise has L states, then $\bar{s}(t)$ can have one of K distributions and $\bar{n}(t)$ can have one of L distributions at any time instant. Thus, $\bar{y}(t)$ can be in one of $K \times L$ states with the distribution:

$$P(\bar{y}(t)|\theta_y) = \int_{\wp} P(\bar{s}(t)|\theta_s)P(\bar{n}(t)|\theta_n)d\bar{s}(t)d\bar{n}(t)$$ (10.17)

where θ_s represents the state of $\bar{s}(t)$, θ_n represents the state of $\bar{n}(t)$, and $\theta_y = (\theta_s, \theta_n)$ represents the corresponding state of $\bar{y}(t)$. For any value of $\bar{y}(t)$ the region of the integral $\wp$ covers all $\bar{s}(t), \bar{n}(t)$ pairs that give rise to that value of $\bar{y}(t)$. The integral in Equation (10.17) has no closed form when $P(\bar{s}(t)|\theta_s)$ and $P(\bar{n}(t)|\theta_n)$ are Gaussian mixtures.

In PMC, it is assumed that the distribution of $P(\bar{y}(t)|\theta_y)$ can also be well modeled by a Gaussian mixture. Therefore, instead of attempting to explicitly solve Equation (10.17), the Gaussian mixture distribution of θ_y is estimated using fabricated examples of noisy speech feature vectors. The fabrication is done by generating examples of clean speech feature vectors and noise feature vectors using the state distributions of θ_s and θ_n, respectively, and combining them using Equations (10.14), (10.15), and (10.16). For computational expediency, it is usually assumed that $P(\bar{y}(t)|\theta_y)$ has exactly as many Gaussian components as $P(\bar{s}(t)|\theta_s)$. The EM algorithm[10] is used to compute the parameters of $P(\bar{y}(t)|\theta_y)$, which is initialized with the parameters of $P(\bar{s}(t)|\theta_s)$. Recognition is now performed with $P(\bar{y}(t)|\theta_y)$.

Maximum Likelihood Linear Regression

In MLLR,[5] model parameters are modified to better fit the noisy test data based on empirical evidence derived from samples of the noisy speech. Specifically, the means of the Gaussians in the state output distributions of the recognizer are modified by a simple transform that results in rotation and translation of the mean vectors (i.e., an affine transform). The parameters of this affine transform are learned from the test data.

Representing the state output probability of the ith state of the HMM for unit U by $P_{U,i}(X)$:

$$P_{U,i}(X) = \sum_k c_{U,i,k} G(X; \mu_{U,i,k}, \Phi_{U,i,k}) \tag{10.18}$$

where $G(X; \mu_{U,i,k}, \Phi_{U,i,k})$ represents a Gaussian with mean $\mu_{U,i,k}$ and variance $\Phi_{U,i,k}$, as in the section "Hidden Markov Models." In MLLR we assume that, in order to best represent the noisy speech, this gets modified to

$$\tilde{P}_{U,i}(Y) = \sum_k c_{U,i,k} G(X; \tilde{\mu}_{U,i,k}, \Phi_{U,i,k}) \tag{10.19}$$

where the relationship between $\tilde{\mu}_{U,i,k}$ and $\mu_{U,i,k}$ is given by the affine transform

$$\tilde{\mu}_{U,i,k} = A\mu_{U,i,k} + B \tag{10.20}$$

Note that in the formulation in Equation (10.20), the same affine parameters A and B are applied to the means of all Gaussians of all states of all the sound units (i.e., they are independent of k, U, and i). In more refined versions of MLLR, a single A and B may be applied only to a subset of the Gaussians in the recognition system, requiring several sets of affine parameters to adapt all the Gaussians in the system.[5]

The matrix A and the vector B are estimated from examples of noisy data, which are called the *adaptation* data. Since the distributions of the states in the HMMs for the sound units must be modified, the transcriptions of the adaptation data are also required. The transcriptions may either be provided to the system, or be obtained through a preliminary attempt at recognizing the noisy speech. The former case is referred to as *supervised* MLLR adaptation of acoustic models. The test data in this case are different from the adaptation data. The latter case is known as *unsupervised* MLLR adaptation. Here the test data and the adaptation data are identical.

The parameters A and B are estimated to maximize the likelihood of the noisy adaptation data computed using HMMs whose state distributions have been modified as in Equation (10.19). Direct maximization of the likelihood is not possible and an EM-based iterative solution is used, which is as follows.

Let W_a be the word sequence, or transcription, corresponding to the adaptation data. Let A^i and B^i represent the estimates for A and B after the ith iteration. Let $Y(t)$ represent the tth vector in the sequence of feature vectors for the adaptation data. Let $\gamma^i_{W_a, j, k}(t)$ represent the *a posteriori* probability that $Y(t)$ was generated by the kth Gaussian of the jth state of the HMM for W_a, computed using A^i and B^i. As mentioned in the section "Hidden Markov Models," $\gamma^i_{W_a, j, k}(t)$ can be computed using the forward-backward (or Baum-Welch) algorithm.[1] The $(i + 1)$th iteration of A and B can be obtained by jointly solving the following equations:

$$\sum_t \sum_j \sum_k \gamma^i_{j,k}(t)\Phi^{-1}_{j,k}Y(t) = \sum_t \sum_j \sum_k \gamma^i_{j,k}(t)\Phi^{-1}_{j,k}(A^{i+1}\mu_{j,k} + B^{i+1}) \quad (10.21)$$

$$\sum_t \sum_j \sum_k \gamma^i_{j,k}(t)\Phi^{-1}_{j,k}(Y(t) - B^{i+1})\mu^T_{j,k} = \sum_t \sum_j \sum_k \gamma^i_{j,k}(t)\Phi^{-1}_{j,k}A^{i+1}\mu_{j,k}\mu^T_{j,k} \quad (10.22)$$

where $\Phi_{j,k}$ and $\mu_{j,k}$ are the covariance matrix and mean, respectively, of the kth Gaussian in the state output distribution of the jth state. Note that the subscript W_a has been dropped for notational simplicity.

The solution for A^{i+1} and B^{i+1} can be significantly simplified if the covariance matrices of all Gaussians are diagonal. Let $a^{i+1}_{j,l}$ represent the element in the jth row and lth column of A^{i+1}, and b^{i+1}_j the jth element of B^{i+1}. Let $\phi_{k,j,l}$ represent the lth diagonal element of $\Phi_{k,j}$. Similarly, let $\mu_{k,j,l}$ represent the lth component of $\mu_{k,j}$. Let $Y(t,l)$ represent the lth component of $Y(t)$.

We can now solve for each row of the matrix A^{i+1} and the corresponding element of B^{i+1} independently of all other terms in them. To solve for the nth row of A^{i+1}, i.e., $a^{i+1}_{n,l}$ $0 < l \le d$, where d is the dimensionality of the feature vectors, and the corresponding element of B^{i+1}, b^{i+1}_n, we define a $d+1 \times d+1$ matrix Q_n such that $q_{n,l,m}$, the element in its lth row and mth column is given by

$$q_{n,l,m} = \sum_t \sum_j \sum_k \gamma^i_{j,k}(t)\phi^{-1}_{k,j,n}\mu_{k,j,l}\mu_{k,j,m} \qquad l, m \ne d$$

$$q_{n,l,d} = \sum_t \sum_j \sum_k \gamma^i_{j,k}(t)\phi^{-1}_{k,j,n}\mu_{k,j,l} \qquad\qquad l \ne d \qquad (10.23)$$

$$q_{n,d,d} = \sum_t \sum_j \sum_k \gamma^i_{j,k}(t)\phi^{-1}_{k,j,n}$$

We define a $d+1$ dimensional vector R_n such that its lth element $r_{n,l}$ is given by

$$r_{n,l} = \sum_t \sum_j \sum_k \gamma^i_{j,k}(t)\phi^{-1}_{k,j,n}Y(t,n)\mu_{k,j,l} \qquad l \ne d$$

$$\qquad\qquad\qquad\qquad\qquad\qquad\qquad\qquad\qquad\qquad (10.24)$$

$$r_{n,d} = \sum_t \sum_j \sum_k \gamma^i_{j,k}(t)\phi^{-1}_{k,j,n}Y(t,n)$$

Solving

$$Q_n Z = R_n \qquad\qquad\qquad\qquad (10.25)$$

results in the $d+1$ dimensional vector $Z = Q^{-1}_n R_n$. The nth row of A^{i+1} and the corresponding elements of B^{i+1} are now given by

$$a_{n,l}^{i+1} = Z_l \quad 0 \leq l \leq d$$
$$b_n^{i+1} = Z_{d+1}$$

$$(10.26)$$

where Z_l represents the lth component of Z. All elements of A^{i+1} and B^{i+1} can now be estimated in this manner. Since this is an iterative algorithm, the affine parameters must be initialized. A^0 is usually initialized as an identity matrix and B^0 with the zero vector.

Typically, one or two iterations of the EM algorithm are sufficient to obtain good estimates of A and B. The means of all the Gaussians in the recognizer are now transformed as in Equation (10.20) and the transformed means are used for recognition.

Maximum *a posteriori* Adaptation

In MAP adaptation,[6,11] model parameters are reestimated based on examples of the noisy data and prior knowledge of the distributions of the parameters. Mathematically, the reestimation of the model parameters can be represented as

$$\hat{\lambda} = \arg\max_\lambda \{ P(Y|\lambda) P(\lambda) \} \tag{10.27}$$

where λ represents an arbitrary set of model parameters and $\hat{\lambda}$ is the MAP estimate of the model parameters that best represent the noisy data Y. $P(Y|\lambda)$ is the likelihood of the noisy data when λ are the model parameters. $P(\lambda)$ represents the *a priori* distribution of model parameters. The nature of this distribution may be best explained by noting that the value of λ that best represents any given noise condition differs from the best value of λ for any other noise condition. If we were to gather the λ values for every possible noise condition and compute their distribution, $P(\lambda)$ would be that distribution. In reality, this distribution can never truly be known and must be approximated. The term within the brackets on the right hand side of Equation (10.27) represents the *a posteriori* distribution of λ. The value of λ at which this term is largest is the MAP estimate for λ.

Typically, the means of the various Gaussians in the HMM are adapted using noisy adaptation data for which transcriptions are available. The Gaussians are assumed to be independent of each other. As a result, the joint *a priori* distribution of the means of all the Gaussians is simply the product of the *a priori* distribution of the individual mean vectors. MAP estimation of all the means can be achieved by simply adapting each of the means independently of all others.

The *a priori* distribution of the mean vectors of any Gaussian is also assumed to be Gaussian. Representing the mean of the kth Gaussian of any state θ as $\mu_{\theta,k}$, the *a priori* distribution of the mean can be represented as

$$P(\mu_{\theta,k}) = \frac{1}{\sqrt{(2\pi)^d \overline{\Phi}_{\theta,k}}} e^{-0.5(\mu_{\theta,k} - \bar{\mu}_{\theta,k})^T \overline{\Phi}_{\theta,k}^{-1}(\mu_{\theta,k} - \bar{\mu}_{\theta,k})} \qquad (10.28)$$

where $\bar{\mu}_{\theta,k}$ and $\overline{\Phi}_{\theta,k}$ are the mean and variance of the *a priori* distribution of $\mu_{\theta,k}$. These parameters must be known beforehand and must be learned in advance. In order to learn $\bar{\mu}_{\theta,k}$ and $\overline{\Phi}_{\theta,k}$, we need well-trained state distributions for the state θ, trained under different noise conditions. Let $\mu_{\theta,k,n}$ represent the mean of the kth Gaussian of θ, under the nth noise condition. Let $\tilde{\mu}_{\theta,k}$ represent the mean of the state when the system has been trained by pooling all the noisy data together. This mean is then identical to the estimate of the mean of the *a priori* distribution of the means:

$$\bar{\mu}_{\theta,k} = \tilde{\mu}_{\theta,k} \qquad (10.29)$$

The variance of the *a priori* distribution of the means can simply be estimated as

$$\overline{\Phi}_{\theta,k} = \frac{1}{N_n} \sum_n (\mu_{\theta,k,n} - \tilde{\mu}_{\theta,k})(\mu_{\theta,k,n} - \tilde{\mu}_{\theta,k})^T \qquad (10.30)$$

where N_n is the number of noise conditions for which $\mu_{\theta,k,n}$ is available. The summation is over all types of noise. The variance given by the above equation can sometimes underrepresent the true variance, especially when the types and number of noise conditions available are small. Therefore, the variance is sometimes estimated as

$$\overline{\Phi}_{\theta,k} = \frac{1}{N_n} \sum_n (\mu_{\theta,k,n} - \tilde{\mu}_{\theta,k,\bar{n}})(\mu_{\theta,k,n} - \tilde{\mu}_{\theta,k,\bar{n}})^T \qquad (10.31)$$

where $\tilde{\mu}_{\theta,k,\bar{n}}$ is the mean of the kth Gaussian estimated using all but the nth noise condition.

The MAP estimate of $\mu_{\theta,k}$ is now obtained in an iterative manner. Let $Y(t)$ represent the tth vector in the sequence of feature vectors for the adaptation data being used to obtain the MAP estimates of the means. Let $\gamma_{\theta,k}^i(t)$ represent the *a posteriori* probability that $Y(t)$ was generated by the kth Gaussian of θ, computed using the MAP estimates of the means that were obtained in the ith iteration. $\gamma_{\theta,k}^i(t)$ can be obtained using the Baum-Welch algorithm. We define the quantities $\eta_{\theta,k}^i$ and $\tilde{\mu}_{\theta,k}^{i+1}$ as

$$\eta_{\theta,k}^i = \sum_t \gamma_{\theta,k}^i(t) \qquad (10.32)$$

$$\tilde{\mu}_{\theta,k}^{i+1} = \eta_{\theta,k}^{-1} \sum_t \gamma_{\theta,k}^i(t) Y(t) \qquad (10.33)$$

$\eta_{\theta,k}^{i}$ is the total expected number of vectors associated with the kth Gaussian of state θ and $\tilde{\mu}_{\theta,k}^{i+1}$ is an estimate for $\mu_{\theta,k}$ based only on the observed adaptation data (without considering the *a priori* distribution). The $(i + 1)$th estimate of $\mu_{\theta,k}$, $\hat{\mu}_{\theta,k}^{i+1}$, is now obtained as

$$\hat{\mu}_{\theta,k}^{i+1} = \eta_{\theta,k}^{i}\overline{\Phi}_{\theta,k}(\Phi_{\theta,k} + \eta_{\theta,k}^{i}\overline{\Phi}_{\theta,k})^{-1}\tilde{\mu}_{\theta,k}^{i+1} + \Phi_{\theta,k}(\Phi_{\theta,k} + \eta_{\theta,k}^{i}\overline{\Phi}_{\theta,k})^{-1}\overline{\mu}_{\theta,k} \quad (10.34)$$

where $\Phi_{\theta,k}$ is the variance of the kth Gaussian of θ. Note that when $\eta_{\theta,k}^{i}$ is very large, $\hat{\mu}_{\theta,k}^{i+1}$ is simply $\tilde{\mu}_{\theta,k}^{i+1}$. When it is very small, $\hat{\mu}_{\theta,k}^{i+1}$ becomes the mean of the *a priori* distribution, $\overline{\mu}_{\theta,k}$.

The algorithm is initialized by setting the initial value $\hat{\mu}_{\theta,k}^{0}$ to the unadapted means. Typically, one or two iterations are sufficient to obtain good estimates of the adapted means. MAP estimates can also be obtained for the variances and mixture weights of the Gaussians. However, most of the improvement in performance is obtained from adapting the means and usually only these are adapted. The interested reader is referred to References 6 and 8 for details of MAP adaptation of variances and mixture weights.

One common problem with MAP estimation as described above is that it is frequently not possible to match the kth Gaussian of a state in the HMM for one noise condition with the kth Gaussian of the corresponding state for a different noise condition, due to variations in indexing in the HMMs for the various noise conditions. As a result, it is not possible to learn $\overline{\mu}_{\theta,k}$ and $\overline{\Phi}_{\theta,k}$, the parameters of the *a priori* distributions of the means. To circumvent this problem, instead of adapting each of the Gaussians in any state θ individually, the mixture Gaussian distribution for θ is collapsed into a single Gaussian. The parameters of this single Gaussian are now given by

$$\mu'_{\theta} = \sum_{k} c_{\theta,k}\mu_{\theta,k}$$

$$\Phi'_{\theta} = \sum_{k} c_{\theta,k}(\Phi_{\theta,k} + \mu_{\theta,k}\mu_{\theta,k}^{T}) - \mu'_{\theta}\mu'_{\theta}^{T} \quad (10.35)$$

The parameters of the *a priori* distribution of μ'_{θ} can be obtained by similarly collapsing the mixture Gaussians in the state distributions of all the noise conditions from which they are learned. MAP adaptation can now be performed for μ'_{θ}. The MAP estimates of the means of the individual Gaussians in the original Gaussian mixture for θ are then obtained as

$$\mu_{\theta,k}^{MAP} = \mu_{\theta,k} + (\mu'_{\theta}^{MAP} - \mu'_{\theta}) \quad (10.36)$$

where μ'_{θ}^{MAP} is the MAP estimate of μ'_{θ}.

Extended Maximum *a posteriori* Adaptation

MAP adaptation has the shortcoming that the parameters of the various Gaussians in the state output distributions are adapted independently of each other. As a result, in order to properly adapt all the Gaussians to the noisy data, it is necessary to have adaptation data that contain vectors representing each of them. In practical situations, however, the amount of adaptation data available is limited and this condition is not usually satisfied, making the adaptation incomplete.

EMAP adaptation[7] tackles this problem by utilizing the correlations between the parameters of the various Gaussians in the adaptation procedure, thereby making it possible to adapt the means of any Gaussian even if there are no adaptation data associated with it.

In EMAP adaptation, the means of all the Gaussians in all the states are concatenated into a single extended mean vector $\underline{\mu}$ as follows:

$$\underline{\mu} = [\mu_{1,1}^T, \mu_{1,2}^T, ..., \mu_{1,K}^T, \mu_{2,1}^T, ..., \mu_{N,K}^T]^T \tag{10.37}$$

where $\mu_{i,j}$ represents the mean of the jth Gaussian in the ith state in the recognizer. N represents the total number of states in the recognizer. We also define an extended covariance matrix $\underline{\Phi}$, as this will be required for the EMAP estimation.

$$\underline{\Phi} = diag[\Phi_{1,1}, \Phi_{1,2}, ..., \Phi_{1,K}, \Phi_{2,1}, ..., \Phi_{N,K}] \tag{10.38}$$

where $\Phi_{i,j}$ represents the covariance matrix of the jth Gaussian in the ith state in the recognizer. $\underline{\Phi}$ is a block diagonal matrix whose diagonal blocks are the various $\Phi_{i,j}$s.

The *a priori* distribution of $\underline{\mu}$ is also assumed to be Gaussian with mean $\underline{\bar{\mu}}$ and variance $\underline{\bar{\Phi}}$. These are learned from the extended mean vectors of HMMs trained under various noise conditions. If we let $\underline{\mu}_n$ represent the extended mean vector for the nth noise condition and $\underline{\tilde{\mu}}$ as the extended mean vector of the HMM trained by pooling all the noisy data together, we have

$$\underline{\bar{\mu}} = \underline{\tilde{\mu}} \tag{10.39}$$

$$\underline{\bar{\Phi}} = \frac{1}{N_n} \sum_n (\underline{\mu}_n - \underline{\tilde{\mu}})(\underline{\mu}_n - \underline{\tilde{\mu}})^T \tag{10.40}$$

where N_n is the number of noise conditions for which $\underline{\mu}_n$ is available. As in the case of MAP, the variance given by Equation (10.40) can underrepresent the true variance. Therefore, Equation (10.40) is sometimes modified to

$$\overline{\Phi} = \frac{1}{N_n}\sum_n (\underline{\mu}_n - \underline{\tilde{\mu}}_{\tilde{n}})(\underline{\mu}_n - \underline{\tilde{\mu}}_{\tilde{n}})^T \qquad (10.41)$$

where $\underline{\tilde{\mu}}_{\tilde{n}}$ represents the extended mean vector of the HMM trained using all but the nth noise condition.

As in the case of MAP, EMAP estimation is also iterative. Let $Y(t)$ represent the tth vector in the sequence of feature vectors for the adaptation data being used to obtain the EMAP estimates of the means. Let $\gamma^i_{\theta,k}(t)$ represent the *a posteriori* probability that $Y(t)$ was generated by the kth Gaussian of state θ, computed using the EMAP estimates of the means that were obtained in the ith iteration. We define the quantities $\eta^i_{\theta,k}$ and $\tilde{\mu}^{i+1}_{\theta,k}$, the extended vector $\underline{\tilde{\mu}}^{i+1}$, and the extended matrix Γ^i as

$$\eta^i_{\theta,k} = \sum_t \gamma^i_{\theta,k}(t) \qquad (10.42)$$

$$\tilde{\mu}^{i+1}_{\theta,k} = \eta^{-1}_{\theta,k}\sum_t \gamma^i_{\theta,k}(t)Y(t) \qquad (10.43)$$

$$\underline{\tilde{\mu}}^{i+1} = [(\tilde{\mu}^{i+1}_{1,1})^T, (\tilde{\mu}^{i+1}_{1,2})^T, ..., (\tilde{\mu}^{i+1}_{1,K})^T, (\tilde{\mu}^{i+1}_{2,1})^T, ..., (\tilde{\mu}^{i+1}_{N,K})^T]^T \qquad (10.44)$$

$$\Gamma^i = \text{diag}[\eta^i_{1,1}I_D, \eta^i_{1,2}I_D, ..., \eta^i_{1,K}I_D, \eta^i_{2,1}I_D, ..., \eta^i_{N,K}I_D] \qquad (10.45)$$

where I_D is a $D \times D$ identity matrix, i.e., a $D \times D$ matrix where all diagonal elements are 1 and all off-diagonal elements are 0. The $(i + 1)$th estimate of the extended mean is now obtained as

$$\underline{\mu}^{i+1} = \overline{\Phi}(\overline{\Phi} + \Gamma^i\overline{\Phi})^{-1}\Gamma^i\underline{\tilde{\mu}}^{i+1} + \Phi(\Phi + \overline{\Phi}\Gamma^i)^{-1}\underline{\tilde{\mu}}_0 \qquad (10.46)$$

Once again, we observe that as the amount of evidence from the adaptation data increases, i.e., as the components of Γ^i get larger, the estimate asymptotically approaches $\underline{\tilde{\mu}}^{i+1}$, the estimate obtained from the data alone without considering the *a priori* distributions.

The algorithm is initialized by setting the initial value μ^0 to the unadapted means. Typically, one or two iterations are sufficient to obtain good EMAP estimates of the means. The adapted means of the individual Gaussians can now be obtained from the adapted extended mean vector. Although EMAP estimates can also be obtained for the variances and the mixture weights of the Gaussians, these involve more detailed procedures and are prone to estimation errors. They are, therefore, not usually performed. Details about these procedures can be found in Reference 8.

One problem with EMAP is that, as in MAP, sometimes the parameters of the *a priori* distribution of the means cannot be uniquely obtained due to noncorrespondence of indices of the Gaussians in the state output distributions of the HMMs for the various noisy conditions from which these parameters are learned. The solution used to tackle this problem is also similar to the one used in MAP: the mixture Gaussian distributions of the various states are collapsed into a single Gaussian and the means of these single Gaussians are adapted. The difference between the adapted and unadapted mean for any state is then added back to the means of each of the Gaussians in the original Gaussian mixture distribution for that state.

A second problem is that the matrix $\bar{\Phi}$ is very large and inverting it may be computationally intractable. Workarounds have been proposed that use only the largest correlations in $\bar{\Phi}$ to reduce the rank of the overall solution. More details of these approaches can be found in Reference 8.

Matched Condition Techniques

In matched condition techniques, recognition is performed using a recognizer that is exactly matched to the noisy test data. This is achieved either by explicitly training the recognizer on data that are similar to the test data being recognized, or by performing recognition based only on those components of the test data that are matched with the recognizer. In this section, we briefly describe *matched condition* and *multistyle* training, which are examples of the former method, and some *missing feature* techniques, which are examples of the latter.

Matched Condition and Multistyle Training

The simplest solution to optimal recognition of noisy speech is to train the recognition system with speech that has been corrupted by noise in an identical manner to the test speech. This is referred to as *matched condition training*. Unfortunately, it is frequently difficult, if not impossible, to obtain sufficient quantities of training data that have been recorded under noise conditions identical to those that are expected in the operating environment. The alternative is to train the system using large quantities of training data from varied noise environments. The diversity of the training data is expected to help account for the noise conditions of the test data even if those precise noise conditions were never seen. This approach is referred to as *multistyle training*[12] and, when sufficient types of noise conditions are available, has also been proved to be highly effective.

Missing Feature Techniques

In missing feature methods, only those components of the test data that match the distributions modeled by the recognizer are used. Mismatched components are deemed to be *unreliable*, or "missing," and are either ignored completely or are discarded and reconstructed on the basis of reliable components. Reliable and unreliable components are usually identified in time-frequency representations or spectrographic representations of the speech signal.

Signal representations used in missing feature analysis typically consist of a sequence of spectral (or log-spectral) vectors derived from the speech signal, one for each analysis segment, or *frame*, in the speech signal. When test data are recorded in a noisy environment, the components of the features that are most affected by the corrupting noise are those for which the local signal-to-noise ratio (SNR) is low. In missing feature methods, the recognizer is usually trained using clean speech. Hence, the judgement of whether or not (or the extent to which) a particular feature component is reliable or unreliable is made on the basis of local SNR, with components less than some specified threshold considered to be unreliable and treated as "missing." Recognition is then performed using an "incomplete" spectrogram of the noisy speech signal that is composed of incomplete log-spectral vectors. Figure 10.4 shows a typical incomplete spectrogram derived from the spectrogram of a noisy speech utterance.

The regions of the spectrogram considered to be unreliable are actually not completely devoid of information regarding the underlying speech signal. If the corrupting noise is additive and uncorrelated to the speech, the value of any unreliable component establishes an upper bound on the value of that component of the underlying clean speech signal. If we represent the jth frequency component of the ith frame of a clean speech signal as $S(i, j)$, and the same component of the signal after corruption by noise as $Y(i, j)$, then $Y(i, j) \geq S(i, j)$. This information can also be utilized by missing feature methods.

There are two approaches to performing recognition with incomplete spectrograms. In the first approach, the recognizer is modified to perform recognition directly with incomplete spectrograms. *Marginalization*[13] and *class-conditional imputation*[14] are examples of such methods. In the second approach, the unreliable regions of the incomplete spectrogram are reconstructed based on the values of the reliable regions and the known statistical properties of clean speech spectrograms. *Covariance-based reconstruction*[15] and *cluster-based reconstruction*[15] are examples of such methods.

The most important step in missing feature methods is that of identifying unreliable regions of the spectrogram. This is a difficult problem which is usually solved by maintaining a running estimate of the spectrum of the noise and using it to estimate the local SNR of the spectrographic elements. Unreliable elements are identified based on the estimated SNR.[16] An alternate approach treats the problem of identifying unreliable elements as one of classification.[17] In the following subsections, we briefly describe these methods. In subsequent sections, we describe marginalization, class-conditional imputation, covariance-based reconstruction, and cluster-based reconstruction.

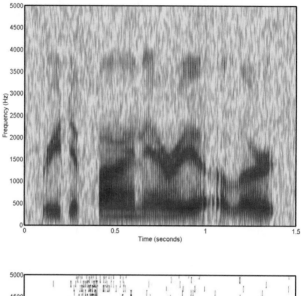

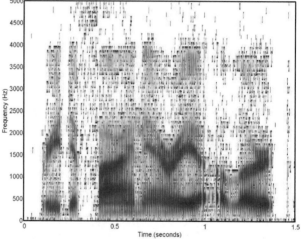

FIGURE 10.4

The top panel shows the spectrographic representation of an utterance that has been corrupted to 15 dB by white noise. The bottom panel shows the same figure when all time-frequency elements with SNR less than 0 dB have been erased. Only the remaining incomplete spectrogram is available for recognition. (From Raj, B., Reconstruction of Incomplete Spectrograms for Robust Speech Recognition, Ph.D. thesis, Carnegie Mellon University, Pittsburgh, PA, 2000. With permission.)

Identifying Unreliable Components

An unreliable component of a spectrogram is generally defined as one with a local SNR that lies below a threshold. The optimal value of this threshold is usually about 0 dB. In practical situations, the unreliable components must be identified without *a priori* knowledge of the true SNR of the spectrographic elements. The matrix of tags that indicates the reliability of components is called a *spectrographic mask.*

Mask Estimation Using Running Noise Estimates

Here a running estimate of the noise spectrum is obtained. This is usually done using the following rule to estimate $N(t, k)$, the magnitude of the kth frequency component of the noise in the tth frame:

$$|N(t, k)| = \begin{cases} (1 - \lambda)|N(t - 1, k)| + \lambda|Y(t, k)| & \text{if } |Y(t, k)| < \beta|N(t, k)| \\ |N(t - 1, k)| & \text{otherwise} \end{cases} \tag{10.47}$$

This is the same rule used in spectral subtraction discussed in Chapter 9. λ usually lies between 0 and 1 and β lies between 2 and 5. The estimated noise spectrum can be used to estimate the SNR of any spectrographic element as

$$SNR(t, k) = \frac{Y(t, k) - N(t, k)}{N(t, k)} \tag{10.48}$$

All elements for which this value lies below a threshold are considered unreliable.[16]

Mask Estimation Using a Binary Classifier

In this method, a classifier is used to identify unreliable elements of the spectrogram. Each element in the spectrogram of the noisy speech is represented by a vector of features. The features are chosen to exploit characteristics of the speech signal that distinguish it from noise. Such features represent information such as the harmonicity and periodicity of the signal, subband energy levels, and spectral contours. The specific details of the features used are beyond the scope of this text and can be found in Reference 17. A classifier is trained using the feature vectors of spectrographic elements from a training corpus of noisy speech for which the identity of the unreliable elements is known beforehand. The feature vectors of the reliable and unreliable elements are used to train two corresponding mixture Gaussian distributions. A spectrographic element of the noisy test speech is identified as reliable if

$$M(X; \Lambda_r) > \alpha M(X; \Lambda_u) \tag{10.49}$$

where X represents the feature vector for that element, $M(X;\Lambda_r)$ is the likelihood of the element computed using the mixture Gaussian for the reliable elements, and $M(X;\Lambda_u)$ is the likelihood computed using the mixture Gaussian for the unreliable elements. α is an empirically determined constant. Additional details of classifier-based mask estimation can be found in References 15 and 17.

Recognition with Incomplete Spectrograms

Recognition with incomplete spectrograms can be performed either by modifying the recognizer to work directly on incomplete spectrograms or by completing the spectrograms prior to recognition by reconstructing the missing regions. In the following sections, we describe two methods from the former category, marginalization and class-conditional imputation, and two methods from the latter category, covariance-based reconstruction and cluster-based reconstruction.

Marginalization

In marginalization,[13] the recognizer is modified to consider only reliable components of log-spectral vectors, and the bounds on the true values of unreliable components implied by their observed noisy values. The recognizer uses these to compute the likelihoods of states. When the Gaussians in state output distributions have diagonal covariance matrices, this is achieved by modifying the manner in which Gaussian likelihoods are computed. The likelihood of the kth Gaussian of state θ for any incomplete log-spectral vector $Y(t)$ is now computed as:

$$P_{\theta,k}(Y(t)) =$$

$$\left(\prod_r \frac{1}{\sqrt{2\pi\sigma_{\theta,k,r}^2}} e^{\frac{-(Y(t,r)-\mu_{\theta,k,r})^2}{2\sigma_{\theta,k,r}^2}} \right) \left(\prod_u \int_{-\infty}^{Y(t,u)} \frac{1}{\sqrt{2\pi\sigma_{\theta,k,u}^2}} e^{\frac{-(X-\mu_{\theta,k,u})^2}{2\sigma_{\theta,k,u}^2}} \, dX \right) \quad (10.50)$$

where $Y(t,i)$ is the ith component of $Y(t)$, the product index r goes over all reliable components of $Y(t)$ and the index u goes over all unreliable components of $Y(t)$. $\mu_{\theta,k,i}$ and $\sigma_{\theta,k,i}^2$ are the mean and variance of the ith dimension of the kth Gaussian of θ. The likelihood of state θ is computed as

$$P_\theta(Y(t)) = \sum_k c_{\theta,k} P_{\theta,k}(Y(t)) \quad (10.51)$$

The modified likelihood given by Equation (10.51) is used to compute state likelihoods during recognition.

Class-Conditional Imputation

In class-conditional imputation,[14] when computing the likelihood of any state θ for an incomplete log-spectral vector during recognition, the unreliable components of that vector are estimated using the mixture Gaussian state distribution of θ. To estimate an unreliable component $Y(t, j)$ of a vector $Y(t)$, we define $P_{\theta, k}(Y(t))$ and $P_{\theta}(Y(t))$ as in Equations (10.49) and (10.50), and $P_{\theta}(k|Y(t))$ and $\hat{S}_{\theta, k}(t, j)$ as

$$P_{\theta}(k|Y(t)) = \frac{P_{\theta, k}(Y(t))}{P_{\theta}(Y(t))} \tag{10.52}$$

$$\hat{S}_{\theta, k}(t, j) = \begin{cases} \mu_{\theta, k, j} & \text{if } \mu_{\theta, k, j} \leq Y(t, j) \\ Y(t, j) & \text{else} \end{cases} \tag{10.53}$$

The unreliable component $Y(t, j)$ is now computed as

$$\hat{Y}_{\theta}(t, j) = \sum_k P_{\theta}(k|Y(t))\hat{S}_{\theta, k}(t, j) \tag{10.54}$$

The estimate $\hat{Y}_{\theta}(t, j)$ is now used in place of the unreliable element $Y(t, j)$ when computing the likelihood of state θ during recognition.

Covariance-Based Reconstruction

In covariance-based reconstruction,[15] unreliable spectrographic elements are estimated based on the values of adjacent reliable elements, using the known covariances between them. These covariances are estimated beforehand from the spectrograms of clean speech utterances. In order to estimate these covariances, it is assumed that the sequence of log-spectral vectors in any spectrogram are the output of a stationary Gaussian random process. Under this assumption, the mean value of any time-frequency component is independent of the time, and the covariance of any two elements of the spectrogram is dependent only on their positions relative to each other, with no reference to where they occur within the spectrogram. The means and covariances can be estimated from the clean speech spectrograms as

$$\mu(k) = \frac{1}{N}\sum_t S(t, k) \tag{10.55}$$

$$C(\tau, k_1, k_2) = \frac{1}{N}\sum_t (S(t, k_1) - \mu(k_1))(S(t + \tau, k_2) - \mu(k_2)) \tag{10.56}$$

where N is the total number of log-spectral vectors in the spectrograms and $S(t, k)$ represents the kth frequency component of the tth log-spectral vector in a spectrogram.

In order to reconstruct the unreliable components of a log-spectral vector $Y(t)$, they are arranged into a vector $U(t)$. All reliable components in the spectrogram that have a high covariance with any of the components of $U(t)$ are arranged into a vector $V(t)$. An estimate of $U(t)$, called $\hat{U}(t)$, is obtained as the value where the Gaussian distribution of $U(t)$, conditioned on $V(t)$ and the bounds on $U(t)$, peaks. This value of $\hat{U}(t)$ is the *bounded MAP* estimate of $U(t)$ and can be estimated using the following iterative procedure for obtaining bounded MAP estimates:

1. Initialize all the components of $\hat{U}(t)$ as $\hat{U}^0(t, k) = U(t, k)$
2. To obtain the $(i + 1)$th estimate of $\hat{U}(t, k)$, construct an extended vector $W(t)$ with all the components of $V(t)$ and all components of the current estimate of $\hat{U}(t, k)$ except the kth element. Construct the mean vector μ_W, and the covariance matrix C_{WW} for $W(t)$ using the means of the components of $W(t)$ and the covariances between its components. Construct the cross-covariance vector $C_{U_k W}$ between $U(t, k)$ and $W(t)$ using the covariances between $U(t, k)$ and the components of $W(t)$. The $(i + 1)$th iteration, $\hat{U}^{i+1}(t, k)$ is now given by

$$
\hat{U}^{i+1}(t, k) = \begin{cases} U(t, k) & \text{if } \mu(k) + C_{U_k W} C_{WW}^{-1}(W(t) - \mu_W) > U(t, k) \\ \mu(k) + C_{U_k W} C_{WW}^{-1}(W(t) - \mu_W) & \text{otherwise} \end{cases} \tag{10.57}
$$

3. Repeat step 2 for all components of $U(t)$ and iterate until $\hat{U}(t)$ does not change any more.

The unreliable components of all the log-spectral vectors in the spectrogram are estimated using the above procedure. Once a complete reconstructed spectrogram is obtained, it can either be used directly for recognition, or other features such as cepstra can be derived from it and used for recognition.

Cluster-Based Reconstruction

In cluster-based reconstruction,[15] the unreliable components of any log-spectral vector are reconstructed based on the reliable components of that vector and the known distribution of the log-spectral vectors of clean speech. This is accomplished by computing a mixture Gaussian distribution from the log-spectral vectors of the spectrograms of a training corpus of clean speech. The Gaussians of this distribution are all assumed to have diagonal covariances. Once the distribution has been computed, a secondary full covariance matrix is also computed that is common across all the Gaussians in the distribution. The distribution and the covariance matrix can both be computed using the EM algorithm.[10]

In order to reconstruct the missing components of any log-spectral vector $Y(t)$, the unreliable and reliable components of the vector are separated out into two vectors $U(t)$ and $R(t)$. A separate estimate of $U(t)$ is obtained for each of the Gaussians in the mixture based on $R(t)$, the mean of that Gaussian and the global covariance matrix. The estimate is obtained using the bounded MAP procedure described in "Covariance-Based Reconstruction." Let us represent the estimate of $U(t)$ obtained for the kth Gaussian as $\hat{U}_k(t)$. We now define the term $P_k(Y(t))$ as:

$$P_k(Y(t)) = \left(\prod_r \frac{1}{\sqrt{2\pi\sigma_{k,r}^2}} e^{\frac{-(Y(t,r)-\mu_{k,r})^2}{2\sigma_{k,r}^2}} \right)\left(\prod_u \int_{-\infty}^{Y(t,u)} \frac{1}{\sqrt{2\pi\sigma_{k,u}^2}} e^{\frac{-(X-\mu_{k,u})^2}{2\sigma_{k,u}^2}} \, dX \right) \quad (10.58)$$

where $\mu_{k,j}$ and $\sigma_{k,j}^2$ represent the mean and variance of the jth dimension in the kth Gaussian. The index r goes over all reliable components of $Y(t)$ and u goes over all unreliable components. We define $P(k|Y(t))$ as

$$P(k|Y(t)) = \frac{P_k(Y(t))}{\sum_k P_k(Y(t))} \quad (10.59)$$

The estimate of the unreliable components of $Y(t)$ is now obtained as

$$\hat{U}(t) = \sum_k P(k|Y(t))\hat{U}_k(t) \quad (10.60)$$

The estimated values of the unreliable elements are now used to reconstruct a complete spectrogram. The reconstructed spectrogram can either be directly used for recognition or can be used to derive other features such as cepstra that can be used for recognition.

Discussion of Relative Merits of the Methods

All methods described in this chapter attempt to recognize noisy speech with an optimal recognizer. In principle, such methods must work better than methods that modify the data.[18] However, most of these methods require access to the parameters of the HMM, which may not be available. For example, commercial recognizers often do not permit the user to access or modify the recognizer components. Model compensation methods usually also require more computation than feature compensation methods and may introduce latencies due to the time taken to adapt the models. When these constraints do not pose a problem, model compensation methods are clearly the methods of choice. In fact, most commercial recognizers include model compensation and adaptation modules that allow the user to adapt the recognizer to the operating noise conditions (and also to the speaker).

In this section, we discuss the relative advantages and disadvantages of the methods described in this chapter. We support our observations with results from various published sources. It must be noted, however, that the various results presented cannot always be compared directly with each other, since they pertain to different tasks of different complexities, over different operating conditions, with different recognizers and databases. This discussion must only be interpreted as representing general trends.

Not all methods described in this chapter are equally effective in all situations. For instance, model decomposition and PMC are only effective if the corrupting transformation is adequately modeled by a linear filter and additive noise. Model decomposition is historically the older method. Its original formulation provided a theoretically sound mechanism for compensating for time-varying noises and was highly effective. Table 10.1 shows some typical early experimental results with this method.

In the original formulation of model decomposition, state output distributions of the compound HMMs were computed using a noise-masking–based relation between the log-spectra of the noisy speech and those of clean speech and noise. This method results in highly accurate models of the distributions of the log spectra of noisy speech.[9] It cannot be translated to derive similar relationships between cepstra, and it becomes necessary to perform recognition with log-spectra. However, recognition systems perform much better when recognition is based on cepstra.[19] PMC is an effective means of performing model decomposition in cepstra-based speech recognizers. Instead of translating the noise-masking relations to the cepstral domain, state out-

TABLE 10.1

Number of Errors Made in Recognizing 300 Words of a Digits Task That Have Been Corrupted by Machine Gun Noise to Various SNRs

Database	+21 dB	+15 dB	+9 dB	+3 dB	−3 dB
Baseline	254	306	377	691	1028
Model Decompensation	6	17	42	81	289

Note: The errors include misrecognition of words and spurious insertions. The recognizer has been trained on the log-spectral vectors of clean speech.

Source: From Varga, A.P. and Moore, R.K., Hidden Markov model decomposition of speech and noise, *Proc. ICASSP90,* 1990, 845–848. © 1990 IEEE.

TABLE 10.2

Word Error Rates for Three Different Test Sets from the DARPA
Database Corrupted to 18 dB by Lynx Helicopter Noise

Test Data	Feb. 89 (%)	Oct. 89 (%)	Feb. 91 (%)	Average (%)
Baseline	38.7	32	33.4	34.7
PMC	7.5	8.1	6.4	7.4

Source: From Gales, M.J.F. and Young, S.J., Robust continuous speech
recognition using parallel model combination, *IEEE Trans. Speech and
Audio Proc.*, 4(5), 352–359, 1996. © 1996 IEEE.

put distributions of the compound HMM are obtained using Monte Carlo
methods or numerical integration. Since recognition is now based on cepstra,
both baseline and noise compensation performance can be expected to be
better. Table 10.2 shows improvements in recognition accuracies obtained on
speech corrupted by Lynx noise by PMC compensation.

When the corrupting transformation cannot be well modeled as a linear
filter and additive noise, neither PMC nor model decomposition can be
expected to be effective. This is the case, e.g., when speech is transmitted
over telephone channels, or compressed using standard codecs.

MLLR, on the other hand, is effective across a wide range of operating
conditions, and is probably the most consistently effective compensation
technique available to date. Table 10.3 shows the improvement in recognition performance obtained on the very difficult "speech in noisy environments" (SPINE)[20] database. No other compensation method was effective
on this database.[21] Even when the linear–filter–additive–noise model is
applicable to the corrupting transformation and PMC can be applied, MLLR
can further improve recognition performance by adapting the state output
distributions computed using PMC. Table 10.4 shows some results that
support this observation.

Like MLLR, MAP uses noisy adaptation data to adapt HMM parameters.
However, unlike MLLR, it also uses *a priori* information about the distributions of the values of the parameters being estimated. MAP attempts to
estimate many more parameters than MLLR, and the degree of adaptation
is greatly dependent on the amount of adaptation data available. As a result,
MLLR is more effective than MAP for small amounts of adaptation data,
whereas MAP is more effective in the reverse situation. Figure 10.5 shows
the behavior of MLLR and MAP with increasing amounts of adaptation
data. Table 10.5 shows typical results with MAP adaptation when sufficient
adaptation data are available. A detailed comparison of MAP and MLLR
techniques for adaptation to new operating conditions can be found in
Reference 22.

MAP attempts to adapt all parameters independently of each other. EMAP,
on the other hand, uses the correlations between parameters to aid adaptation and is expected to require less data than MAP. Consequently, for any
given amount of noisy adaptation data, EMAP can be expected to perform

TABLE 10.3

Error Rates for SPINE1 Data before and after
MLLR Adaptation

Feature	Baseline (%)	MLLR (%)
WMFC	35.1	33.3
PLP	38.0	34.8
LMFC	47.4	40.1

Note: Results are shown for recognition systems
trained with three different types of features.
WMFC, PLP, and LMFC stand for "wide-band
mel-frequency cepstra," "perceptual linear prediction cepstra," and "lowpass filtered mel-frequency cepstra," respectively, three variants of
the basic cepstral parameters.

Source: From Singh, R., Seltzer, M., Raj, B., and Stern,
R.M., Speech in noisy environments: Robust automatic
segmentation, feature extraction, and hypothesis combination, *Proc. ICASSP2001*, 273–276, 2001. © 2001 IEEE.

TABLE 10.4

Word Error Rate on the DARPA Spoke-5 Data

	No Adapt (%)	MLLR (%)
Baseline	17.4	12.1
PMC	10.6	8.6

Note: The first and second rows represent baseline
models and models that have been compensated using PMC, respectively. The columns show
the type of adaptation performed on these models. Baseline performance, performance after
MLLR adaptation, performance after PMC compensation, and performance with MLLR adaptation of PMC-compensated models are shown.

Source: From Woodland, P.C., Gales, M.J.F., and Pye,
D., Improving environmental robustness in large vo-
cabulary speech recognition, *Proc. ICASSP96*, 65–68,
1996. © 1996 IEEE.

better than MAP. Table 10.6 shows some results to this effect. In general,
however, the full potential improvement of EMAP is not achieved due to
the difficulty in accurately estimating the large number of *a priori* parameters
required by it. Several solutions have been proposed to reduce the number
of parameters required by EMAP to improve its performance.[8] Unsupervised
adaptation, where the transcriptions of the adaptation data are obtained by
recognition, is not usually as effective for MAP or EMAP as it is for MLLR.

Matched training is not always possible, since sufficient training data that
have been recorded under identical recording conditions to the noisy test
data are rarely available. Under such conditions, multistyle training is

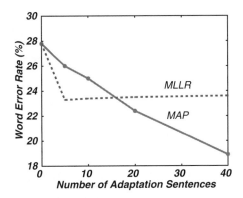

FIGURE 10.5
The effect of increasing amounts of adaptation data on recognition accuracy. Data from the DARPA *Wall Street Journal* (WSJ) database were used in this experiment. (From Doh, S.-J., Enhancements to Transformation-Based Speaker Adaptation: Principal Component and Inter-Class Maximum Likelihood Linear Regression, Ph.D. thesis, Carnegie Mellon University, Pittsburgh, PA, 2000. With permission.)

TABLE 10.5

Recognition Accuracy with and without MAP Adaptation on Two Sets of Data

	No Adaptation (%)	MAP (%)
Set1, single mic	52.8	91.0
Set1, 6 mics	74.9	98.3
Set2, single mic	40.4	87.4
Set2, 6 mics	57.3	96.3

Note: Each set has two subsets, one in which only a single microphone in used, and a second where an array of six microphones has been used to improve the SNR of the signal.

Source: From Omologo, M., Svaizer, P., and Matassoni, M., Environmental conditions and acoustic transduction in hands-free speech recognition, *Speech Communication*, 25, 75–95, 1998. © 1998 by Elsevier Science. With permission.

frequently done. Table 10.7 shows recognition accuracies obtained on data recorded over a Crown PZM microphone, using matched, mismatched, and multistyle trained recognizers. For Crown PZM data multistyle training is superior to training with clean speech only. In Table 10.7, the performance with multistyle training is worse than that with a fully matched recognizer. However, in many practical situations, the performances are comparable.

Missing feature methods have not been extensively evaluated, since they are relatively new. In pilot experiments, they have demonstrated great potential for improving the noise robustness of recognition systems. Of the missing feature methods, marginalization and class-conditional imputation work by

TABLE 10.6

Relative Improvements in Word Error Rate Due to MAP and EMAP Adaptation of Means

	Error Rate (%)	Improvement (%)
Baseline (no adaptation)	7.16	—
MAP adaptation of means	6.47	9.6
EMAP adaptation of means	6.28	12.3

Note: In this experiment the recognition system was adapted to speakers, rather than to noise conditions. The DARPA H2 database was used for the experiment. Common correction terms were estimated for groups of states in order to reduce the total number of parameters to be estimated.

Source: From Zavaliagkos, G., Maximum *a posteriori* Adaptation Techniques for Speech Recognition, Ph.D. thesis, Northeastern University, Boston, 1995. With permission.

TABLE 10.7

Recognition Accuracies Obtained with Matched, Mismatched, and Multistyle Trained Models on Clean Speech (Recorded over a Close-Talking Microphone), and Speech Recorded over an Open Desktop Crown PZM Microphone

	CLSTK (%)	CRPZM (%)
CLSTK	85.3	18.6
CRPZM	36.9	76.5
Multistyle	78.5	67.9

Source: From Acero, A., Acoustic and Environmental Robustness in Automatic Speech Recognition, Ph.D. thesis, Carnegie Mellon University, Pittsburgh, PA, 1990. With permission.

directly modifying the recognizer. For these methods, recognition must be performed using log-spectral vectors. Since these methods utilize the optimal recognizer for the incomplete spectrograms of noisy speech, they outperform methods that reconstruct unreliable elements when recognition is performed with log-spectra. However, reconstruction methods such as covariance-based reconstruction and cluster-based reconstruction result in complete spectrograms from which cepstra can be derived. Recognition results obtained with these cepstra can be significantly superior to those obtained with marginalization or class-conditional imputation. Figure 10.6 shows recognition accuracies obtained using all four missing-feature methods described in this chapter.

Apart from the selected methods described in this chapter, several other important techniques within the categories mentioned have been described in the literature. Most of these methods are similar to, or based on, the techniques described here. For example, variations to PMC have been

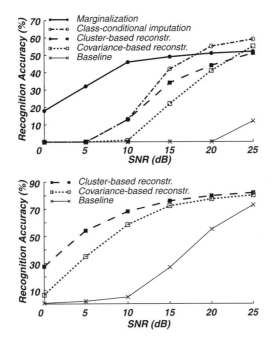

FIGURE 10.6

Recognition accuracy of various missing feature methods on speech corrupted by white noise to various SNRs. The top panel represents accuracies obtained when recognition is performed with log-spectral vectors. The bottom panel shows recognition accuracies obtained using cepstra. In all experiments, unreliable components of spectrograms have been identified using a classifier. (From Raj, B., Seltzer, M., and Stern R.M., Robust speech recognition using missing features: the case for restoring missing input features, *Proc. CRAC Workshop*, Aalborg, Denmark, 2001. With permission.)

proposed that attempt to make it more effective or faster.[9] Other model decomposition techniques use Taylor series expansions to model the state output distributions of the compound HMMs.[23] Several variants of MLLR have been proposed that use multiple sets of affine parameters for adaptation (e.g., see Reference 24). Other methods have combined MLLR and MAP to obtain MAP estimates of MLLR parameters.[25] Still other methods deal with robust estimation of MLLR parameters based on principal component analysis of model parameters.[26] EMAP-style variants of MLLR utilize the correlations between different groups of model parameters to estimate multiple sets of affine parameters for MLLR.[26] Several variants of the MAP and EMAP algorithms have also been proposed that attempt to adapt the various HMM parameters more robustly (e.g., see Reference 8). However, although there exist many noise robustness techniques as described in this chapter and elsewhere, no single method is expected to be equally effective in all noise conditions. The actual technique used must always be tailored to the expected operating conditions and the techniques described can at best be used as points of departure in doing so.

Acknowledgments

We thank Sam-Joo Doh of Speechworks Inc. and Bent Schmidt-Nielsen of Mitsubishi Electric Research Labs for their insightful suggestions, which have improved the correctness and quality of several sections of this chapter. We also thank Gillian Davis for inviting us to write this chapter.

Preparation of this report was sponsored by the Space and Naval Warfare Systems Center, San Diego, under Grant No. N66001-99-1-8905. The content of the information in this publication does not necessarily reflect the position or the policy of the U.S. Government, and no official endorsement should be inferred.

References

1. Juang, B.H., Levinson, S.E., and Sondhi, M.M., Maximum likelihood estimation for multivariate mixture observations of Markov chains, *IEEE Trans. Info. Theory*, IT-32(2), 307–309, 1986.
2. Brand, M., Structure discovery in conditional probability models via an entropic prior and parameter extinction, *Neural Comput.*, 11, 1155–1183, 1999.
3. Viterbi, A.J., Error bounds for convolutional codes and an asymptotically optimal decoding algorithm, *IEEE Trans. Info. Theory*, IT-13, 260–269, 1967.
4. Varga, A.P. and Moore, R.K., Hidden Markov model decomposition of speech and noise, *Proc. ICASSP90*, 1990, 845–848.
5. Leggetter, C.J. and Woodland, P.C., Speaker adaptation of HMMs using linear regression, *Techn. Rep. CUED/F-INFENG/ TR. 181*, Cambridge University, Cambridge, U.K., 1994.
6. Gauvain, J.-L. and Lee C.-H., Maximum *a posteriori* estimation for multivariate Gaussian mixture observations of Markov chains, *IEEE Trans. on Speech and Audio Processing*, 2(2), 291–298, 1994.
7. Lasry, M.J. and Stern, R.M., A posteriori estimation of correlated jointly Gaussian mean vectors, *IEEE Trans. on Pattern Anal. and Mach. Intel.*, 6, 530–535, 1984.
8. Zavaliagkos, G., Maximum *a posteriori* Adaptation Techniques for Speech Recognition, Ph.D. thesis, Northeastern University, Boston, 1995.
9. Gales, M.J.F., Model-Based Techniques for Noise Robust Speech Recognition, Ph.D. thesis, Cambridge University, Cambridge, U.K., 1996.
10. Dempster, A.P., Laird, N.M., and Rubin, D.B., Maximum likelihood from incomplete data via the EM algorithm (with discussion), *J. Royal Stat. Soc., Series B*, 39, 1–38, 1977.
11. Duda, R.O., Hart, P.E., and Stork, D.G., *Pattern Classification*, 2nd ed., John Wiley & Sons, New York, 2000.
12. Lippmann, R.P., Martin E.A., and Paul, D.B., Multi-style training for robust isolated-word speech recognition, *Proc. ICASSP87*, 1987, 705–708.
13. Cooke, M.P., Morris, A., and Green, P.D., Missing data techniques for robust speech recognition, *Proc. ICASSP97*, 1997, 863–866.

14. Josifovski, L., Cooke, M., Green, P., and Vizinho, A., State based imputation of missing data for robust speech recognition and speech enhancement, *Proc. Eurospeech99*, 1999, 2837–2840.

15. Raj, B., Reconstruction of Incomplete Spectrograms for Robust Speech Recognition, Ph.D. thesis, Carnegie Mellon University, Pittsburgh, PA, 2000.

16. Vizinho, A., Green, P., Cooke, M., and Josifovski, L., Missing data theory, spectral subtraction and signal-to-noise estimation for robust ASR: An integrated study, *Proc. Eurospeech99*, 1999, 2407–2410.

17. Seltzer, M., Automatic Detection of Corrupted Speech Features for Robust Speech Recognition, Master's thesis, Carnegie Mellon University, Pittsburgh, PA, 2000.

18. Moreno, P.J., Speech Recognition in Noisy Environments, Ph.D. thesis, Carnegie Mellon University, Pittsburgh, PA, 1996.

19. Davis, S. and Mermelstein, P., Comparison of parametric representation for monosyllabic word recognition in continuously spoken sentences, *IEEE Trans. on Acoustics, Speech, and Signal Proc.*, 28(4), 357–366, 1980.

20. Singh, R. et al., Speech in noisy environments: Robust automatic segmentation, feature extraction, and hypothesis combination, *Proc. ICASSP2001*, 2001, 273–276.

21. Hansen, J.H.L. et al., Robust speech recognition in noise: An evaluation using the SPINE corpus, *Proc. Eurospeech2001*, 2001, 905–908.

22. Fischer, A. and Stahl, V., Database and online adaptation for improved speech recognition in car environments, *Proc. ICASSP99*, 1999, pp. 445–449.

23. Kim, D.Y., Kim, N.S., and Un, C.K., Model-based approach for robust speech recognition in noisy environments with multiple noise sources, *Proc. Eurospeech97*, 1997, 1123–1126.

24. Gales, M.J.F., The generation and use of regression class trees for MLLR adaptation, *Tech. Rep. CUED/F-INFENG/TR.263*, Cambridge University, Cambridge, U.K., 1996.

25. Chesta, C., Siohan, O., and Lee, C.-H., Maximum *a posteriori* linear regression for hidden Markov model adaptation, *Proc. Eurospeech99*, 1999, pp. 211–214.

26. Doh, S.-J., Enhancements to Transformation-Based Speaker Adaptation: Principal Component and Inter-Class Maximum Likelihood Linear Regression, Ph.D. thesis, Carnegie Mellon University, Pittsburgh, PA, 2000.

11

Noise and Voice Quality in VoIP Environments*

Dennis Hardman

CONTENTS

---

* Agilent Technologies, Inc. makes no warranty as to the accuracy or completeness of the material presented here and hereby disclaims any responsibility.

Introduction

In many ways, noise reduction techniques in a Voice over Internet Protocol
(VoIP) environment mirror those used in traditional voice transmission systems. In other words, as long as public switched telephone network (PSTN)
technologies are used in tandem with VoIP technologies — and in almost
all cases, they are — PSTN-like noise reduction is required. However, when
voice signals are encoded, packetized, and transmitted (even for part of the
voice path) across a VoIP network, other network behaviors and impairments
come into play that might or might not be adequately handled by traditional
telephony noise reduction and cancellation techniques. Across a VoIP network, voice signals are encoded in new ways and are transported from point-
to-point across networks designed for non-real-time traffic. In addition, VoIP
networks are often not subject to historical and, until relatively recently (due
to the deregulation of the late 1990s), regulatory standards and constraints.[1]
As a result, an interesting and challenging host of noise sources emerge.
Noise reduction in VoIP networks must take these new sources into account.

Noise is *any* interfering sound. In the context of VoIP, a broader definition
is perhaps required. Noise can be more generally defined as *distortion*. In
other words, noise can be thought of as any undesirable characteristic that
degrades the signal of interest. Given this definition, in a VoIP environment
there are two types of distortion: additive and subtractive (see the section
"Additive vs. Subtractive Distortion"). And along with the signal distortion

described here (which clearly affects *sound quality*), VoIP network behavior can impact *conversational quality* in ways not often seen in most PSTNs.

Strictly speaking, VoIP does *not* introduce any "new" sources of noise or distortion that do not already exist in one form or another on other communications networks. For example, IP networks have always exhibited packet loss and jitter (delay variation). PSTNs produce analog channel noise and echo and always have. Quantization distortion, attenuation/level problems, low-bit rate codec distortion, and so on have all existed for some time. It is the relatively new and unique combination of real-time voice with nonrealtime data network behavior, *and* the interworking of VoIP with traditional PSTNs that create the new challenges of voice and conversational quality. Because of this, noise *avoidance* in a VoIP environment is as important as noise *reduction*.

This chapter briefly describes VoIP technologies and deployments, introduces in more detail the signal distortion and conversational quality impairments that VoIP exhibits, and discusses some of the techniques being used to ameliorate these impairments. Finally, this chapter provides an examination of measurement techniques that target the unique VoIP environment. Note that this chapter approaches noise reduction (and avoidance) from a *system* point of view. Detailed descriptions of network components, processes, or noise reduction techniques can be found in the reference material or in other parts of this book.

VoIP Overview

VoIP refers to an expanding family of voice processing and transport technologies that seek to take advantage of existing data network infrastructures. VoIP networks promise to reduce the cost of local and long distance telephone calls for individuals and businesses alike, and they have the potential to provide unique new services and hasten computer–telephony integration. Relative to traditional telephone networks and data communications networks, VoIP is still in its infancy. But as voice and data service providers look for new ways to improve service offerings while increasing profits and reducing costs, VoIP stands a good chance of becoming one of the most important voice processing and transport technologies in the communications industry. To be widely accepted and deployed, however, VoIP must address several significant challenges. One of these challenges is matching the signal and conversational quality that is consistently delivered by PSTNs and to which telephone customers have become accustomed. Related to the challenge of achieving acceptable sound and conversational quality is the technical challenge of integrating and interworking VoIP with existing voice networks.[2]

With regard to voice signal quality, one of the primary differences between the PSTN and the VoIP network is that the PSTN provides a dedicated voice channel of consistent bandwidth for each voice call, whereas a VoIP network provides best-effort voice packet delivery consistent with IP network behavior. Another way of looking at this difference is that PSTN voice channels are designed with the voice signal in mind (i.e., they have just the right amount of bandwidth and the right frequency response to minimally support a conversational quality voice signal). IP networks, on the other hand, were never really designed for real-time, dedicated bandwidth applications like voice. This difference affects virtually all aspects of noise and distortion avoidance for VoIP implementations and VoIP/PSTN integration. Another interesting difference is the fact that PSTNs provide call setup and management intelligence in the core of the network (via SS7 signaling and central office processing), whereas VoIP networks have pushed this intelligence to the edge of the network where it resides in VoIP endpoints such as personal computers or IP/ethernet telephones. This can also impact voice quality, because the network core is no longer as tightly controlled or regulated.

This section provides a basic overview of VoIP implementations and technologies, showing where noise and distortion issues can arise. Douskalis[2] and Minolli and Minolli[3] provide more detailed information about the technologies, implementations, and measurement techniques for VoIP.

VoIP Implementations

In its most basic and generic form, a VoIP network consists of user endpoints (e.g., telephone, fax, modem, VoIP computer terminal) connected to media gateways which, in turn, are connected to the IP signaling and media transport network. This basic architecture is shown in Figure 11.1.

Because a VoIP network must provide ubiquitous call access, it is almost certainly connected to and integrated with various other voice transport networks including cellular, integrated services digital network (ISDN), PSTN, and proprietary enterprise data and voice networks. Depending on the VoIP protocols and equipment used, other devices can be deployed and implementations can become quite complex. Please note that in Figure 11.1 and in the remainder of this chapter, PSTN generally refers to any analog voice circuit ranging from an analog telephone connected to the analog foreign exchange station (FXS) port of a VoIP gateway or router to an analog telephone connected to a service provider's local loop and central office.

There are various places in a typical VoIP and/or VoIP/PSTN implementation that can cause, or make worse, noise and signal distortion. Figure 11.2 identifies the main sources of voice signal impairment: IP network behavior and processing, VoIP network processing, and PSTN/VoIP integration. This chapter focuses primarily on these three sources of distortion. Remember, however, that PSTN-specific impairments can and do affect VoIP signal

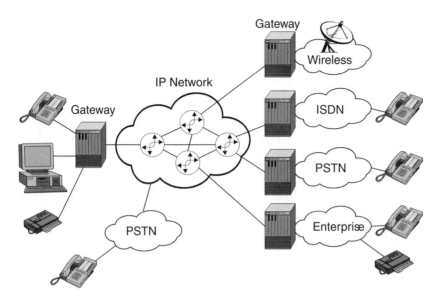

FIGURE 11.1
Basic VoIP implementation.

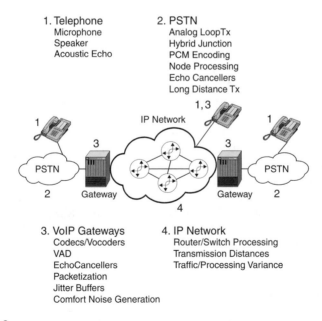

FIGURE 11.2
Sources of voice quality impairments. (Copyright 2001, Agilent Technologies, Inc. Reproduced with permission.)

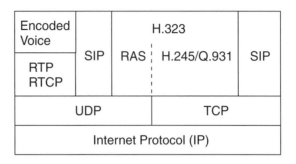

FIGURE 11.3
VoIP protocol stack(s).

quality when VoIP networks and PSTNs are involved in the same voice
signal path.

VoIP Protocols

An increasing number of VoIP protocols provide the signaling, call services,
audio/video stream transport and, in some cases, quality of service needed
to successfully place and answer VoIP telephone calls. Two of the more
commonly implemented protocols are the complex and feature-rich ITU-T
Recommendation H.323 and the simpler Session Initiation Protocol (SIP).
These protocols share a common basic protocol stack as shown in Figure 11.3.
    Whether the VoIP protocol is H.323 or SIP, the protocol stack has some
common characteristics. IP is carried over the physical and network transport layers. User datagram protocol (UDP) and transport control protocol
(TCP) are encapsulated into IP packets. VoIP-specific protocol packets are
encapsulated into UDP or TCP depending on the particular signaling function. Digitized and packetized voice handled by real-time transport protocol
(RTP) and real-time transport control protocol (RTCP) is encapsulated into
UDP datagrams. In the context of voice signal distortion, it is the
IP/UDP/RTP portion of the stack that is the most interesting. Aspects of the
VoIP signaling stack (H.323, SIP) also affect voice signal quality to some
extent because characteristics of the voice channel are often defined by the
signaling process when calls are set up. Finally, RTCP plays a role in maintaining the quality of a VoIP call, because it can be used to gather information
about delay, jitter, and packet loss.

General Noise/Distortion Issues in VoIP

Before exploring VoIP-specific distortion issues and how they are dealt with,
a few basic concepts should be introduced. Although these concepts are not

necessarily limited to VoIP applications, they do, however, affect VoIP signal quality. In fact, early VoIP network designers have all too often failed to meet existing standards for voice quality that apply to any voice network regardless of the underlying technology.[4]

General Telephony Impairments

As described in the previous section, VoIP networks almost always interface with some aspect of the PSTN. This means that most PSTN impairments can impact voice and conversation quality on interconnected VoIP networks. For example:[5]

- Signal level is arguably *the* most important factor affecting perceived voice quality. Clearly, if signal levels are too low, users cannot understand what is said, and if levels are too high, clipping (distortion) can occur.

- Circuit noise and background noise have many sources from both the analog and digital portions of a telephony network. Since much of this noise is outside the voice band, it can cause some problems for VoIP vocoders if not eliminated via adaptive noise filters or other techniques.[6]

- Sidetone is in fact a form of intentional echo that occurs at the telephone set. It is designed into telephone sets so that users can regulate their own voice levels and receive the necessary feedback that the circuit over which they are speaking is still "alive." A similar phenomenon is addressed in VoIP networks in which voice activity detectors (also called silence suppressors) are used. In this case, artificial background noise is actually injected into the voice circuit during silent periods between speech utterances to provide feedback that the circuit is still active.

- Attenuation and group delay distortion are impairments that are dependent on the frequency characteristics of a particular voice channel. Similar to analog circuit noise, attenuation, and group delay distortion can cause unpredictable effects when coupled with low-bit rate perceptual codecs used in VoIP.[6]

- Absolute delay is the time it takes for a voice signal to travel from talker to listener, and delay values typical of PSTNs (tens of milliseconds) have little effect on perceived voice quality if there is no echo or if echo is adequately controlled. However, due to signal processing, VoIP networks introduce unavoidable delays of 50 ms and above which can expose echo (as described below) and affect conversational quality.

- Talker and listener echo can be problematic in traditional PSTNs and have been around for many years. In most situations, this echo is not perceptible because it returns to the talker/listener too quickly to be distinguished from regular speech. However, when larger end-to-end delays are introduced by VoIP processing, existing PSTN echo can become a real problem.

- Quantizing and nonlinear distortion occurs in digital systems when an analog signal is encoded into a digital bit stream. The difference between the original analog signal and that which is recovered after quantizing is called quantizing distortion or quantizing noise. High-quality PCM encoders used in PSTNs exhibit a predictable level of quantization noise and can, therefore, be dealt with in a relatively straightforward way. However, this assumption cannot be carried into the VoIP domain because voice-band codecs (vocoders) operate on a different premise and produce nonlinear distortion. Thus, in VoIP environments, quantization noise cannot always be measured or eliminated in the same way.[6]

Because such PSTN impairments as described above can have an unpredictable effect on voice signals processed and transported across VoIP networks, aggressive noise reduction on circuits known to interface with VoIP networks should probably be employed.

Additive vs. Subtractive Distortion

All voice transmission systems are subject to the effects of both additive distortion (e.g., circuit noise, background noise) and subtractive distortion (e.g., transient signal loss, severe attenuation). For VoIP systems, however, these types of distortion are even more significant. Because perceptual codecs play such an important role in VoIP applications (as described in "VoIP Processing"), noise added to the voice signal prior to encoding can have unpredictable effects depending on whether the noise has frequency components within the voice band or not and depending on the type of encoding used. In VoIP, traditional subtractive distortion such as excessive attenuation is now accompanied by the effects of packet loss where discrete portions of the encoded voice signal simply disappear. Again, due to the use of low-bit rate codecs to preserve network bandwidth, this packet loss can be particularly disruptive. An equally interesting and related source of distortion is error concealment in which subtractive distortion such as packet loss is actually compensated for by *intentional* additive distortion in the form of predictive packet insertion.[7]

Nonlinearity and Time Variance

Two of the primary differences between a PSTN or PSTN-like voice channel and a VoIP voice channel are the conditions of time variance and linearity. For the most part, a PSTN voice channel is linear and time invariant (LTI).

(A voice channel is more or less linear if the voice waveform that enters the system is reproduced at the receiving end. A voice channel is time invariant if, once it is set up, its transmission characteristics normally do not change over time.) A VoIP voice channel, on the other hand, is often nonlinear and time variant, a condition that makes noise reduction in a VoIP environment particularly challenging. For example, the end-to-end delay of the digital encoding/decoding scheme of a VoIP channel can change during a single telephone call (time variance), resulting in changes in sound and conversational quality. Modern VoIP codecs (as described in "VoIP Processing") encode and decode voice signals in nonlinear ways, because they strive primarily to preserve the subjective sound quality of a given voice signal rather than the objective audio waveform. Depending on how these codecs are implemented (and depending on other network conditions such as packet loss), significant levels of distortion can be introduced to the voice signal.

Human Perception's Role

It is very difficult to separate the quantification of voice quality (i.e., the evaluation or measurement of noise and distortion) from the subjective experience of the human talker and listener. Voice quality can really only be judged relative to the situation being assessed and the human experience of it.[8] Voice circuit designers know that the physiology of the human ear and the psychology of human perception must be taken into account when designing voice processing and transmission systems and, therefore, when detecting and avoiding distortion. Digital signal processing (DSP) and voice processing design efforts increasingly concern themselves with only those parts of the voice signal likely to be perceived.[9] This selective processing ultimately reduces transmission bandwidth requirements, benefiting those who must implement VoIP systems in bandwidth-limited situations. Therefore, noise reduction and avoidance in a VoIP environment often concerns itself only with the perceptually important aspects of noise and distortion.

Obviously, the human ear can detect only those auditory signals within a finite frequency and loudness range. However, cognitive aspects of human perception play an important role in network design. For example, humans adapt to very brief auditory drop-outs without losing the meaning or content of a spoken phrase. Human listeners will perceive a particular voice sample as having worse quality if a burst of distortion occurs at the end of the sample as opposed to at the beginning of the sample.[8,10] In addition, listeners' expectation and mood can also affect their assessment of voice quality. These and other aspects of human perception play a role in noise reduction in VoIP.

Listening Quality vs. Conversational Quality

As mentioned in "VoIP Overview," two of the biggest challenges facing VoIP systems are listening/sound quality and conversational quality. These two

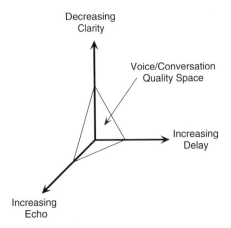

FIGURE 11.4
Voice quality metrics. (Copyright 2000, Agilent Technologies, Inc. Reproduced with permission.)

types of quality are related because end users often do not make a conscious distinction between them. However, the distinction between the two should be preserved. Clearly, listening/sound quality is directly impacted by noise or other types of distortion. It is also clear that a distorted voice signal will negatively impact a telephone conversation. But several telephony phenomena, further exacerbated by VoIP processing, affect the character of voice conversations without really affecting sound quality at all. These phenomena include end-to-end and round-trip network delay, delay variance (jitter), and echo. Delay and echo will be covered along with sound quality (also called clarity) in the next section.

Primary VoIP Quality Metrics

In VoIP environments, three elements (shown in Figure 11.4) emerge as the primary factors affecting voice listening/sound and conversation quality.

Clarity and delay can be thought of as orthogonal in that they normally do not directly affect each other. Echo, on the other hand, affects perceived clarity and, in many cases, can be made more perceptible (and annoying) by increasing delay. Although Figure 11.4 shows a rough relationship between clarity, delay, and echo, a strict mathematical relationship does not exist. Suffice it to say, however, when clarity is good, delay is short, and echo is reduced, overall voice quality is improved. Often, trade-offs must be made between these parameters. For example, to decrease delay, VoIP designers can use less complex encoding schemes, but the clarity of the voice signal can suffer (i.e., coding distortion can increase).

Clarity

Clarity generically refers to a voice signal's fidelity, clearness, lack of distortion, and intelligibility. This is primarily a sound quality metric where the presence of noise and distortion plays the most significant role. Clarity is a very subjective metric and is challenging to measure, particularly in VoIP applications. Traditionally, the clarity of a voice signal or voice channel has been measured subjectively according to ITU-T Recommendation P.800 resulting in a mean opinion score (MOS). MOS values can range from 1 to 5 with 5 being the best possible score. MOS and other more economical and objective measurement techniques that take into account human perception and physiology are described later in "Measuring Noise and Distortion in a VoIP Environment." In a VoIP environment, clarity problems are often caused by packet loss, uncontrolled jitter, and analog circuit noise. Clarity is also significantly impacted by the codecs used on the voice channel.

Delay and Delay Variance (Jitter)

End-to-end delay is the time it takes a voice signal to travel from talker to listener. This voice signal delay is the additive result of VoIP/IP network processing and packet transport. Delay affects the quality of a conversation without affecting the actual sound of the voice signal — delay does not introduce noise or distortion into the voice channel. When end-to-end delay reaches about 250 ms, participants in a telephone conversation begin to notice its effects. For example, conversation seems "cold" and participants start to compensate. Between 300 and 500 ms, normal conversation is difficult. End-to-end delay above 500 ms can make normal conversations impossible. In PSTNs, end-to-end delay is typically under 10 ms. In VoIP networks, however, an unavoidable lower limit on end-to-end delay can be as much as 50 to 100 ms because of codec operations such as packetization and compression.

There is, however, one aspect of delay that has the potential to cause voice signal distortion, and that is delay variance (or jitter). Jitter is the variation in individual voice packet arrival times at voice gateways. For data networks, jitter is less of a problem because arriving packets can be buffered for longer periods of time. For real-time applications such as voice, however, some jitter can be tolerated, but more stringent upper limits must be imposed. When packets arrive outside this upper limit, the packets are discarded or ignored causing what amounts to packet loss. Packet loss directly affects voice signal distortion (described in more detail in "Packet Transmission"), and it must be controlled or managed in VoIP systems to reduce its negative effect.

Echo

Echo is the sound of the talker's voice returning to the talker's ear. Echo, like delay, influences conversational quality more than it does sound quality. However, echo can significantly affect a talker's *perception* of sound quality in much the same way an interrupting burst of noise affects a listener's perception of sound quality. In the context of VoIP, echo (which often already exists on the PSTN but is rarely noticed) is made more noticeable by the unavoidable delay caused by VoIP processing. The causes and solutions to

VoIP-exposed echo will be covered later in "VoIP/PSTN Hybrid Network Implementations."

Specific Noise/Distortion Issues in VoIP

This section describes some of the sources of (and solutions to) noise and distortion that are either created or made significantly worse by VoIP technologies and implementations.

Packet Transmission

When voice is introduced into networks not originally designed for real-time audio transmission, normal network behaviors can suddenly become the source of significant voice quality impairments. Clearly, the idea of voice carried over data networks is not new. Asynchronous transfer mode (ATM) networks provide services, protocols, and quality of service (QoS) processes designed specifically for this application. Frame relay networks have come a long way with regard to QoS and voice over frame relay (VoFR) services. However, IP packets can be encapsulated into a broad range of wide area network (WAN) data network protocols, not all employing robust QoS and voice-handling capabilities. A typical voice path could involve a number of these protocols (carrying VoIP packets). And although there have been significant advances in IP QoS, the very fact that IP was designed as a data network protocol implies that there will be voice quality problems associated with otherwise normal data network behavior.

This section begins with a brief description of physical layer bit errors and data link layer frame/cell loss and then describes the two primary packet-based causes of distortion on a VoIP network: packet loss and jitter. A brief description of IP QoS follows.

Layer 1 Bit Stream Errors/Layer 2 Frame or Cell Loss

Bits and bytes can be errored or lost at the physical layer of the open system interconnection (OSI) data communications stack. Bit error rates, if below those expected of normally operating T1, E1, DS3, or 10/100 base — T ethernet networks will not affect the sound of a voice signal in any significant way (although a single errored sample can produce an audible click or pop). In fact, if bit error rates become high enough to be truly disruptive, chances are the integrity of the call itself is at risk. ITU-T recommendation G.821 defines levels of bit error rates for specific media and distance specifications. It is beyond the scope of this chapter to describe the details of bit errors and bit error rates. However, with regard to the effect bit errors can have on VoIP applications, the following can be said:[11]

- In telephony applications, bit errors generally come in bursts and are usually caused by clock synchronization problems, electrical disturbances, and physical layer processing problems.
- Intuitively, one might conclude that evenly distributed low-bit error rates would have little effect on overall voice quality. However, voice applications may routinely discard any IP packet that has even one error, particularly TCP packets. If packet sizes are large, the resulting packet loss can be debilitating.
- UDP, the portion of the VoIP stack that contains encoded voice, can be configured to tolerate bit errors. This characteristic is configured in the operating system and can reduce the packet loss associated with small numbers of bit error.

Frame or cell errors or loss at the OSI data link layer can also have a significant impact on the clarity of voice traffic carried by protocols higher in the stack. Frame relay summarily discards errored frames and relies on transport layer processes for retransmission, thus increasing jitter and ultimately increasing packet loss. ATM discards cells when QoS or traffic shaping processes are triggered to maintain agreed upon traffic levels, relying on upper level protocols to recover or retransmit lost data. Typically, if cell or frame loss at layer 2 is a problem, mere signal quality at the VoIP application layer will be the least of a VoIP implementer's worries. Call and channel reliability is the more significant issue. The good news is that, for the most part, layer 2 data protocols often provide error correction and run over very robust physical layers.

IP Packet Loss

By its very nature, IP is an unreliable networking protocol. In its most basic (and ubiquitous) form, IP makes no delivery, reliability, flow control, or error recovery guarantees and can, as a result, lose or duplicate packets or deliver them out of order.[3] IP assumes that higher layer protocols or applications will detect and handle any of these problems. Obviously, this kind of network behavior can be problematic for real-time VoIP. When an IP packet carrying digitized voice is lost, the voice signal will be distorted. Before describing the kinds of distortion packet loss can create, it is useful to briefly describe the causes of packet loss:[11]

- *Packet damage:* Many applications will discard incoming packets when presented with one that has been damaged. An example of packet damage is bit errors due to circuit noise or equipment malfunction.

- *Network congestion, buffer overflow, and IP routing:* Perhaps the largest cause of packet loss is packet discard due to network congestion. When a particular network component receives too many packets at one time, its receive buffers overflow causing packets to be discarded. IP networks also deal with network congestion by rerouting traffic to less congested network paths, but this can increase delay and jitter.

Typically, when packets are intentionally discarded due to damage or congestion, networking applications will retransmit the data. This can cause duplicate packets to be sent, can result in packets arriving too late to be used, or can cause packets to be received in the incorrect order. For nonreal-time applications, this kind of network behavior is not catastrophic — in fact, it is expected. However, late or misordered packets can have, from a VoIP standpoint, the same effect as lost packets.

Determining the effect packet loss has on voice signal distortion is a complex task and depends on several variables. Fundamentally, lost packets mean lost voice information resulting in audible dropouts, pops, and clicks. Generally speaking, more packet loss means more distortion. However, the location in the packet stream at which packet loss occurs, the type of codec used (and its bit rate, packet size, compression algorithms, and error concealment methods), and the amount of jitter on the network all contribute to just how much (and how perceptible) the distortion will be. In "VoIP Processing," codec type, packet loss rates, and jitter will be related to specific distortion measures. However, a few general thoughts are presented here:

- With regard to human perception, there is a difference between a steady-state and widely distributed packet loss rate and bursty packet loss. One might expect a steady state of annoyingly perceptible distortion would be more disruptive than an occasional burst of distortion. In addition, the location of the burst affects perceived voice signal quality as well. For example, in a 60-s call, packet loss bursts toward the end of the call are perceived to be more disruptive than those that occur near the beginning of the call.

- Low-bit rate, perceptual codecs exhibit more distortion for a given packet loss percentage than waveform codecs. For example, G.711, a waveform preserving, linear codec, encodes the most voice information (no compression, maximum number of bits for each voice sample) as compared to most other codecs. Therefore, when a G.711 codec is being used, packet loss has less effect on perceived quality than with other codecs. On the other hand, perceptual codecs (G.729, G.723, G.721) encode and decode based on perceptual relevance using compression to reduce the number of bits needed. Experimental evidence shows that lost packets can have a larger impact on the voice signal in this case.

Jitter (Varying Packet Delay)

One of the primary causes of practical packet loss is varying packet delay (jitter) that is not accounted for by network components such as VoIP gateways. In an ideal network, each voice packet arrives at its destination with the same end-to-end delay. This would allow the receiving gateway to assemble and play out the voice packets once they started arriving. As long as the end-to-end delay does not exceed about 120 to 180 ms, end users will report no conversational impairment. However, real IP networks can and do deliver voice packets with varying end-to-end delay due to multiplexer and switch operations, queues, routing changes, congestion, and other network behavior.[11] For example, when a series of voice packets arrive at the destination 50, 58, 43, 89, 104, and 66 ms, respectively, after each was sent, the receiving device can have problems reassembling and playing out the voice signal unless a process is in place to account for this jitter. To account for jitter, jitter buffers are implemented in voice gateways. Jitter buffers are described in more detail in "VoIP Processing."

The key point to remember is that *jitter does not sound like anything to the end user* unless it is bad enough that packets arrive too late to be used. This late arrival time results in a situation that for all practical purposes is the same as packet loss.

Solutions to Packet Loss and Jitter — QoS

There are various ways that the negative effects of packet loss and jitter can be avoided or even eliminated. Since many VoIP calls will span WANs as well as local area networks (LANs), the QoS methods mentioned next involve aspects of both WAN and LAN networking technologies. Other solutions to packet loss and jitter involve specific VoIP processing. Examples of QoS solutions include:[12]

- Overprovisioning involves making sure that the network has much more bandwidth capacity than it needs, thus ensuring that VoIP traffic is never subject to congestion or other causes of packet loss and jitter. This, however, is not practical for large telephony service providers.
- ATM and frame relay both provide QoS support, with ATM having the most robust and extensive capabilities (and often the most expensive), particularly with regard to cell/packet loss and jitter.
- IP type of service (TOS) and filtering provides basic QoS and is built into the IP protocol. However, this method requires specific router configurations and may be unsuitable for larger networks.
- Integrated services and resource reservation protocol (RSVP) permit a terminal or voice gateway to request a specific IP quality of service. However, limited packet loss and jitter control is offered.

- Differential services (including multiprotocol label switching (MPLS)) is a relatively new technology that offers both packet loss control and jitter control.

In addition to the more general QoS methods mentioned above, packet loss and jitter can be dealt with by VoIP processing such as codec error concealment in which lost packets are replaced, optimum codec packet size, and intelligent jitter buffer configuration.

VoIP Processing

In addition to the voice sound quality problems caused by voice packet delivery and processing (i.e., IP network performance) discussed in the previous section, voice quality is impacted by processes that are very specific to VoIP gateways and other VoIP equipment. Although some of these processes are used to solve quality issues of one sort or another, they themselves can introduce voice distortion or conversation impairments.

Codec Characteristics and Performance

Perhaps the most important factor with regard to voice signal quality in a VoIP environment is the voice codec (coder/decoder) implemented in VoIP gateways/routers, IP telephones, and other VoIP terminals. In fact, it is the voice codec (along with the initial quality of the signal being encoded/decoded) that defines the best possible voice quality that can be delivered. In other words, the quality of a voice signal will never be better than what a particular codec can deliver under optimum conditions, although it can certainly be worse due to conditions such as background noise or packet loss.[4]

Codec Description

Codecs digitize and packetize voice signals prior to their transmission across an IP network. Some codecs also compress the voice signal to preserve network bandwidth. Voice codecs are implemented in software and/or hardware and are often rated according to the following parameters:[9]

- Bit rate is a measure of the compression achieved by the codec.
- Delay is a measure of the amount of time a codec requires to process incoming speech signals. This processing delay is a portion of the overall end-to-end delay experienced by a voice packet.
- Complexity is an indication of a codec's cost and processing power.
- Quality is a measure of how speech ultimately sounds to a listener.

Clearly, trade-offs must be considered when deciding which codecs to use in a given VoIP network or device. For example, in situations where bandwidth is at a premium, low-bit rate codecs may be preferred at the expense of some signal quality. In other situations, voice quality must be preserved resulting in higher complexity, cost, and bandwidth requirements.

For telephony applications, there are three categories of codecs:[7]

- Waveform codecs are the most common type and are used ubiquitously in most PSTNs. These codecs seek to reproduce the analog signal waveform at the receiving end of the call and generally introduce the least amount of distortion and noise. They also require the highest amount of bandwidth. ITU-T's G.711 is the most common waveform codec.

- Vocoders (also called source codecs) do not seek to reproduce the analog signal waveform, but instead seek to reproduce the subjective sound of the voice signal. Vocoders are targeted strictly at voice signals, use less bits to encode the voice signal (thus, requiring less bandwidth), and are generally believed to be marginally suitable for telephony applications (although they have been and are used in some VoIP environments).

- Hybrid codecs are the most commonly used codecs in VoIP networks. Hybrid codecs meld the best characteristics of both waveform codecs and vocoders and also operate at very low bit rates.

Both vocoders and hybrid codecs seek, to a lesser or greater extent, to encode the perceptually relevant characteristics of a voice signal with the ultimate goal of producing good voice quality using less bandwidth than the waveform codec. Because of this, the analog voice waveform is not always reproduced. Traditionally, when the analog waveform is altered from its original shape, this is thought to represent either additive or subtractive distortion. All codecs introduce some level of distortion (e.g., quantization distortion). Whether this waveform distortion results in a degraded voice signal depends on the quality of the codec and other network conditions. It can also depend on whether the codec uses noise shaping techniques to reduce the amount of perceptual noise that is actually encoded, or error concealment to reduce the negative effects of packet loss.

Codecs, Bit Rates, Packet Loss, Jitter, and Voice Quality

One generalization that can be made is that lower bit rate codecs introduce more perceptually relevant distortion (i.e., lower voice signal quality) than waveform codecs operating at higher bit rates.[4,13] Figure 11.5 shows measurement results in which an MOS prediction algorithm — perceptual analysis measurement system (PAMS) listening quality (Ylq) described in "Objective/Predictive Testing" — was used to evaluate the speech quality produced by four different codecs. As bit rates decrease, so too does voice quality (i.e., perceptually relevant distortion increases).

Network conditions such as packet loss and jitter also affect the voice quality produced by a specific codec. It is very difficult to accurately quantify

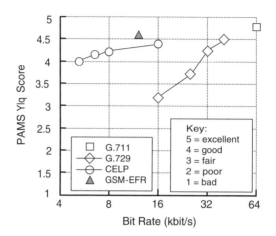

FIGURE 11.5
Listening quality vs. codec bit rate. (Courtesy of Psytechnics, Inc.)

the effect packet loss and jitter will have on a particular voice signal passing through a specific codec. Predictably, as packet loss and/or jitter increases, so does signal distortion. But whether that distortion will have a significant impact on perceived quality depends on the type and location of the packet loss, whether jitter buffers are adequately compensating for varying packet arrival time, and on error concealment methods used by the codec.

Figure 11.6 shows the results of "distortion" measurements made on an ITU-T G.729 codec as packet loss percentages were increased.[14] The decrease of perceived signal quality as packet loss increases is consistent with other experimental results as well as with the experience of VoIP system implement-

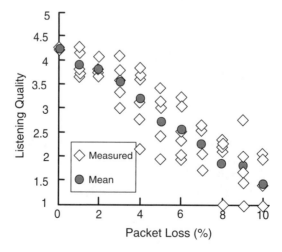

FIGURE 11.6
Listening quality vs. packet loss. (Courtesy of Psytechnics, Inc.)

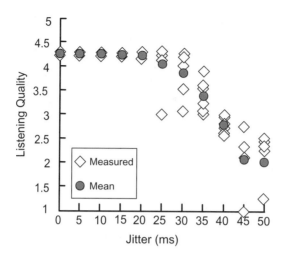

FIGURE 11.7
Listening quality vs. jitter. (Courtesy of Psytechnics, Inc.)

ers. Note, however, that at a specific packet loss percentage, the measured listening quality spans a broad range. These types of results are also shown to be true for other codec types under different experimental conditions.[13]

Jitter can also affect the signal quality produced by codecs. Figure 11.7 shows the same G.729 codec measured with increasing amounts of packet jitter. Again, the spread of measured listening quality is relatively broad, but the general trend is downward at higher jitter values and is consistent with other experimental results.[13]

Jitter Buffers

As described in "Primary VoIP Quality Metrics," jitter (packet arrival variance) does not *sound* like anything as long as receiving VoIP equipment or processes can handle it. Receiving VoIP devices handle jitter by implementing a jitter buffer that can smooth out the packet delay variance so real-time applications can work properly. Basically, a jitter buffer delays the playout of individual arriving voice packets until enough of them have arrived to play out contiguous speech. This implies that jitter buffers add delay to the system.

There are two types of jitter buffers:[4]

- Static buffers provide a fixed length playout delay and any packet that arrives late is discarded. This playout delay is usually configurable, but the underlying network must exhibit a predictable jitter for static buffers to be effective.

- Dynamic jitter buffers are more sophisticated in that they can adjust the playout delay based on the jitter exhibited by previous packets. This provides an automatic balancing act between avoiding lost packets and adding too much delay to incoming packets.

Jitter buffers can impact voice quality in a number of ways. For example, although most dynamic jitter buffers adjust their playout delay during periods of silence, they can and sometimes do adjust during speech utterances, causing momentary distortion. Another example is simply a misconfigured static buffer that does not account for larger jitter values on a network or introduces too much delay. Again, delay and the potential for packet loss must be balanced.

Voice Activity Detection/Comfort Noise Generation

To use bandwidth more efficiently, VoIP networks employ functionality referred to as silence suppression or voice activity detection. A voice activity detector (VAD) is a component of a voice gateway or terminal that suppresses the packetization of voice signals between individual speech utterances (i.e., during the silent periods) in a voice conversation. VADs generally operate on the send side of a gateway, and can often adapt to varying levels of noise vs. voice. Since human conversations are essentially half duplex in the long term, the use of a VAD can realize approximately 50% reduction in bandwidth requirements over an aggregation of channels.

Although a VAD's performance does not affect voice signal quality directly, if it is not operating correctly, it can certainly decrease the intelligibility of voice signals and overall conversation quality. Excessive front-end clipping (FEC), for example, can make it difficult to understand what is said. Excessive hold-over time (HOT) can reduce network efficiency, and too little hold-over time can cause speech utterances to "feel" choppy and unconnected.

Complementary to the transmit-side VAD, a comfort noise generator (CNG) is a receive-side device. During periods of transmit silence, when no packets are sent, the receiver has a choice of what to present to the listener. Muting the channel (playing absolutely nothing) gives the listener the unpleasant impression that the line has gone dead. A receive-side CNG generates a local noise signal that it presents to the listener during silent periods. The match between the generated noise and the "true" background noise determines the quality of the CNG.

VoIP/PSTN Hybrid Network Implementations

Thus far, noise and distortion sources have been discussed with regard to IP network behavior or VoIP-specific processing. Another important source of signal distortion and conversational quality degradation is the interoperation between a VoIP network and the PSTN.

Level/Loss Plans

PSTNs are designed with specific signal level, gain, and loss characteristics depending on where in the network the signal is measured and the type of equipment across which a signal passes.[15] VoIP networks, however, do not always adhere to specific loss plans. When voice signals pass from the PSTN to a VoIP network and back to the PSTN, they may have been attenuated and then amplified resulting in an increased noise floor. Other problems such as clipping can occur. Automatic gain control (AGC) is increasingly being used in VoIP gateways, but AGC can create noise problems of its own.[4]

Transcoding/Multiple Encoding

In a pure VoIP network, a single codec type can be used at each end of a given voice call so the number of times a voice signal is encoded and decoded is limited to one and the coding and encoding schemes are compatible. This would limit the unavoidable codec and quantization noise introduced into the system and would keep end-to-end delay at a minimum. Although this represents perhaps an optimum network design, it is not always practical.

Because a given voice call will likely traverse multiple VoIP systems or VoIP/PSTN hybrid networks, it is more common for multiple codecs to be used and for voice signals to be encoded and decoded multiple times.[16] In these situations, codec and quantization distortion accumulates (often in nonlinear ways), attenuation distortion is multiplied, and idle channel noise is added to the signal at each coding stage. In addition to multiple codec processing, codecs of different bit rates might be used on a single voice signal.

In general, when a voice signal is processed by multiple codecs (particularly of different types) along a single voice path, that voice signal's clarity cannot be better than that produced by the "worst" codec. The quality may, in fact, be noticeably worse if two or more low-bit rate codecs are used. In addition, because many codecs distort speech in nonlinear ways, the order in which they encode/decode speech will affect sound quality. Finally, end-to-end delay can increase significantly when more than one encoding and decoding process is in the voice path, resulting in increased echo perception and causing severe conversational quality problems.

Echo and Echo Cancellation

In most cases, echo is caused by an electrical mismatch between analog telephony devices and transmission media in a portion of the network called the tail circuit.[15] Specifically, this electrical mismatch occurs in a device called a hybrid that provides the junction between an analog four-wire ear and mouth (E&M) trunk line or digital transmission channel and an analog two-wire foreign exchange office (FXO) line. The hybrid separates send-path and receive-path signals so they can be carried on separate pairs of wires or transmission channels. Because the methods used to separate send signals from receive signals are often not ideal, some of the received signal leaks onto the send path and is perceived as echo. Another cause of echo can be acoustic coupling problems (called acoustic echo) between a telephone's speaker and microphone, for example, the hands-free set of a speaker telephone, PC terminal, or cellular telephone. Both types of echo are present on

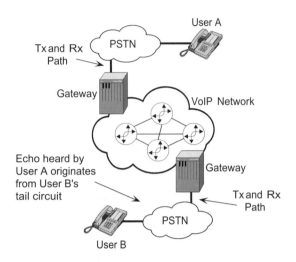

FIGURE 11.8
Perceived echo's origination point. (Copyright 1999, Agilent Technologies, Inc. Reproduced with permission.)

many PSTN networks, but because they are received at the talker's ear so quickly (under 30 ms), they are perceived as sidetone or not perceived at all.

When a VoIP segment is introduced into the voice path, existing (and usually unnoticed) echo can suddenly become perceptible. It can be assumed that any echo generated from a near-end hybrid will still return to the talker too quickly to be perceived. However, far-end echo will be subjected to the unavoidable round-trip delay introduced into the voice path by VoP network processing causing existing echo originating from the far-end analog tail circuit to become perceptible and even annoying to end users.

Although not a source of voice signal distortion in the sense that the transmitted signal is degraded in some way, echo can definitely affect a talker's *perception* of call quality and disrupt *conversational quality*. It can even be argued that, as perceived by the talker, the sound of returning echo combined with the sound of the talker's voice constitutes a *distorted* voice signal.

To deal with unwanted echo, functional components known as "echo cancellers" are deployed in the local exchange, the VoIP gateway, or the VoIP terminal (e.g., PC, IP telephone), usually as close as possible to the tail circuit that generates the echo. Referring to Figure 11.8, an echo canceller next to the hybrid on User B's side of the network "faces out" at User B and cancels the echo of User A's voice that would otherwise be heard by User A.

Modern echo cancellers form a mathematical model of the tail circuit they monitor, and then use this model (along with representations of the signal likely to be echoed, e.g., User A's voice) to estimate the expected echo. This estimated echo is then subtracted from the speech originating on the tail circuit side of the echo canceller (User B's voice). Thus, normal speech is allowed to pass through the echo canceller, but echoes of received speech are removed. An interesting characteristic of most modern echo cancellers is their ability to "adapt" to signal and tail circuit conditions. In other words, at the start of a voice call, echo cancellers take some finite time to converge on the echo estimate that will be subtracted from far-end speech signals. For example, at the beginning of a VoIP telephone call that terminates through an analog tail circuit, echo may be perceptible but quickly diminishes as the echo canceller converges. Echo cancellers are often designed and configured to expect echo within a specific time window (echo delay) and within a specific level range (echo return loss). If the echo signal does not fit within these parameters, the echo canceller can contribute to perceived signal distortion by failing to remove the echo or by converging on inaccurate echo estimates. Voice circuits, particularly when VoIP components are used, must be intelligently designed to exhibit the correct echo return loss and echo delay.

An interesting point of failure (or poor performance) for many echo cancellers is when the talker at the far-end interrupts the near-end talker (a condition known as "double-talk"). Echo cancellers work with the assumption of a linear and time-invariant tail circuit. Double-talk, however, causes the tail circuit to appear to be nonlinear, resulting in echo canceller divergence (in other words, its echo estimate becomes more *inaccurate*). In this case, the interrupting speech can become distorted.

Inconsistent QoS Implementation across Networks

As mentioned in the section "Solutions to Packet Loss and Jitter — QoS," IP QoS can be an effective solution to packet loss and jitter, two important causes of noise and distortion in VoIP environments. However, because VoIP networks interoperate with PSTNs and other voice transport systems, QoS mechanisms must be defined on an end-to-end basis, requiring sufficient network resources to be provided *throughout* the voice path. This is not an overwhelming issue for an enterprise network or a single ISP environment where all resources can be administered through one network manager. But it is almost impossible to administer when multiple ISPs or service providers are involved, as is the case in virtually every national or international long distance call. In addition, this fulfillment of QoS assumes that all equipment in the network is equally capable of identifying voice traffic and of providing the required network resources. Although progress is being made on this front, end-to-end QoS is still the exception rather than the rule in today's IP networks because standards for many of these mechanisms have not been finalized and implemented by equipment manufacturers.

Measuring Noise and Distortion in a VoIP Environment

To reduce or avoid noise and distortion in a VoIP network (or any network for that matter), it is important to be able to characterize it or measure it in some way. Traditionally, voice signal quality testing techniques involved comparing waveforms on a screen and measuring signal-to-noise ratio (SNR) and total harmonic distortion (THD) among others. These and other linear measurements are useful only in certain cases, because they assume that changes to the voice waveform represent unwanted signal distortion. These testing methods also assume that telephony circuits are essentially linear and time invariant. With VoIP and other voice-over-packet networks, particularly when they use low-bit rate speech codecs such as G.729 and G.723.1, neither waveform preservation nor circuit linearity can be assumed. Because of these conditions, specialized testing methods are often used.

VoIP Network Measurement Concepts

Before describing some of the more common measurement techniques used in VoIP, general measurement concepts need to be covered. Although some of these concepts apply equally well in other telephony environments, they are particularly important in VoIP.

Passive Monitoring vs. Active/Intrusive Testing

VoIP network testing is similar to other data and telecommunications testing in that it consists of passive monitoring and active/intrusive testing:

- Passive monitoring is a testing method in which the test device or process "listens" to some aspect of the voice traffic (digital or analog) to gather statistics and perform various types of analysis. Passive monitoring is nonintrusive and does not affect voice traffic or network behavior. It is often used in digital environments in which information encapsulated in frames, cells, or packets can be used to alert test personnel of a problem, or can be analyzed later to determine problem causes and identify traffic trends. Strictly speaking, subjective testing such as MOS (described later) can also be considered passive monitoring. Passive monitoring is often coordinated from 24×7 network operations centers (NOC), and is performed by those tasked with keeping an installed network up and running.

- Active/intrusive testing usually consists of injecting traffic of some type onto the voice channel and analyzing either the effect the traffic has on the channel or the effect the channel has on the traffic. This "energetic" approach to noise and distortion testing usually requires more sophisticated test equipment and software capable of emulating VoIP processes. Active testing is often performed by those responsible for new VoIP devices and software who do their work in research and development labs. Active testing is also useful when isolating the causes of noise and distortion.

Call Setup, Call Completion, and Services Testing

An important area of VoIP operations and performance that must be tested involves the signaling that occurs to establish, maintain, and disconnect VoIP telephone calls. Metrics include percentages of call success/completion, call services validation, call setup times, and so on. This aspect of VoIP operations has little direct effect on voice signal quality. However, "negotiations" occur during some call setup processes between VoIP entities, which can result in a noise or distortion baseline. For example, SIP signaling protocols negotiate codecs and other channel characteristics. Protocol analyzers that can deliver data stream decodes are often used for this type of testing.

Packet Performance Testing

Given the impact packet loss and jitter have on a voice signal carried across a VoIP network, it is clear that packet delivery performance must be tested. Test methods can range from monitoring actual IP traffic to find evidence of packet loss and jitter, to injecting into the network under test-specific packet streams with specific transmission and payload characteristics. Data communications test solutions that provide VoIP decodes, RTP and RTCP monitoring, and general IP traffic analysis capabilities represent perhaps the best ways to measure packet performance in a VoIP environment.

Sound Quality, Distortion, and Noise Testing

VoIP testing must also include a direct measure of sound quality, noise, and distortion. Although VoIP signaling and packet performance are often measured at network interfaces within the VoIP network itself, sound quality measurements are performed from the perspective of the end user of the telephony system. In other words, the quality of the signal received at the telephone set is what must be measured because this is what the user of the system will experience. Test devices that can transmit, receive, and analyze actual voice signals are preferred, although some testing methods use voice-like signals that emulate the frequency characteristics of voice.

Subjective Testing

Because of the subjective nature of voice signal quality, and because traditional audio measures are not always useful, new methods have been developed to evaluate voice clarity in a voice-over-packet environment. Early methods included mean opinion score (MOS), based on the ITU-T P.800 recommendation. This method requires that relatively large numbers of human listeners rate voice quality as part of a controlled and well-defined test process. The advantage of this method is that clarity evaluations are derived directly from the individuals who experience a voice call. Another advantage is the statistical validity provided by numerous evaluators.

However, MOS evaluations can be very expensive, difficult to repeat when new telephony products need to be tested, and time consuming. Because of this, software- or hardware-based predictive methods have been developed to provide objective and repeatable measurement results.

Objective/Predictive Testing

In recent years, algorithms have been developed that can predict MOS results, avoiding some of the disadvantages of full-blown MOS testing. To be successful, these algorithms must evaluate the quality of voice signals in much the same way that nonlinear codecs encode and decode audio signals. That is, they evaluate whether a particular voice signal is distorted with regard to what a human listener would find annoying or distracting. Typically, these algorithms compare "clean" test signals (either actual voice signals or special voice-like signals) to more or less distorted versions of the same signal (having passed through some communications system). Using complex weighting methods that take into account what is perceptually important, the physiology of the human ear, and cognitive factors related to what human listeners are likely to notice, these algorithms provide a qualitative score that often maps closely to MOS. Two very important clarity algorithms in use today are:

- Perceptual evaluation of speech quality (PESQ) is based on the ITU-T P.862 standard that defines the algorithms used to compare reference speech samples with test samples to measure quality degradation due to distortion. PESQ replaces a previous perceptual quality algorithm called perceptual speech quality measure (PSQM), which was based on P.861.
- PAMS is an algorithm developed and licensed by Psytechnics, Inc. that compares speech-like samples to obtain listening effort and listening quality scores.[17]

Both PESQ and PAMS produce MOS-like scores as well as high-resolution disturbance values and error surfaces that allow testers to identify distinct distortion events including packet loss, transient noise spikes, and VoIP processing problems such as VAD front-end clipping. Figure 11.9 shows PESQ measurement results in an implementation produced by Agilent Technologies, Inc.

Another approach to predicting perceived voice quality involves passively monitoring IP traffic to determine packet loss, jitter, and error burst characteristics. These metrics can then be analyzed mathematically in conjunction with known VoIP network characteristics such as delay and codec type, and human cognitive factors to ultimately produce a MOS estimation. Nonintrusive measurement techniques of this sort can be embedded into VoIP equip-

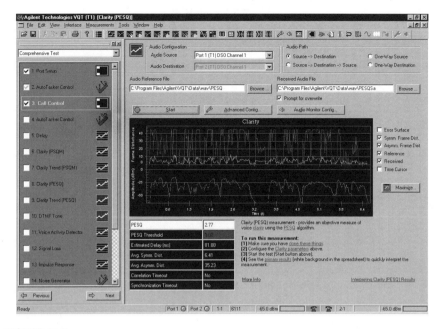

FIGURE 11.9
PESQ measurement results. (Copyright 2001, Agilent Technologies, Inc. Reproduced with permission.)

ment or test equipment with relative ease, and they can provide perceptually relevant distortion measures without producing additional network traffic.[10]

Conclusions

Reducing noise and distortion in a VoIP environment requires not only an understanding of traditional voice signal characteristics, processing, and transmission, but also an understanding of IP network behavior, VoIP-specific processing, and the interaction between emerging VoIP systems and existing telephony infrastructures (i.e., the PSTN). Noise reduction, therefore, involves making design and implementation decisions that balance desired voice quality with network capacity and cost.

Acknowledgments

The author would like to thank John Anderson, Joe Haver, Stefan Pracht, and Ron Sexton of Agilent Technologies, Inc., and Richard Reynolds of

Psytechnics, Inc., for their input and research assistance, and Fran Beckman of Agilent Technologies, Inc., for editing and organizational input.

References

1. McPhillips, E., *Voice over IP Equipment and Service Markets*, HPL-IRI-1999-005, Hewlett-Packard, 1999, 3.
2. Douskalis, B., *IP Telephony: The Integration of Robust VoIP Services*, Prentice-Hall, Upper Saddle River, NJ, 2000.
3. Minolli, D. and Minolli, E., *Delivering Voice over IP Networks*, John Wiley & Sons, New York, 1998.
4. Reynolds, R.J.B. and Rix, A.W., Quality VoIP — An engineering challenge, *BT Tech. J.*, 19(2), 23–32, 2001.
5. Effect of Transmission Impairments, ITU-T Recommendation P.11, March 1993.
6. Bellemy, J.C., *Digital Telephony*, 3rd ed., John Wiley & Sons, New York, 2000, chap. 3.
7. Collins, D., *Carrier Grade Voice Over IP*, McGraw-Hill, San Francisco, 2001, chap. 3.
8. Moller, S., *Assessment and Prediction of Speech Quality in Telecommunications*, Kluwer Academic Publishers, Boston, 2000, 116–117.
9. Madisetti, V.K. and Williams, D.B., Eds., *The Digital Signal Processing Handbook*, CRC Press, Boca Raton, FL, 1998, chap. 45.
10. Clark, A., *Passive Monitoring for Voice over IP Gateways*, Telecommunications Industry Association, TR41.4-01-02-068, February 2001.
11. Siegel, E.D., *Designing Quality of Service — Solutions for the Enterprise*, John Wiley & Sons, New York, 2000, chap. 4.
12. Siegel, E.D., *Designing Quality of Service — Solutions for the Enterprise*, John Wiley & Sons, New York, 2000, chap. 7.
13. Berger, J. et al., *Report of 1st ETSI VoIP Speech Quality Test Event*, ETSI TC STC #10 (Tiphon #22), March 2001.
14. Reynolds, R., *IP, Voice, Quality, and Their Relationships*, informal Psytechnics, Inc. report, June 2001.
15. Green, J.H., *The Irwin Handbook of Telecommunications*, 3rd ed., Irwin, Chicago, 1997, chap. 2.
16. Subjective Performance Assessment of Telephone-Band and Wideband Digital Codecs, ITU-T Recommendation P.830, February 1996.
17. Rix, A.W. and Hollier, M.P., *The Perceptual Analysis Measurement System for Robust End-to-End Speech Quality Assessment*, IEEE ICASSP, June 2000.

12

Noise Canceling Headsets for Speech Communication

Lars Håkansson, Sven Johansson, Mattias Dahl, Per Sjösten, and Ingvar Claesson

CONTENTS

Introduction

Headsets for speech communication are used in a wide range of applications. The basic idea is to allow hands-free speech communication, leaving both hands available for other tasks. One typical headset application is aircraft pilot communication. The pilot must be able to communicate with personnel on the ground and at the same time use both hands to control the aircraft.

A communication headset usually consists of a pair of headphones and a microphone attached with an adjustable boom. Headphone design varies considerably between different manufacturers and models. In its simplest form, the headphone has an open construction providing little or no attenuation of the environmental noise. In headsets designed for noisy environments, the headphones are mounted in ear cups with cushions that provide some attenuation.

The microphone is designed primarily to pick up the speech signal, but if the headset is used in a noisy environment, the background noise will also be picked up and transmitted together with the speech. As a consequence, speech intelligibility at the receive end will be reduced, possibly to zero. To increase the speech-to-noise ratio (SNR), it is common to use a directional microphone that has a lower sensitivity to sound incident from other directions than the frontal direction. In addition to this, the microphone electronics are usually equipped with a gate function that completely shuts off the microphone signal if its level drops below a threshold value. The purpose of the gate is to open the channel for transmission only when a speech signal is present.

Headsets are frequently used in noisy environments where they suffer from problems of speech intelligibility. Even if an ear cup–type headset is used, the attenuation is relatively poor for low frequencies. Low-frequency noise has a masking effect on speech, which significantly reduces the speech intelligibility.[1,2] Several cases have been reported in which the sound level of the communication signal was increased to hazardous levels by the user to overcome this low-frequency masking effect.[1,2] Ear exposure to the communication system resulted in hearing damage, such as hearing loss, tinnitus, and hyperacusis.

Passive Headsets

This section discusses headsets based on traditional passive ear defenders. Basic theory for passive ear defenders is introduced and practical issues that influence the performance of passive headsets are presented.

Traditional passive methods to attenuate noise employ barriers to block sound transmission and sound-absorbing materials to absorb the sound energy.[2] A passive ear defender — circumaural or closed-back headset — is based on a rigid ear cup containing sound-absorbing material.[2,3] Principally two ear cups are sealed to the users head via cushions by a spring band over the head. Passive ear defenders may be equipped with loudspeakers and boom- or throat-mounted microphone to provide one- or two-way communication. A closed back passive headset for two-way communication is shown in Figure 12.1.

The transmission ratio, $p_i(f)/p_e(f)$, for a passive headset at the frequency f[Hz] is given by:[3]

$$\frac{p_i(f)}{p_e(f)} = \frac{K_a}{K_a + K_c - (2\pi f)^2 M + j2\pi f R} \tag{12.1}$$

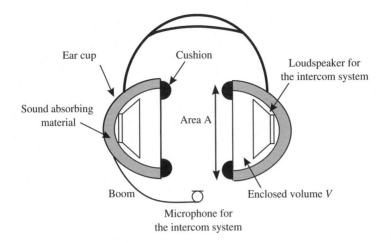

FIGURE 12.1
A closed-back passive headset for two-way communication.

where $p_i(f)$ is the sound pressure inside the ear cup, $p_e(f)$ is the external sound pressure acting on the ear cup, M is the mass of the rigid ear cup, R is the damping in the cushion, and K_c is the mechanical stiffness of the cushion. The mechanical stiffness of the air inside the headset K_a is given by:

$$K_a = \frac{A^2 c_0^2 \rho_0}{V} \quad [N/m] \tag{12.2}$$

Here c_0 and ρ_0 are the speed of sound in air and the density of air at normal temperature and pressure, respectively. A is the area of the plane surface enclosed by the external curvature of the headset cup where it is attached to the cushion and V is the volume of air enclosed by the headset cup. Based on parameters originating from Shaw et al.,[3] the frequency response of the transmission ratios, $p_i(f)/p_e(f)$, for two well-designed passive headsets with different values of cushion damping have been calculated and are shown in Figure 12.2.

A passive headset produces good attenuation of noise at frequencies above its eigenfrequency, as Figure 12.2 illustrates. To maximize the attenuation of noise at lower frequencies, the parameters of the headset need to be chosen appropriately. From Equation (12.1), it follows that the transmission ratio below the eigenfrequency is given approximately by $K_a/(K_a + K_c)$. Hence to reduce transmission of noise at low frequencies, the mechanical stiffness of the air inside the headset K_a should be small, while the mechanical stiffness of the cushion K_c should be large. From Equation (12.2), it follows that small K_a can be achieved by increasing the air volume inside the headset cup and decreasing the area A, of the plane surface enclosed by its external curvature. Both approaches to reducing K_a, however, have practical limits. Since a

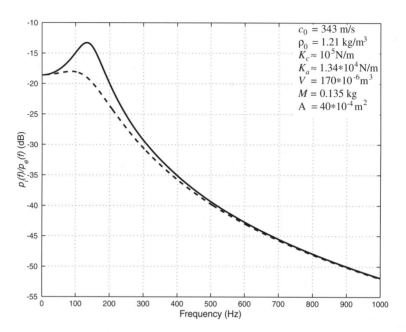

FIGURE 12.2
The transmission ratio between the internal and external sound pressure for two well-designed passive headsets; solid line $R = 70$ Ns/m and dashed line $R = 140$ Ns/m. The eigenfrequencies of the two headsets shown in this figure are at frequencies of about 100 and 130 Hz.

headset cup must often be designed to fit most people, the minimum area of the plane surface of the headset cup is limited by the fact that the ear cup must fit most ears. Furthermore, increasing the air volume inside the headset cup must also be limited so that the headset does not become too bulky and uncomfortable to wear.

As mentioned above, the mechanical stiffness of the cushion K_c is also a factor that influences the low-frequency performance of the headset. By selecting a cushion material with larger mechanical stiffness, the headset should theoretically produce greater low-frequency attenuation. However, in practice, the stiffness acting on the ear cup due to the cushion is limited by the layer of flesh underlying the cushion. Furthermore, as the stiffness of the cushion increases, not only does the headset become more uncomfortable to wear but its ability to provide good sealing around the ear also decreases. As a consequence of this, air leakage around the ear increases which prevents further improvement in the low-frequency noise attenuation being realized.

As discussed above, a passive headset designed to provide good low-frequency noise attenuation is likely to cause the wearer considerable discomfort due to its size, weight, and cushion stiffness. Active noise control techniques do not have the same limitations and have been proven to be very successful at improving the low-frequency attenuation achievable with headsets while at the same time allowing them to be comfortable to wear.[4]

Active Noise Control Headsets

As discussed above, bulky headsets are required in order to attenuate low-frequency noise through traditional passive methods.[3] Active approaches can complement these passive methods, and by combining these two approaches, high noise attenuation over a wide frequency range is made possible. Indeed, attenuation can be achieved over the entire audible frequency range from 30 Hz up to 20 kHz.[4,5]

This section discusses active control techniques for headset applications. Both analog and digital controllers are introduced as well as their combination. Analog feedback controllers are covered first, followed by a discussion of digital control algorithms. Finally, the combination of analog feedback controllers and digital controllers, which may be of either feedback or feedforward type, are discussed.

Active noise control is based on the principle of destructive interference between two sound fields, one sound field originating from the primary noise source, e.g., an engine, and the other generated by a secondary sound source such as a loudspeaker.[6] The loudspeaker produces a sound field of equal amplitude and opposite phase — 180° out of phase — to the unwanted sound field. The accuracy of the amplitude and phase of the generated sound field, the antisound, determine the noise attenuation achievable.

Active noise control works best on low-frequency sounds where the acoustic wavelengths are large compared to the space in which the noise is to be attenuated. In such a case, the antisound is approximately 180° out of phase in the whole space.[6] In general, the closed cavity within the ear cup of a headset and the eardrum is small compared to the wavelengths of sounds for which passive noise cancellation is poor and active techniques are of interest. Using such active control methods, attenuation of noise at frequencies below approximately 1 kHz by up to 20 dB has been achieved.[4,5] Such active control systems have been based on analog and/or digital techniques,[4,6] and both approaches are discussed in the following sections.

Analog Active Noise Control Headsets

Today, most commercial active headsets are based on analog feedback control technology. This type of headset typically includes a loudspeaker, an error microphone, and an analog control unit. The error microphone is generally placed as close as possible to the ear canal, since the objective of the active control is principally to minimize the perceived sound pressure.

The sound pressure under control, $p_c(f)$, inside an analog hearing protector can be written as:

$$p_c(f) = \frac{p_i(f)}{1 + KC(f)H(f)} \tag{12.3}$$

where $p_i(f)$ is the sound pressure inside the analog hearing protector without control, K is the amplifier gain, $H(f)$ is the frequency function of the compensation filter, and $C(f)$ is the frequency function of the control path, i.e., the transfer path comprising the loudspeaker, headset cavity, and error microphone. By letting the amplifier gain K assume large values, the magnitude of the denominator in Equation (12.3) becomes large and the sound pressure under control approaches zero.

In practice, however, the performance of an active feedback control system is limited by closed-loop stability requirements, i.e., the Nyquist stability criterion.[7] Physical paths, such as the electro-acoustic response of the loudspeaker and the acoustic path from the loudspeaker to the microphone, introduce time delay due to propagation time, and this will introduce increasing phase shift with frequency and thus limit the performance of the control system. As the net phase shift in the electro-acoustic response of the loudspeaker and the acoustic path from the loudspeaker to the microphone approaches 180°, the feedback becomes positive and the magnitude of the open loop frequency response $KC(f)H(f)$ for the feedback control system must be less than one in order to remain stable. Thus, the frequency range of the control system where the loop gain $|KC(f)H(f)|$ can be large is usually upper limited by the frequency where the net phase shift in the electro-acoustic response of the loudspeaker and the acoustic path from the loudspeaker to the microphone approaches 180°.[7] By using a compensation filter to provide phase lag compensation, the low-frequency loop gain of the feedback control system may be increased as the phase-lag filter attenuates the high-frequency gain. In this way the gain margin (i.e., the maximum factor by which the open loop frequency response for the feedback control system can be amplified without the feedback control system becoming unstable[6,7]) of the open loop frequency response for the feedback control system can be improved, and the phase shift added by the compensation filter can be minimized.[6,7]

Although different compensation filter designs have been reported,[8,9] they have not been described in detail for commercial reasons. It is clear, however, that since the cavity enclosed by the headset is likely to vary between different users, it is important to ensure that the design of the controller used in the active headset is robust to such variations.[6] Robustness of digital controllers regarding variations in the control path is discussed in "Digital Active Noise Control Headsets."

To enable radio communication via the headset loudspeaker of an analog active noise control communications headset, the communication signal may be injected between the error microphone and the compensation filter, as shown in Figure 12.3. This results in a sound pressure under control, $p_c(f)$, given by:[8]

$$p_c(f) = \frac{p_i(f)}{1 + KC(f)H(f)} + \frac{KC(f)H(f)p_s(f)}{1 + KC(f)H(f)} \tag{12.4}$$

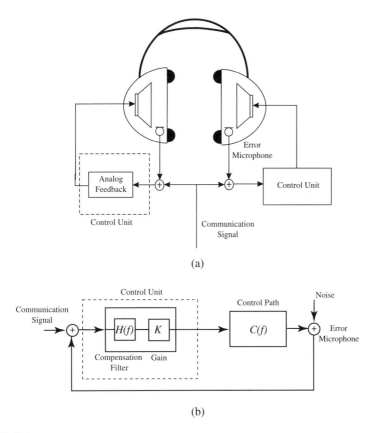

(a)

(b)

FIGURE 12.3
(a) Analog active noise control headset for two-way communication and (b) the corresponding block diagram of the feedback control system with communication signal injection.

Here $p_s(f)$ is the sound pressure due to the communication signal, $p_i(f)$ is the sound pressure inside the analog hearing protector without control, K is the amplifier gain, $H(f)$ is the frequency function of the compensation filter, and $C(f)$ is the frequency function of the control path. If the amplifier gain K is chosen to be large to produce a large loop gain, $|KC(f)H(f)| \gg 1$, the sound pressure under control, $p_c(f)$, is given approximately by:

$$p_c(f) \approx \varepsilon + p_s(f), \quad \text{where } |\varepsilon| \ll 1 \tag{12.5}$$

Hence, the influence of the feedback control on the communication signal $p_s(f)$ is reduced significantly, and, in addition, the distortion generally introduced in this signal by filtering caused by the loudspeaker and headset cavity is also reduced.

Closed-back headsets can be uncomfortable to wear, especially if the requirements are such that they have to be worn continuously for considerable periods of time. For such a headset, heat building up in the acoustic cavity and the pressure to maintain the acoustic seal around the ear necessary to provide passive attenuation can cause substantial discomfort. In contrast, an open-backed headset design, such as the headset type typically used in combination with portable tape and compact disk (CD) players, offers a more comfortable headset solution. This type of headset design does not, however, provide any passive high-frequency attenuation. In addition, it allows large variability in the acoustic path between the loudspeaker and error microphone.[4,10] Thus, the performance of an active open-backed headset typically falls short of that achievable with an active closed-back headset.[4,10]

Analog active noise control headsets typically produce an attenuation of about 20 dB at 100 to 200 Hz, which falls to zero below approximately 30 Hz and above approximately 1 kHz.[5] Higher attenuation of narrowband noise components may be achieved with nonadaptive analog active noise control headsets by using a more sharply tuned compensation filter.[6] However, as narrowband noise components may be time variable and are likely to differ between different environments, an adaptive controller is preferable. Since it is difficult and expensive to implement adaptive controllers that are analog, digital controllers tend to be used for the control of nonstationary and stationary narrowband noise.[6]

Digital Active Noise Control Headsets

This section covers digital active noise control headsets based on both adaptive feedforward and adaptive feedback control algorithms for active noise control headsets. The discussions of both types of algorithms are based on the well-known filtered-x least mean squares (FXLMS) algorithm. Since this algorithm was originally defined for feedforward control applications, this section begins by introducing active noise control headsets of the feedforward type.

Feedforward Control Systems

Feedforward control systems are theoretically more robust than feedback control systems. Feedback controllers are generally designed based on a model of the system to be controlled.[7] Variation in the control path may cause the feedback to become positive and lead to instability of the control system, i.e., the Nyquist stability criterion is violated.[7] In contrast to feedback control systems, a feedforward system is not based on a feedback control signal that may introduce this positive feedback and thereby instability.

In contrast to the feedback systems discussed above and in "Feedback Control Systems," feedforward control systems rely on the availability of a reference signal that contains information about the frequency content of the noise to be controlled.[11] The attenuation achievable is related to the amount of information about the noise to be controlled in the reference signal. The reference signal is processed by an adaptive digital control system prior to feeding the loudspeaker. For the control of broadband noise, a broadband reference signal can be provided to the digital controller by a microphone mounted on the exterior of the ear cup. For reducing tonal noise, e.g., generated by engines and propellers, the reference microphone can be replaced by a nonacoustic reference sensor, e.g., a tachometer or an optical or inductive sensor.[11,12] In this case, the periodic reference signals can be produced internally within the digital controller by using the output signal from the non-acoustic reference sensor.

There are several advantages in using nonacoustic sensors. For example, the reference signals based on such sensors will contain only the tonal components that are desired to be controlled and the properties of the reference signals, i.e., frequency and signal power, are known. With reference signals generated in this manner, the adaptive control becomes extremely selective. It is possible to determine which frequencies are to be controlled and which are not. Compared with a reference microphone, a nonacoustic sensor usually results in a reference signal with a lower noise level, resulting in higher performance.[6,11] In addition, undesired acoustic feedback from the loudspeaker to the reference sensor, which can cause instability, is also eliminated. In this case, the controller is purely feedforward, i.e., its performance is completely unaffected by the action of the loudspeaker.

Acoustic feedback, experienced when using a reference microphone, can be reduced by electronic techniques. However, this approach requires the control system to have a complicated structure in order to compensate for the acoustic feedback.[6,11] The acoustic feedback problem is particularly important for open-back headsets, where significant coupling between the loudspeaker and the reference microphone is likely to be present. For stationary noise, i.e., noise whose statistical properties are time invariant, a convenient estimate of the maximum noise suppression achievable by an active feedforward noise control system in decibels (dB) is given by $-10\log_{10}(1-\gamma_{xd}^2(f))$,[6,11] where $\gamma_{xd}^2(f)$ is the coherence function[6] between the uncontrolled noise and the reference signal, which is a measure of the linear relationship between them. For a coherence $\gamma_{xd}^2(f)$ of 0.99, an attenuation of potentially 20 dB may be achieved.

In broadband active feedforward control it is important that the causality condition is fulfilled, i.e., the delay introduced by the controller plus the control path does not exceed the acoustic delay from the reference microphone to the error microphone.[6,11] Consequently, in a headset application, feedforward control of broadband noise requires that the reference microphone be positioned such that it picks up the acoustic noise sufficiently in advance of its arrival at the ear to allow time for the processing of the noise signal and for it to be fed to the loudspeaker.

To fulfill the causality condition, a reference microphone mounted on an adjustable boom attached to the headset may be used. By manually tuning the microphone boom to point toward the noise source, the acoustic noise can be arranged to arrive at the error microphone in advance of its arrival at the ear and hence the causality condition to be fulfilled. However, as the headset user moves around, the direction of the microphone boom has to be continuously adjusted to point toward the noise source in order to continue to fulfill the causality condition. Unfortunately, both this requirement for continual manual adjustment of the boom, as well as the need for the boom microphone in the first place, may cause an active headset based on a digital controller for broadband applications to be impractical to use. If the causality constraint is *not* fulfilled, the system can efficiently only reduce more deterministic noise, e.g., tonal noise, for which it is always possible to find a correlation.

We will continue our discussion on feedforward active noise control headsets by introducing some important feedforward adaptive control algorithms suitable for active noise control headsets.

The human ear responds mainly to the mean square value of the pressure it perceives. Consequently, the "quantity" or "cost" function that most adaptive active control systems are designed to minimize is the mean square value of the error microphone signal, which is proportional to the acoustic energy.[6,11]

A digital feedforward control system is illustrated in Figure 12.4. In both broadband and narrowband applications, the control filter is commonly based on a transversal filter, i.e., finite impulse response (FIR) filter, steered by the well-known FXLMS algorithm. This algorithm is developed from the least mean squares (LMS) algorithm and is based on a gradient search method that relies on the optimization technique known as the *method of steepest descent*.[6,11] The FXLMS algorithm is given by:[6,11]

$$y(n) = w^T(n)x(n)$$

$$e(n) = d(n) + y_C(n) \tag{12.6}$$

$$w(n+1) = w(n) - \mu x_{\hat{C}}(n)e(n)$$

where μ is the adaptation step size and

$$x_{\hat{C}}(n) = [\sum_{i=0}^{I-1} \hat{c}(i)x(n-i), \ \ldots \ , \sum_{i=0}^{I-1} \hat{c}(i)x(n-i-M+1)]^T \tag{12.7}$$

is the filtered reference signal vector, which usually is produced by filtering the reference signal $x(n)$ with an FIR-filter estimate $\hat{c}(i)$, $i \in \{0,1,\ldots,I-1\}$ of the physical path between the loudspeaker and error microphone (i.e., the control path), and M is the length of the adaptive FIR filter, $y(n)$ is the output signal from the control filter, $w(n) = [w_0(n),\ldots,w_{M-1}(n)]^T$ is the control filter weight vector, $x(n) = [x(n),\ldots,x(n-M+1)]^T$ is the reference signal vector, $d(n)$ is the noise to be controlled, $e(n)$ is the estimation error, i.e., error microphone signal, and $y_C(n)$ is the output of the control path.

In practice, the elements in the filtered reference signal vector $x_{\hat{C}}(n)$ are produced by filtering the reference signal, $x(n)$, with an FIR-filter estimate

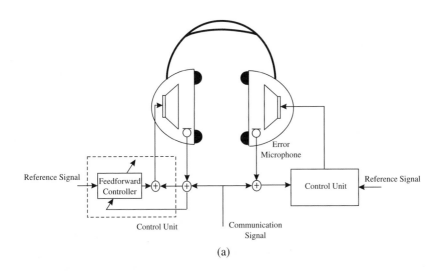

(a)

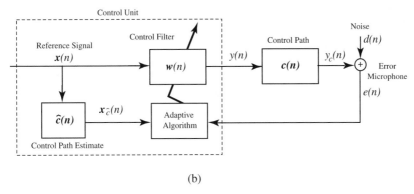

(b)

FIGURE 12.4
(a) Feedforward digital active noise control headset for two-way communication and (b) the corresponding block diagram of the adaptive feedforward control system.

of the control path $\hat{c}(i)$, $i \in \{0,1,\ldots,I-1\}$ and the product of this filtered reference signal vector and the estimation error usually produces the FXLMS algorithm's gradient estimate, i.e., $x_{\hat{c}}(n)e(n)$. However, the gradient estimate in the FXLMS algorithm's weight vector adjustment algorithm is by definition based on a filtered reference signal that is produced by filtering the reference signal $x(n)$ with the actual impulse response of the control path.[6,11] As a result, the filtered reference signal vector will be an approximation, and differences between the estimate of the control path and the true control path influence both the stability properties and the convergence rate of the algorithm.[6,11] Differences between the estimate of the control path and the actual control path will influence the gradient estimate used in the algorithm, i.e., $x_{\hat{c}}(n)e(n)$, and this will cause the algorithm to adjust its coefficient vector in a direction that is biased compared with the direction of steepest descent.[6,11]

The algorithm is robust, however, to errors in the estimate of the control path.[6,11] For example, in the case of narrowband reference signals, the algorithm will converge even for phase errors in the estimate of the control path of up to 90° provided that the step size μ is sufficiently small.[6,11] Furthermore, phase errors smaller than 45° will have only a minor influence on the algorithm convergence rate.[6,11]

In order to ensure stable action of the FXLMS algorithm, it has been found that the step size μ should be selected according to:

$$0 < \mu < \frac{2}{(\Delta + M) E[x_{\hat{c}}^2(n)]} \tag{12.8}$$

where $E[x_{\hat{c}}^2(n)]$ is the power of the filtered reference signal and Δ is the number of samples corresponding to the overall delay in the control path.[6,11] In practice, however, the power of a reference signal obtained by, for example, a reference microphone, might be time varying. If the reference signal has a time-varying power, it follows from Equation (12.8) that the upper limit for the step size μ is time varying. Variations in reference signal power influence the performance, e.g., the stability and convergence speed, of the FXLMS algorithm.[6,11] A common way to improve the performance of the FXLMS algorithm, regarding variations in the power of the reference signal, is to replace the fixed step size μ with a time varying step size μ(n) in the FXLMS algorithm (Equation [12.6]) according to:

$$\mu(n) = \frac{\mu_0}{\varepsilon + M \hat{E}[x_{\hat{c}}^2(n)]} \tag{12.9}$$

Here μ_0 is a step-size parameter typically less than two, $\hat{E}[x_{\hat{c}}^2(n)]$ is an estimate of the power of the filtered reference signal, and ε is a small positive number added in order to avoid division by zero if $\hat{E}[x_{\hat{c}}^2(n)] = 0$. By using the time-varying step size given by Equation (12.9) in the FXLMS algorithm, the normalized FXLMS algorithm is obtained.[11]

The mean power of the filtered reference signal vector can be updated according to different update laws.[11] One recursive update law for estimating the signal power is given by[11]

$$\hat{E}[x_{\hat{c}}^2(n+1)] = \hat{E}[x_{\hat{c}}^2(n)] + \frac{x_{\hat{c}}^2(n+1) - x_{\hat{c}}^2(n-L+1)}{L} \tag{12.10}$$

where L is the block length.

The stability and convergence properties of the FXLMS algorithm are related to errors in the estimate of the control path. One efficient way to improve the robustness to errors in the estimate of the forward path is to use the leaky FXLMS algorithm that is defined by[6,11]

$$w(n+1) = \gamma w(n) - \mu x_{\hat{c}}(n)e(n) \tag{12.11}$$

where γ is a real positive leakage factor, $0 < \gamma < 1$.

Some applications of the FXLMS algorithm require long-control FIR filters, for example, where tonal components that are close in frequency are to be controlled. Long-control FIR filters result in slow convergence of the adaptive algorithm and a considerable computational burden.[11] To improve the convergence speed and reduce the computational burden in such narrowband control applications, the complex FXLMS algorithm described below may be used.

The capacity of an adaptive control system to handle tonal components that are close in frequency depends on the structure of the controller, i.e., how the multiple frequencies are processed. Suitable controllers are generally based on either a single-filter structure or a parallel-filter structure using several filters.[11] The single-filter structure is based on a composite reference signal containing all frequencies to be controlled. For tonal components that are close in frequency, a long-control FIR filter is required resulting in slow convergence of the adaptive algorithm as mentioned above.[11] For the parallel-filter structure, each frequency component is individually processed. This enables shorter filters, and thereby better convergence performance, to be achieved. If possible, therefore, the parallel structure rather than the single-filter structure should be used to achieve efficient and robust control of frequencies that are close together.[12] An example of such a noise field is the beating sound produced by propellers rotating at slightly difference speeds.

An alternative approach to the FIR-based control system for controlling tonal noise is a system based on complex arithmetic.[12] Here each frequency is controlled by an adaptive complex weight. The complex FXLMS algorithm is based on a recursive weight adjustment which is made for each tone required to be controlled, i.e., for each frequency f to be controlled the complex adaptive weight $w_f(n)$ is updated according to[12]

$$w_f(n+1) = w_f(n) - \mu_f x_f^*(n)\hat{C}_f^* e(n) \tag{12.12}$$

where $x_f(n)$ is a complex scalar reference signal at the frequency f, $\hat{C}_f$ is a complex control path estimate corresponding to the frequency f, $(\cdot)^*$ denotes the complex conjugate, μ_f is the adaptation step size at the frequency f, and $e(n)$ is the broadband error microphone signal.[11,12] The output signal from the parallel adaptive filter is produced by $y(n) = \sum_{f \in F} \mathcal{R}\{w_f(n)x_f(n)\}$, where F is the set of controlled frequencies and $\mathcal{R}\{\cdot\}$ denotes the real part of the complex quantity $w_f(n)x_f(n)$. In a practical implementation, $\mathcal{R}\{\cdot\}$ implies that only the real part is evaluated.

For the same reason as in the case of the FXLMS algorithm of reducing susceptibility to reference signal power variations, it can be important to introduce a normalized version of the complex FXLMS algorithm by using a time-varying adaptation step size

$$\mu_f(n) = \frac{\mu_0}{\varepsilon + \hat{E}[\,|\,x_f(n)\,|^2\,]\,|\,\hat{C}_f\,|^2} \qquad (12.13)$$

in Equation (12.12). As in the case of the normalized FXLMS algorithm, the estimate of mean power of the reference signal $\hat{E}[\,|\,x_f(n)\,|^2\,]$ may be updated according to different update laws.[11]

Feedback Control Systems

As discussed above, feedforward control systems rely on the existence of some prior knowledge of the noise to be controlled. This knowledge is provided by a reference signal that drives the control loudspeaker through the controller. Generally speaking, the ideal active controller is of this feedforward type provided, of course, that a reference signal highly correlated with the undesired acoustic noise is available.[6,11]

Active feedforward control is typically well suited to applications where it is simple and practical to obtain a reference signal of the noise requiring cancellation. Such a situation is typically found in helicopter and aircraft cockpits, and consequently the active headsets used in these environments are often of the feedforward type. In some headset applications, however, the generation of a suitable reference signal may be impractical or too costly. For example, there might be a large number of different noise sources with reference signals produced by tachometers or optical or inductive sensors, all of which have to be fed to the headset controller.

In such a situation, the use of feedback rather than feedforward control, for which no reference signal is required, has an obvious advantage. Furthermore, although feedforward control systems are theoretically more robust than feedback control systems (see "Feedforward Control Systems"), the performance of the feedforward controller is highly dependent on the *quality* of the reference signal, and in many cases a feedback system may perform equally well or better than a system with feedforward control.

The performance of a feedback controller in broadband applications is largely determined by the delay in the feedback loop. To obtain high performance, a small delay in the feedback loop is required.[6] This delay affects the length of the prediction interval of the controller, i.e., how far into the future the controller has to produce an estimate of the error signal.[6] Due to the inherent delay in digital controllers associated with their processing time, A/D- and D/A-conversion processes, analog anti-aliasing, reconstruction filtering, and analog feedback controllers, as discussed in "Active Noise Control Headsets," are usually preferred for use in broadband applications. For example, the hardware cost of a fast digital controller that introduces a

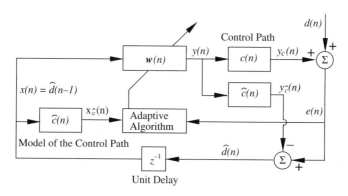

FIGURE 12.5
Block diagram of internal model control (IMC) controller based on an adaptive control FIR filter steered by the FXLMS.

delay comparable to or below that of an analog system makes an analog controller a more cost efficient solution for broadband control problems.[6]

However, in narrowband applications the performance of a feedback controller is less sensitive to delays in the control loop, since narrowband signals exhibit more deterministic behavior. Furthermore, as narrowband noise components may be time variable and are likely to differ between different environments, adaptive control is often required. Consequently, in the case of nonstationary narrowband noise, a digital feedback controller, which can provide adaptive control more easily and cost effectively than an analog controller, is preferable. For example, at the error microphone of an active noise control headset an adaptive digital feedback controller may provide up to 20 dB more attenuation of the narrowband noise than can be achieved with an analog controller.[6] An active noise control headset for use in a variety of environments involving both broadband and narrowband noise is thus likely to be one which involves both an adaptive digital feedback controller and an analog feedback controller.[6]

An adaptive digital feedback controller suitable for use in active noise control headsets is the internal model control (IMC) controller based on an adaptive control FIR filter steered by the FXLMS algorithm.[6] In Figure 12.5, a block diagram of this adaptive IMC controller is shown. The adaptive IMC controller algorithm is obtained by adding the two equations

$$\hat{d}(n) = e(n) - \sum_{i=0}^{I-1} \hat{c}(i)y(n-i)$$

$$x(n) = \hat{d}(n-1)$$

(12.14)

to those defining the FXLMS algorithm (see Equation [12.6]). Here, $\hat{d}(n)$ is an estimate of the noise to be controlled, $\hat{c}(i), i \in \{0, 1, \ldots, I-1\}$ is an FIR-filter estimate of the control path between the loudspeaker and error microphone,

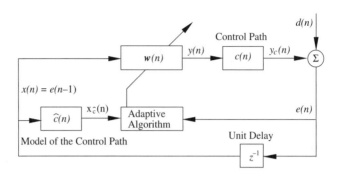

FIGURE 12.6
Block diagram of the feedback FXLMS controller.

$y(n)$ is the output signal from the control filter, and $e(n)$ is the error microphone signal. The relationship between the delayed estimate of the noise to be controlled $\hat{d}(n-1)$ and the reference signal $x(n)$ describes the fact that we are dealing with an adaptive digital filter in a feedback control system.

This type of controller is based on feedback cancellation; it uses an estimate of the control path to cancel the feedback.[6] If the controller is provided with a good estimate of the control path, it will act as an adaptive feedforward predictor.[6] However, differences between the estimate of the control path and the actual control path reduce the stable region of operation of the adaptive control system and this algorithm may suddenly become unstable.[6] Using the leaky FXLMS algorithm (Equation [12.11]) instead of the FXLMS algorithm (Equation [12.6]) in the IMC controller improves robustness to differences between the estimate of the control path and the actual control path.[6] This is an important issue, especially when it comes to open-backed headsets, as the control path may be subject to large variations in this type of headset, as discussed in "Analog Active Noise Control Headsets."[10]

Another adaptive feedback controller that might be suitable for narrowband applications is the feedback FXLMS algorithm. A block diagram of such a controller is shown in Figure 12.6. This algorithm is defined by adding Equation (12.15)[13]

$$x(n) = e(n-1) \tag{12.15}$$

to the equations defining the FXLMS algorithm (see Equation [12.6]). Here the relation between the delayed error signal $e(n-1)$ and the reference signal $x(n)$ describes the fact that we are dealing with an adaptive digital filter in a feedback control system. As in the case of the IMC controller (see Equation [12.14]), a leakage factor in the weight adjustment equation — the leaky FXLMS algorithm (Equation [12.11]) is used instead of the FXLMS algorithm (Equation [12.6]) in the feedback FXLMS algorithm — improves the robustness of the algorithm to differences between the estimate of the control path and the actual control path.[13] In contrast to the IMC controller (Equation [12.14]), this algorithm does *not* rely on cancellation of the feedback path.

Since the feedback FXLMS algorithm does not rely on cancellation of the feedback path, it might be argued that this algorithm is likely to be favorable compared with the IMC controller in circumstances where the control path shows large variability.

As mentioned in "Analog Active Noise Control Headsets," analog active noise control headsets typically produce an attenuation of about 20 dB at 100 to 200 Hz that falls to zero below approximately 30 Hz and above approximately 1 kHz. Combining such analog feedback control with the digital feedback controllers described above will result in broadband attenuation of the noise due to the former and further reduction of nonstationary narrowband noise components due to the latter. Thus, using digital feedback control in combination with analog feedback control is likely to result in the best overall performance active noise headsets.[6]

In such a combined system, the digital adaptive feedback controller is implemented as an outer control loop of the analog feedback control system.[6] When such a digital feedback loop is used as part of a headset communication system, degradation of the communication signal may occur since the digital feedback controller is likely to affect the communication signal. Selecting the adaptation step size μ sufficiently small, and thereby prohibiting the adaptive filter from tracking the nonstationary speech signal, might allow a communication signal of sufficient quality to be achieved while simultaneously maintaining the attenuation of the narrowband noise.

Hybrid Active Noise Control Headsets

An active control system that combines feedforward and feedback techniques is usually called a *hybrid* active control system.[11] Such a system can provide narrowband as well as broadband noise attenuation and can be combined with open- or closed-back headsets. The principle of a hybrid active noise control headset is illustrated in Figure 12.7. Hybrid active noise control can be used to improve the noise attenuation achievable within an environment which has dominant low-frequency tonal noise embedded in broadband noise. Such environmental noise dominates that found in the interior cabins of propeller aircraft and helicopters. Compared with an analog feedback controller giving broadband noise attenuation, an adaptive digital feedforward controller is likely to produce greater attenuation of the nonstationary tonal noise components.

A hybrid system can be based on either digital technology or a combination of analog and digital technologies. The performance of a closed-back hybrid headset based on a nonadaptive analog feedback controller combined with a digital adaptive feedforward controller is shown in Figure 12.8(a).[14] In this figure the dominant low-frequency tones correspond to the fundamental blade passage frequency, i.e., rotational speed of the propeller axis multiplied with the number of the propeller blades, and the harmonics of the main rotor blade passage frequency of the helicopter. If only a passive commercial headset is used, limited low-frequency noise attenuation can be obtained, as

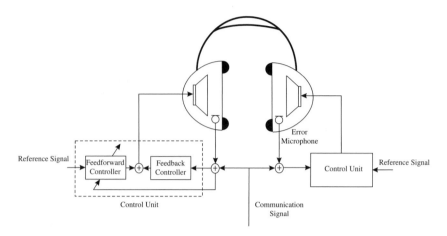

FIGURE 12.7
The principle of a closed-back hybrid headset based on digital feedforward and analog feedback active noise control techniques with communication signal injection.

illustrated in Figure 12.8(b). By combining the passive headset with analog feedback active noise control, an improved overall low-frequency performance can be obtained, as shown in Figure 12.8(c). A combination of feedforward and feedback control results in significant attenuation of both the tonal components and low-frequency broadband noise, as illustrated in Figure 12.8(a).

Speech Enhancement for Headset Applications

Thus far, this chapter has focused on passive and active control of the noise inside the ear cups of headsets. Although these techniques reduce the environmental noise for the headset wearer, in the case of a communications headset, the noise picked up by the intercom microphone remains and reduces communication quality. This section focuses on two techniques for reducing the amount of background noise picked up by the intercom microphone and transmitted with the speech: spectral subtraction and a new in-ear technique.

Spectral Subtraction

Spectral subtraction is a broadband noise reduction method well suited for use in speech communication systems in severe noise situations such as intercom systems in boats, motorcycles, helicopters, and aircraft.[15] It is an efficient and robust background noise reduction technique that can be used in combination with the conventional active noise control techniques

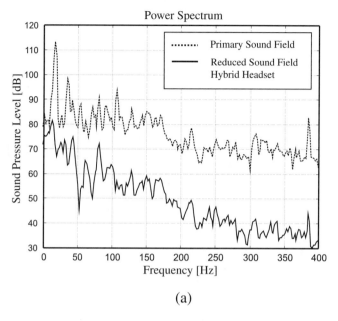

(a)

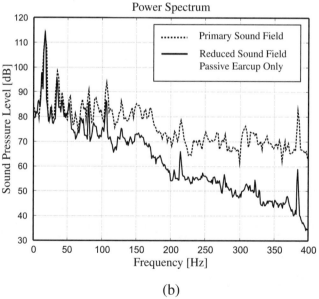

(b)

FIGURE 12.8
In (a–c), the sound pressure level of the interior helicopter cabin noise — AS332 "Super Puma" MKII helicopter — outside the ear cups is shown with a dashed line. (a) Solid line; reduced sound pressure level inside the ear cups after the *hybrid headset* has been switched on; (b) solid line; reduced sound pressure level inside the ear cups when only *passive* damping is applied.

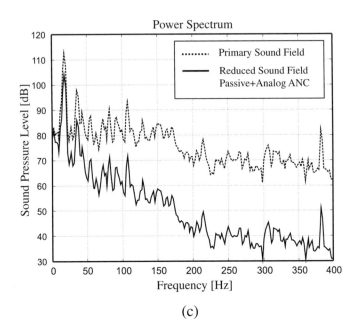

(c)

FIGURE 12.8 *Continued.*
(c) Solid line; reduced sound pressure level inside the ear cups after the *analog feedback* controller has been switched on.

discussed earlier in this chapter. Spectral subtraction is based on a fast Fourier transform subtraction technique.[15] Two steps are required to remove noise: Step 1 is a noise level estimation step based on data gathered during speech inactive periods, i.e., collection of information about the type of noise to be removed. Step 2 is a spectral subtraction step involving subtraction of the noise from the speech using the estimated noise level from Step 1. Figure 12.9 illustrates schematically this spectral subtraction technique.

The SNR improvement achievable with spectral subtraction techniques is normally substantial. From experience, a rule of thumb is that the SNR improvement in headset applications achievable is of the same order of magnitude as the SNR before reduction. One disadvantage of spectral subtraction is, however, that sometimes background distortion of the processed signal may occur in the form of musical tones.[15] Such distortion, if heard at all, depends on the type of spectral subtraction scheme used, the level of noise reduction achieved, and the degree to which the background noise is nonstationary.[16] Spectral subtraction is commonly used on the transmit side of the communication channel, but the technique can be used on the receive side as well.

In-Ear Microphone

A common approach to achieving good SNR in an ordinary communication headset is to mount the microphone on a boom close to the mouth. In severe

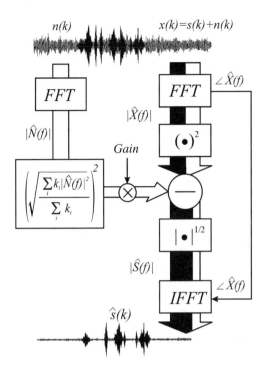

FIGURE 12.9
The spectral subtraction principle. The white parts of the arrows indicate the noise content and the black parts of the arrows indicate the desired speech. The upper waveform shows the original noise sequence and the lower waveform shows the spectral subtraction cleaned sequence. The left branch shows the noise estimation process, and the right branch the spectral subtraction stage.

noise environments, where sound pressure levels are high, alternative microphone techniques can be used to improve performance. One approach is to replace the ordinary communication microphone with a more sensitive microphone mounted in a foam plug that is placed in the auditory canal (auditory meatus). The foam plug itself can provide substantial reduction of the noise picked up by the microphone. Further noise reduction can be achieved by using a miniature active noise control system or by using an additional passive or active headset together with the in-ear microphone. Figure 12.10 shows an active noise control headset that uses such an in-ear microphone.

Many of the techniques for improving noise cancellation with ordinary headsets, passive and active or a combination thereof, which were covered earlier in this chapter, can also be applied to a headset incorporating an in-ear microphone. For example, as in the case of a passive headset, both the choice of sound-absorbing earplug material, as well as the plug-to-auditory-canal fit, are important noise-damping factors. Combining an in-ear microphone with an active noise control headset can be useful in various situations, civil as well as military, in order to enhance the speech intelligibility of the intercom system.[17]

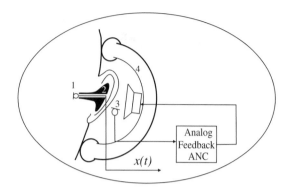

FIGURE 12.10
The ear microphone (1) is mounted in a plug (2). In this case, the ear microphone is combined with an analog active noise control (ANC) feedback headset (4) using a separate error microphone (3).

Conclusions

Headsets for speech communication are used in a wide range of applications. A communication headset usually consists of a pair of headphones and a microphone attached to the headset with an adjustable boom. In its simplest form, the headset has an open construction with little or no attenuation of the environmental noise. In headsets designed for noisy environments, the headphones are mounted in ear cups with cushions that provide some attenuation. The microphone is primarily designed to pick up the speech signal, but if the headset is used in a noisy environment, the background noise will also be picked up and transmitted together with the speech.

Passive headsets produce good attenuation of noise above their eigenfrequency, typically in the order of 40 dB above 500 Hz. Analog feedback active noise control techniques for use in headset applications have received considerable attention. These techniques have proved to be very successful at improving attenuation at frequencies below 1000 Hz by up to 20 dB and at the same time enabling comfortable noise canceling headsets to be designed. For the control of narrowband noise both digital feedback and digital feedforward controllers enable further attenuation to be achieved compared with analog feedback controllers. Digital feedback and feedforward controllers may be used in combination with analog feedback controllers or on their own.

In headset applications such as intercom systems, the background noise picked up by the boom microphone will be transmitted together with the speech. In order to enhance the speech transmitted by such systems, spectral subtraction is typically used. It is an efficient and robust broadband background noise reduction technique, which will not interfere with conventional active noise control systems used to improve the low-frequency attenuation of the headphone. Alternatively, the use of a headset microphone boom with spectral subtraction may be replaced by in-ear microphone techniques.

References

1. Backteman, O., Köhler, J., and Sjöberg, L., Infrasound-tutorial and review, *J. Low Freq. Noise Vibration*, 2(4), 176–210, 1983.
2. Beranek, L.L., *Acoustics*, 4th ed., American Institute of Physics, New York, 1993.
3. Shaw, E.A.G. and Thiessen, G.J., Acoustics of circumaural ear phones, *J. Acoustical Soc. Am.*, 34, 1233–1246, 1962.
4. Casali, J.G. and Robinson, G.S., Narrow-band digital active noise reduction in a siren-cancelling headset: real-ear and acoustical manikin insertion loss, *Noise Control Eng. J.*, 42(3), 101–115, 1994.
5. Crabtree, R.B. and Rylands, J.M., Benefits of active noise reduction to noise exposure in high-risk environments, *Proc. Inter. Noise'92*, 1992, 295–298.
6. Elliott, S.J., *Signal Processing for Active Control*, Academic Press, London, 2001.
7. Franklin, G.F., Powell, J.D., and Emami-Naeini, A., *Feedback Control of Dynamic Systems*, 3rd ed., Addison-Wesley, Reading, MA, 1994.
8. Carme, C., A new filtering method by feedback for A.N.C. at the ear, *Proc. Inter. Noise '88. Institute of Noise Control Engineering*, 1988, 1083–1086.
9. Bai, M. and Lee, D., Implementation of an active headset by using the H_∞ robust control theory, *J. Acoustical Soc. Am.*, 102, 2184–2190, 1997.
10. Rafaely, B., Garcia-Bonito, J., and Elliott, S.J., Feedback control of sound in headrest, *Proc. Active '97*, 1997, 445–456.
11. Kuo, S.M. and Morgan, D.R., *Active Noise Control Systems*, John Wiley & Sons, New York, 1996.
12. Johansson, S. et al., Comparison of multiple- and single-reference mimo active noise control approaches using data measured in a Dornier 328 aircraft, *Int. J. Acoustics Vibrations*, 5(2), 77–88, 2000.
13. Claesson, I. and Håkansson, L., Adaptive active control of machine-tool vibration in a lathe, *Int. J. Acoustics Vibrations*, 3(4), 155–162, 1998.
14. Winberg, M. et al., A new passive/active hybrid headset for a helicopter application, *Int. J. Acoustics Vibrations*, 4(2), 51–58, 1999.
15. Boll, S.F., Suppression of acoustic noise in speech using spectral subtraction, *IEEE Transactions on Acoustics, Speech, and Signal Processing*, ASSP-27(2), 113–120, 1979.
16. Gustafsson, H., Nordholm, S., and Claesson, I., Spectral subtraction, truly linear convolution and a spectrum dependent adaptive averaging method, *IEEE Transactions on Acoustics, Speech, and Signal Processing*, 2001.
17. Westerlund, N., Dahl, M., and Claesson, I., In-ear microphone equalization exploiting an active noise control headset, 30th International Congress and Exhibition on Noise Control Engineering, *Proc. Inter. Noise 2001*, 2001.

13

Acoustic Crosstalk Reduction in Loudspeaker-Based Virtual Audio Systems

Darren B. Ward

CONTENTS

Introduction

In everyday life, a person with normal hearing makes use of his or her directional hearing ability to help process information. For example, at a large social gathering where many people may be speaking simultaneously, a listener can focus on a single conversation without too much difficulty. This so-called "cocktail party phenomenon" is mainly due to the fact that we have two ears and are able to process information based on its origin in space.

The aim of virtual, or three-dimensional (3D), audio systems is to give a listener the impression of being immersed in a virtual acoustic environment. Such a system could provide the audio component of a virtual reality system. Besides its obvious use in entertainment, virtual audio has been found to greatly increase recognition in situations where personnel must monitor several voice communications channels simultaneously.[1] This is achieved by placing different voice channels at different locations in virtual space around the user.

In teleconferencing applications with multiple participants, 3D audio systems can be used to place the voice channels of different participants at different points in space around the listener, in particular, to match the position of the participant's video image on the screen. For example, the voice of the participant whose image appears at the top left of the listener's screen would come from above and to the left of the listener, and so on.

There are two basic approaches to virtual audio: binaural and soundfield. The aim of binaural systems is to reproduce only at the listener's ears the acoustic pressures that would result if the listener were physically present in the environment being virtualized. Binaural implicitly assumes that the listener is at a specific position in the virtual environment. The aim of soundfield systems is to reproduce within a large region of space the complete acoustic soundfield that would exist in the environment being virtualized. Any listener whose head is within this controlled region would therefore be physically located within a local acoustic environment that is identical to the virtual acoustic environment. One can think of the distinction between the two approaches as follows: soundfield attempts to recreate an acoustic event itself, whereas binaural merely attempts to reproduce the impression of the acoustic event. By its nature, binaural is generally suitable only for a single listener, whereas several listeners may use soundfield simultaneously (providing the reproduction region is large enough). Binaural can be produced using headphones or a small number of loudspeakers, whereas soundfield typically requires a large number of loudspeakers. In this chapter, we will focus on single-user binaural systems.

Binaural Processing

Consider a sound source at a point in space to a listener's left. When the sound arrives at the listener's ears there are three general effects. First, the sound arrives at the left ear before the right ear; this is called the interaural time difference. The sound will also be louder at the left ear compared to the right because of the shadowing effect of the listener's head; this is called the interaural intensity difference. Finally, the listener's external ear, head, and torso modify the sound before it arrives at the eardrums; this results in acoustic filtering of the sound.

For a sound source at a given point in space, there is a pair of acoustic transfer functions (TFs) from the source to the listener's ears. These are referred to as head-related transfer functions (HRTF). Each person has an individual set of HRTFs, determined by the size and shape of his or her head, torso, etc. These can be measured by placing microphones in the listener's ears and generating a known test signal from a point in space.

Repeating this for all points in space around a listener gives the complete set of HRTFs, parameterized by the source position. A set of HRTFs measured using a mannequin at the Massachusetts Institute of Technology (MIT) is available in the public domain,[2] and it can be used to design and implement 3D audio systems. A mono-audio channel can be positioned at a particular point in space by filtering it with the appropriate HRTFs (a process known as binaural spatialization) and then delivering these binaural signals to the listener's ears. For a review of binaural spatializers, refer to Reference 3. The remainder of this chapter will consider issues that arise when one attempts to deliver binaural signals to a listener using loudspeakers.

Crosstalk Cancellation Systems

In delivering binaural audio to a single listener, the aim is to control the acoustic pressure at each of the listener's ears as accurately as possible. This is trivial using headphones. If, however, loudspeakers are used to deliver the binaural signals, then additional signal processing is required. If the binaural signals are simply played through their respective loudspeakers, then the "crosstalk" signal that arrives at each ear from the opposite loudspeaker must be removed. Adaptive filter techniques could be used to minimize the binaural signal arriving at the opposite-side ear,[4] but this would require a microphone located close to the ear to provide a feedback path. If no feedback of the acoustic signal at the listener's ear is available (as will be the case most often in practice), one must assume a model for the transmission and use a prefiltering structure. This structure is called a crosstalk cancellation system (CCS) and was first introduced in the 1960s by researchers at Bell Laboratories.[5] With recent developments in digital signal processing hardware making its implementation more practical, there has been renewed interest in CCS over the past few years.

The general N-loudspeaker CCS is shown in Figure 13.1, and is described by the following linear system:

$$\begin{bmatrix} \hat{B}_L(f) \\ \hat{B}_R(f) \end{bmatrix} = \begin{bmatrix} A_{1,L}(f) & A_{2,L}(f) & \cdots & A_{N,L}(f) \\ A_{1,R}(f) & A_{2,R}(f) & \cdots & A_{N,R}(f) \end{bmatrix} \begin{bmatrix} H_1(f) & H_{N+1}(f) \\ H_2(f) & H_{N+2}(f) \\ \vdots & \vdots \\ H_N(f) & H_{2N}(f) \end{bmatrix} \begin{bmatrix} B_L(f) \\ B_R(f) \end{bmatrix} \quad (13.1)$$

$$\hat{\mathbf{b}}(f) = \mathbf{A}(f)\,\mathbf{H}(f)\,\mathbf{b}(f)$$

where $A_{1,L}(f)$ is the acoustic TF from loudspeaker l_1 to the left ear (and similarly for the other loudspeaker/ear combinations), $B_L(f)$ is the desired binaural signal to be delivered to the left ear (and similarly for the right ear),

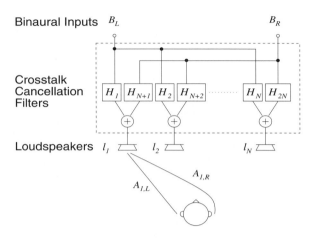

FIGURE 13.1
Block diagram of a general crosstalk cancellation system.

and $\hat{B}_L(f)$ is the actual signal received at the ear. The aim is to design the CCS filters $H_n(f)$, $n = 1,\ldots,2N$, to equalize the response from the binaural inputs to the ears. The solution that minimizes the least squares error is $\mathbf{H}(f) = \mathbf{A}^+(f)$ where $\mathbf{A}^+(f)$ denotes the pseudo-inverse of $\mathbf{A}(f)$. If the TF matrix $\mathbf{A}(f)$ was known, then designing the CCS filters is a trivial problem. However, the TF matrix will never be known in practice for a variety of reasons.

One of the main reasons for uncertainty in the acoustic TF matrix is that a 3D audio system will typically be used in a real room in which reverberation is present. Reverberation causes filtering of the sound source, and this filtering varies significantly from room to room, from point to point within the same room, and even with temperature. Although the reverberant TF from loudspeaker to ear could theoretically be measured and included in the design process, this is impractical for any CCS that is to be generically applicable. It has also been shown that attempting to equalize a reverberant acoustic channel results in an extremely nonrobust system.[6] Hence, in practice, a CCS is typically designed by considering direct-path acoustic TFs only (the effect of reverberation on such a design has been considered in Reference 7).

Minor variations in the acoustic TF matrix will also occur due to individualized HRTF and unknown loudspeaker TFs (although theoretically these could also be measured and included in the design process).

A more fundamental problem with designing the CCS filters is that, even if the TF matrix were known exactly, as soon as the listener moves slightly the equalizer ceases to be effective. Consider the following example. Assume that a CCS is designed with two loudspeakers placed at angles of ±30° relative to the center of the listener's head at a distance of 1 m, and the CCS filters are designed to equalize the acoustic TFs with the head in a specific position. Using this CCS, the resulting ear responses for the left binaural signal are shown in Figure 13.2 when the head moves laterally by 2 cm (dashed) and 4 cm (dotted). These simulations were performed using a

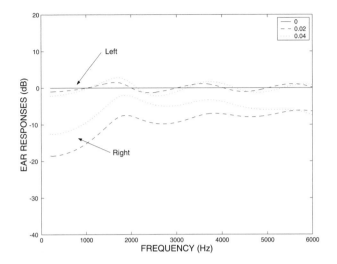

FIGURE 13.2
Effect of head movement on the performance of a CCS using loudspeaker angles of ±30°. Ear responses for the left binaural input are shown for the design position (solid), and when the head moves sideways by 2 cm (dashed) and 4 cm (dotted).

spherical head model[8] both for design of the CCS filters and for the resulting simulated ear responses; reverberation was not included in the simulation. Ideally, the left ear response should be unity and the right ear response should be zero, and indeed this is the case when the head is in the design position (denoted by the solid line). However, as the head moves away from the design position, the ear responses differ dramatically from their ideal values. The design position is often referred to as the "sweet spot."

In designing a CCS, one is therefore faced with the fact that the acoustic TFs between loudspeakers and ears can never be known exactly. The typical approach to deal with this is to assume a model for the acoustic TFs, and then design the CCS filters to equalize the corresponding modeled TF matrix. We will refer to this modeled matrix as the design TF matrix $A_0(f)$. The aim is to design the CCS filters so that they give good performance, not only for the design TF matrix $A_0(f)$, but also for slight unknown variations. Moreover, when the listener moves the filters must be updated to reflect a new design TF matrix for the new position of the listener's head.

In order to deal with the robustness problem, one would like to know how robust the equalizer is to unknown perturbation of the design TF matrix. One measure for this robustness is the condition number of $A_0(f)$, defined as the ratio of its largest and smallest eigenvalues. For a robust system the condition number will be small (ideally unity), and for a nonrobust system the condition number will be large. Hence, one should attempt to choose a design TF matrix with small condition number, since this will ensure that an equalizer will provide reasonable performance even if the actual TF matrix differs from the design matrix. As an example, the condition number

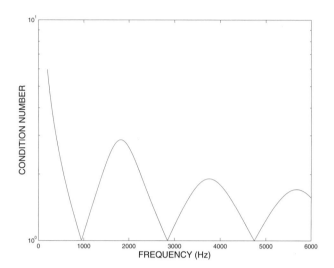

FIGURE 13.3
Condition number of the acoustic TF matrix used to design the CCS filters in Figure 13.2.

for the design matrix used in Figure 13.2 is shown in Figure 13.3. Notice that at the frequencies at which the responses in Figure 13.2 differ most from their ideal (e.g., around 1900 Hz, 3800 Hz), the condition number in Figure 13.3 has a peak.

A small amount of head movement can be tolerated by a robust system. However, if the head moves more than a few centimeters from its design position, then it is necessary to use external means to track the head position and update the CCS filters accordingly. The design TF matrix should therefore be explicitly parameterized by the head position, so that as the head moves, the design matrix and the corresponding equalization filters can be similarly updated.

Techniques to Improve Robustness

It was shown in the previous section that the major problem with loudspeaker-based 3D audio systems is that they lack robustness. In this section, we will consider several techniques that have been proposed to make the CCS perform better in practical situations.

Loudspeaker Geometry

The robustness of the linear system in Equation (13.1) to perturbation is dependent on the conditioning of the design TF matrix $\mathbf{A}_0(f)$. This in turn

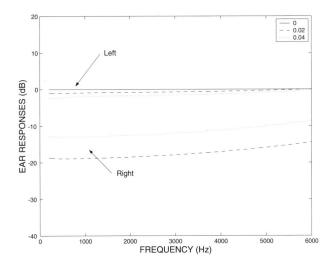

FIGURE 13.4

Effect of head movement on the performance of a CCS using loudspeaker angles of ±5°. Ear responses for the left binaural input are shown for the design position (solid) and when the head moves sideways by 2 cm (dashed) and 4 cm (dotted).

is determined (among other things) by the position of the loudspeakers. Hence, one of the simplest and most effective ways to increase the robustness of 3D audio systems is through judicious positioning of the loudspeakers. For a CCS using two loudspeakers, it has been shown[9] that robustness can be parameterized by the interaural path difference Δ, defined as the added distance from each loudspeaker to the opposite-side ear compared with the distance to the same-side ear. At each frequency there is a certain value of Δ at which the condition number of the design TF matrix is a minimum. The optimum solution for the interaural path difference (optimum in the sense of minimizing the condition number) is for the loudspeaker spacing to vary inversely with frequency. For a wideband audio system, this is impractical. An effective approach, however, has been to use closely spaced loudspeakers[10] or to have different pairs of loudspeakers that are used at different frequencies. Figure 13.4 shows the resulting ear responses for the left binaural signal using two loudspeakers located 1 m from the head center at angles of ±5°. Comparing these results with those of Figure 13.2, notice that even if the head moves 4 cm from its design position, the crosstalk signal (i.e., the right ear signal) is attenuated by at least 10 dB for all frequencies up to around 5 kHz. It has recently been shown that asymmetric loudspeaker positions can also improve the robustness to head movement.[11]

Additional Equalizer Constraints

Rather than choosing loudspeaker positions to give a well-conditioned design TF matrix and then designing the filters to exactly equalize this response, one can also design the CCS filters to attempt to provide reasonable equalization for a family of design TF matrices. In other words, the aim is to design the CCS filters to explicitly increase the size of the sweet spot. One approach is to use a multiple-point control technique.[12] The idea is to create a larger sweet spot by equalizing the response at multiple points in the vicinity of the modeled ear positions. An alternative is to impose derivative constraints on the response at the ears, by constraining the partial derivatives of the TF to be zero at the modeled ear positions.[13]

Reducing the Effect of Reverberation

The effect of reverberation on a CCS that is designed only to equalize the direct-path TFs has recently been investigated.[7] As one would intuitively expect, if the ratio of direct to reverberant energy were quite low, the per-formance of the CCS degrades. One possible way to improve the direct to reverberant energy ratio is to use multiple loudspeakers in an attempt to make the sound radiated by the loudspeakers more directional. Consider the following example. The expected ear responses for the left binaural signal were calculated using the results in Reference 7 in a room with dimensions of $6.4 \times 5 \times 4$ m and a reverberation time of 0.2 s. The loudspeakers were placed symmetrically about the line bisecting the listener's head at a distance of 1 m. No HRTF effects were included. Results are shown in Figure 13.5 using different numbers of loudspeakers; in each case the distance between adjacent loudspeakers was 0.175 m. As the number of loudspeakers increases, one observes two effects from these plots: (1) the response at the left ear (which should ideally be unity) becomes flatter over the entire fre-quency band and (2) the response at the right ear (which should ideally be zero) is reduced, most noticeably at low frequencies.

Visual Information

There is one situation in which no CCS can be effective. This corresponds to the case where the listener has moved far enough away from the design position that the left ear is now where the right ear should be, or vice versa. In this case, additional information must be used to provide a good model for the acoustic TF matrix. One promising approach has been to use a video camera to track the position of the listener's head and then adaptively update the CCS filters as the listener moves.[14] This requires that the acoustic TF matrix is parameterized by the head position so that an appropriate design matrix can be used to design new equalization filters whenever the head moves. It has also been suggested that visual processing systems could help obtain user-specific HRTFs (for example, through matching pinna shapes from a library),[14] although the practicality of such a system has yet to be proven.

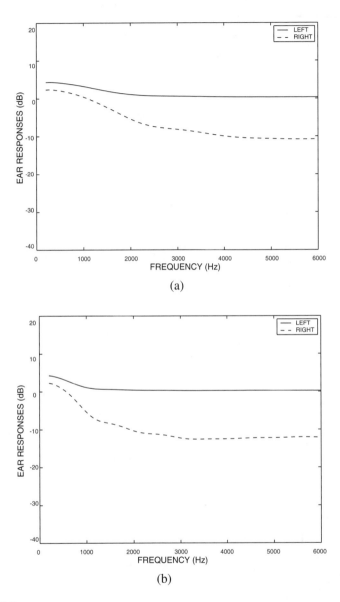

(a)

(b)

FIGURE 13.5
Effect of reverberation on the performance of a CCS using (a) two loudspeakers and (b) three loudspeakers.

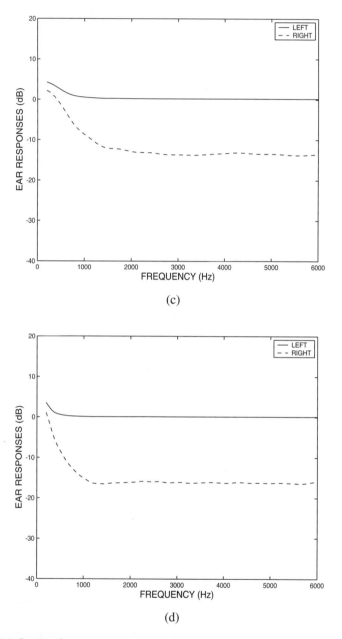

(c)

(d)

FIGURE 13.5 *Continued.*
Effect of reverberation on the performance of a CCS using (c) four loudspeakers and (d) eight loudspeakers.

Discussion and Conclusions

Acoustic crosstalk cancellation is a signal processing technique that is used to deliver binaural signals to a single listener using loudspeakers. It is well known that the main problem with a CCS is that its performance is critically dependent on how much the actual acoustic transfer functions between loudspeakers and ears differ from the modeled transfer functions used to design the CCS. In this chapter, we have presented an overview of this problem, along with techniques that have been proposed to create a CCS that is more robust to modeling errors. It was shown that loudspeaker geometry plays a critical role in determining the system performance. Specifically, using two closely spaced loudspeakers results in a system that performs reasonably well even if the listener moves by a few centimeters from the design position (see Figure 13.4). However, such a system has poor low-frequency performance in a reverberant environment (see Figure 13.5[a]). If one has more than two loudspeakers available, then performance can be significantly improved. Based on these observations, a CCS that is both robust to head movement and provides reasonable performance in a reverberant environment is shown in Figure 13.6. The center two loudspeakers (l_2 and l_3) are used over all frequencies, whereas the outside two loudspeakers (l_1 and l_4) are used only at frequencies below 3 kHz. Thus, at low frequencies, the system uses an unequally spaced array of four loudspeakers, and at high frequencies, two closely spaced loudspeakers are used. The performance of this system is shown in Figure 13.7, and demonstrates that

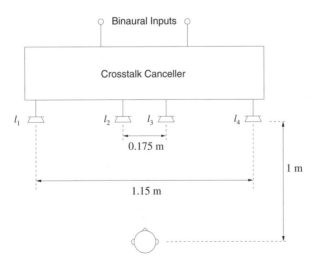

FIGURE 13.6
Geometry for robust CCS. Below 3 kHz all four loudspeakers are used, whereas above 3 kHz only the center two loudspeakers are used.

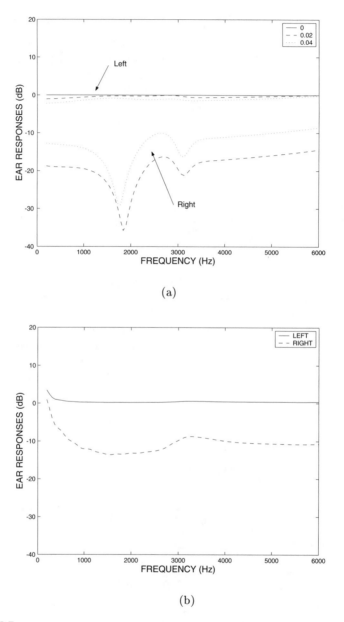

(a)

(b)

FIGURE 13.7
Performance of robust CCS: (a) effect of head movement and (b) effect of reverberation.

it achieves reasonable performance when (a) the head moves, or (b) reverberation is present.

Regardless of how robust it is, if the listener moves more than a few centimeters from the design position, then any CCS is doomed to fail. In this case, the 3D virtual acoustic environment will collapse and sounds will appear to come from one of the loudspeakers. In future virtual audio systems, it is inevitable that tracking of the listener's head will be required, most likely through visual tracking. One can envisage future systems that will also update the virtual acoustic field as the listener's head turns. This will require very efficient algorithms for both head tracking and CCS filter implementation if latency is not to cause serious difficulties for the listener. Incorporating visual tracking with robust design techniques should result in loudspeaker-based virtual audio systems that provide very good performance in most environments.

References

1. Begault, D.R. and Erbe, T., Multichannel spatial auditory display for speech communication, *J. Audio Eng. Soc.*, 42, 819–826, 1994.
2. Gardner, W.G. and Martin, K.D., HRTF measurements of a KEMAR, *J. Acoust. Soc. Am.*, 97(6), 3907–3908, 1995.
3. Chen, J., 3D audio and virtual acoustical environment synthesis, in *Acoustic Signal Processing for Telecommunication*, Gay, S. and Benesty, J., eds., Kluwer Academic Publishers, Boston, 2000, chap. 13.
4. Nelson, P.A., Hamada, H., and Elliot, S.J., Adaptive inverse filters for stereophonic sound reproduction, *IEEE Trans. Signal Processing*, 40(7), 1621–1632, 1992.
5. Atal, B.S. and Schroeder, M.R., Apparent sound source translator, U.S. Patent 3,236,949, February 1966.
6. Radlovic, B.D., Williamson, R.C., and Kennedy, R.A., Equalization in an acoustic reverberant environment: robustness results, *IEEE Trans. Speech Audio Processing*, 8(3), 311–319, 2000.
7. Ward, D.B., On the performance of acoustic crosstalk cancellation in a reverberant environment, *J. Acoust. Soc. Am.*, 110(2), 1195–1198, 2001.
8. Brown, C.P. and Duda, R.O., A structural model for binaural sound synthesis, *IEEE Trans. Speech Audio Processing*, 6(5), 476–488, 1998.
9. Ward, D.B. and Elko, G.W., Effect of loudspeaker position on the robustness of acoustic crosstalk cancellation, *IEEE Signal Processing Lett.*, 6(5), 106–108, 1999.
10. Kirkeby, O., Nelson, P.A., and Hamada, H., The stereo dipole — a virtual source imaging system using two closely spaced loudspeakers, *J. Audio Eng. Soc.*, 46(5), 387–395, 1998.
11. Ward, D.B. and Elko, G.W., A new robust system for 3D audio using loudspeakers, in *Proc. IEEE Int. Conf. Acoust., Speech, Signal Processing (ICASSP-2000)*, Vol. 2, 781–784, Istanbul, Turkey, June 2000.
12. Abe, K. et al., Sound field reproduction by controlling the transfer functions from the source to multiple points in close proximity, *IEICE Trans. Fundamentals Electronics Commun. Comp. Sci.*, E80-A(3), 574–581, 1997.

13. Asano, F., Suzuki, Y., and Sone, T., Sound equalization using derivative constraints, *Acustica*, 82, 311–320, 1996.
14. Kyriakakis, C., Fundamental and technological limitations of immersive audio systems, *Proc. IEEE*, 86(5), 941–951, 1998.

14

Interference in Telephone Circuits

George Keratiotis, Larry Lind, Minesh Patel, John W. Cook,
and Pete Whelan

CONTENTS

Introduction

The local or access part of the telecommunications network is defined as the part of the network that connects customers to the local exchange (or central office). Despite the significant modernization of the trunk and junction parts of the network with the introduction of optical fiber systems, the local network is still based on copper. This can be attributed to the significant amount of capital investment required to replace millions of installed twisted pair lines.

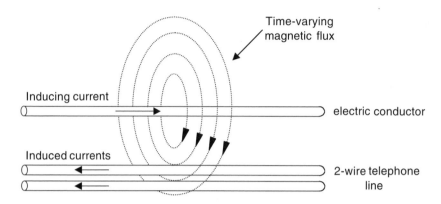

FIGURE 14.1
Electromagnetic coupling.

Since twisted pair lines cannot be easily replaced, an important task for the network operator is the maintenance of the network and the preservation of the quality of the electrical signal as it propagates over the copper lines. One of the measures of the quality of a twisted pair line is its ability to prevent electromagnetic interference from corrupting the information signal. However, due to cable degradation over the years, this ability is reduced, making the twisted pair line susceptible to external sources of interference.

A common source of interference induced on local telephone loops is interference resulting in cases where the twisted pairs are routed in parallel and in proximity to active electrical conductors. When an alternating field is associated with the electrical conductor (as shown in Figure 14.1), electromagnetic coupling occurs and voltages are induced into each wire of the cable. The two wires in the pair shown in Figure 14.1 are not twisted to simplify the illustration.

Let V_1 and V_2 be the voltages induced into each wire of the pair with respect to earth. These voltages will not in general be the same because their values depend partly on the spacing between the wires and the interfering electrical conductor. The average value of these voltages at a time instant is known as common mode (CM), or longitudinal, interference and their difference is known as differential mode (DM), or transverse, interference. The DM level is generally very small compared to the CM level.

On a twisted pair, the information signal (speech or data) is represented by the difference between the wire voltages, and hence it is the DM interference that corrupts the information signal. Ideally, due to the twisting, very little interference should appear as a DM voltage. The interference rejection properties of a twisted pair are described by its balance, which is defined as:

$$\text{Balance (dB)} = 20 \log_{10} \frac{\text{CM}_{\text{interf}}(\text{V})}{\text{DM}_{\text{interf}}(\text{V})} \tag{14.1}$$

TABLE 14.1

Network Cable Balance Figures

Balance (dB)	Category
> 70	Good
65–70	Reasonable
60–65	Poor
< 60	Bad

Source: Ruddock, B., *N&S Field Engineering Unit,* Issue 4, British Telecommunications, 1999.

The smaller the balance, the greater the conversion of the CM interference to DM interference and hence the greater the degradation of the information signal. Some typical balance values are given in Table 14.1.

In practice, some lines are not well balanced due to a number of factors like manufacturing impairments, cable deterioration, insulation imperfection, and poor joints. In such cases, significant DM interference is added to the information signal. Typically, a DM noise voltage would be a few millivolts but it can also be a few hundreds of millivolts, depending on the severity of the interference and the balance of the circuit.[1]

In general, the balance of the cable reduces as the frequency of the interfering signal increases, and therefore higher frequency interference can be more easily converted to a differential signal and thus affect the telecommunication signals. An example of such interference is the radio frequency interference (RFI) due to commercial AM or amateur radio signals which affect mostly overhead (aerial) service cables. Generally, the radio signal is picked up by the wire, which acts as an antenna, and may affect the voice band services if it is demodulated by nonlinear components, such as varistors, transistors, and diodes in the telephone set and from corroded connections and terminations. RFI is one of the limiting factors for other services that operate at higher frequencies, such as those based on the digital subscriber line (DSL) technology.[2]

Examples of Interference

One of the most common types of low-frequency interference encountered on telephone circuits is periodic interference, which originates from either a three-phase or single-phase power line (also known as power line interference). A three-phase power line has the same current in each phase conductor and as these are placed at a phase angle of 120° relative to each other, the sum of the currents at any instant is zero and their inductive effects on a parallel telephone line cancel out. However, in practice, this precise balance

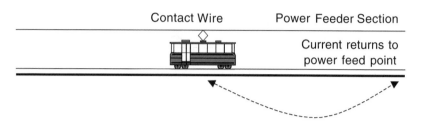

FIGURE 14.2
AC electrified railway.

of the phase conductors is seldom met and a small out-of-balance earth-return current circulates in the system, causing an external magnetic field and interference on nearby cables.

A typical example of a single-phase system is the one used in AC electrified railways. These systems generate a lot of harmonics due to the power conversion/motor drive controls. The current is supplied to the trains through an overhead contact wire and it returns to the feeding station partly through the traction rails, the return conductors (if fitted), and through the earth (Figure 14.2). Since this is an inherently unbalanced system compared to three-phase power distribution, voltages may be induced on nearby communication lines by magnetic induction from the contact wire current.

A second type of low-frequency interference arises from agricultural electrified fences (Figure 14.3). This type of fencing has seen rapid growth recently, mostly due to economic reasons. The fence is energized by means of very short, high-voltage (up to 8 kV; see Reference 3) pulses, which are injected onto the wires by an electronic fence energizer unit. This energizer feeds a pulse which has duration of a few milliseconds with a repetition rate of approximately 1 s.[3]

The interference occurs when leakage currents flow from the energized fence wire to earth. Vegetation growing adjacent to the fence makes contact with the energized wire and provides a path to earth. This path is also provided by poor insulation at the points where the wires attach to the fence posts.

The voltage pulse, the associated current pulse, and any pulsed earth potential all contribute to interference in telephone circuits. As the induced voltage from such a system depends on the amount of leakage, the measured CM voltage varies greatly from a few tens of volts up to 300 volts. These high-voltage pulses cause very loud "clicks" on a telephone handset, distort transmitted data/fax messages, and in some cases cut off the line.[1]

The rest of the chapter further investigates the problem of power line interference, the most frequent type of interference. Additional information on low-frequency interference can also be found in the International Telecommunication Union (ITU) recommendation ITU-T K.10. More detailed information and methods of suppressing electric fence interference can be found in References 4 and 5.

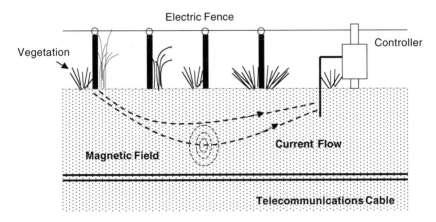

FIGURE 14.3
Electric fence interference.

Psophometric Weighting Characteristic

A method for rating the subjective effects of the interference on voice band telephony is to use the so-called psophometric weighting characteristic shown in Figure 14.4. This is an ITU-T directive (ITU-T O.41) and its attenuation vs. frequency curve conforms closely to the response of the human ear to sounds emitted from a typical telephone handset.

As can be seen from Figure 14.4, if the power line interference were a pure 50 Hz (UK frequency) sinusoid, its subjective effect would be minimal, since the telephone set attenuates this tone by more than 60 dB. However, in practice, the periodic interference is rich in harmonics due to the opera-

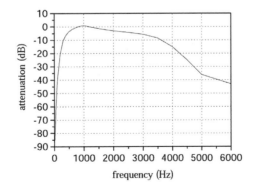

FIGURE 14.4
Telephone circuit psophometric weighting characteristic.

tion of nonlinear apparatus, such as transformers, DC motors, rectifying equipment, and pumps. When these harmonics extend inside the voice-frequency bandwidth (300 to 3400 Hz) of the telecommunication lines, they cause significant degradation of the information signal and give rise to signaling problems.

A Typical Example of Power Line Interference

Figures 14.5(a) and (b) show two cycles of a typical CM–DM pair of power line interference signals as measured on a telephone line in Newcastle, U.K. The DM signal has been filtered with a low-pass filter (LPF) to remove the frequencies outside the voice band. The source of the interference was a nearby electrified railroad line. The shape of the signals reveals the existence of higher-order harmonics.

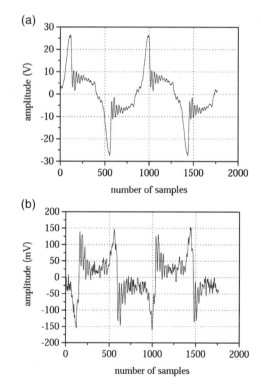

FIGURE 14.5
(a) CM and (b) DM interference (sampled at 44.1 kHz) as captured on a telephone line.

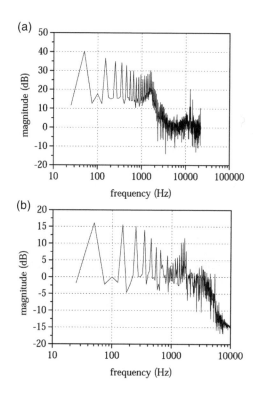

FIGURE 14.6
Spectrum of the CM interference (a) and the DM interference (b).

In this type of interference, most of the signal's energy is distributed in the odd harmonics of 50 Hz. Figures 14.6(a) and (b) show the spectrum of the CM and DM waveforms. As can be seen, the DM interference signal has significant energy inside the voice-frequency bandwidth (300 to 3400 Hz) of the telecommunication lines and causes degradation in telephony and fax operation and signaling problems as well.[6] Similar measurements have been reported by Hierman.[7]

Traditional Noise Reduction Techniques

There are a number of techniques available to a telecommunications company for the suppression of such types of low-frequency DM interference. Ideally, the CM noise component also should be attenuated, since it might be converted to DM (thus audible interference) beyond the remote unit (point of demarcation between the telephone company network and the customer premises) due to the poor balance of the cable and/or customer equipment.

These techniques can be summarized as follows:

1. *Magnetic Screening:*[8] The magnetic screening is achieved by periodically connecting points of the cable screen or a spare pair to earth. By doing so, a counter electromagnetic field (emf) is induced that reduces the longitudinal flow of current in the signal pairs. The effectiveness of the technique is variable and its deployment is economic only in cases where the number of affected customers is large. Only the CM noise component is affected with this method.

2. *Induction Neutralizing Transformers (INT):*[8] The principle of operation is the same as for magnetic screening. The counter emf is induced by routing the longitudinal current flow through a transformer formed by using a spare pair in the cable, which is earthed at both ends of the network. The INT introduces some inductance on the pair, which is not a problem for speech band frequencies but may be prohibitive for DSL-type services. Only the CM noise component is affected with this method.

3. *Line Filters:*[8] These operate like the INT but are specially designed for operation at the customer end of the line. They provide very good CM attenuation but no DM attenuation at all. Also, the filter has a low-pass type of characteristic for the differential signals, an undesirable effect for services that make use of higher frequencies.

4. *Digital Access Carrier Systems (DACS) and Pair-Gain Systems:*[9] This system provides noise immunity by using digital transmission. The method has two major disadvantages: (a) there is a maximum reach limit and (b) there is no attenuation of the CM interference.

The disadvantage of the first three methods is that they cannot provide any DM interference suppression so noise would still be audible on a telephone handset that is using the affected line. On the other hand, the employment of DACS is not cost effective, since a CM suppression method has to be used as well.

Problem Modeling

The basic difficulty in the power-line interference problem is the co-existence of speech and noise in the same frequency band. A direct solution would be the insertion of a fixed comb filter for the attenuation of all the odd harmonics of 50-Hz noise. The distortion introduced on the speech signal could be minimized by reducing the bandwidth of the notches. However, actual noise recordings have revealed that there is a frequency drift present on the

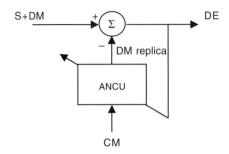

FIGURE 14.7
The adaptive noise cancellation concept.

harmonics, which is time varying and depends on a number of factors such as site location and train movement. In such a case, the effectiveness of a fixed filtering approach is significantly reduced.

A better approach is to use the well-known adaptive noise cancellation set-up, which was originally proposed by Widrow et al.[10] As shown in Figure 14.7, the information signal S is transmitted over the line with the interference signal DM superimposed on it. The idea behind the noise suppression method is to create an artificial replica of the DM noise signal and then subtract it from the incoming S + DM signal. To account for the time-varying nature of the interference, an adaptive system has to be employed. Let the differential voltage after the subtraction be called the differential error (DE) signal. Ideally, this should be equal to the information signal S. Since the CM signal is highly correlated with the DM noise component, it can be used by an adaptive noise cancellation unit (ANCU) to create an output that is as close as possible to the DM noise signal. This output is subtracted from the input S + DM to suppress the interference signal DM. The residual signal DE is used by the ANCU in order to increase the efficiency of the suppression. The adaptation should take place only in the absence of the information signal, so that the true error between the incoming interference signal and its replica is used to improve the suppression procedure.

In the majority of applications, a finite impulse response (FIR) filter is assumed to model the transfer function between the two noise components. This is because of the simple implementation of the FIR filter and the well-understood properties of the associated adaptive algorithms. The CM noise suppression can be achieved in a similar manner as shown in Figure 14.8. This is the typical linear-prediction mode of operation, where an adaptive FIR filter is used to replicate the incoming CM noise component using the common mode error (CE) samples.

Various adaptive algorithms, such as the ones discussed in Chapter 1, can be used for the CM and DM noise suppression. The particular characteristics of the case under consideration impose tight constraints on the possible adaptive algorithms that might be used, since there is a need for a cost-efficient and robust approach. The least mean square (LMS) adaptive algo-

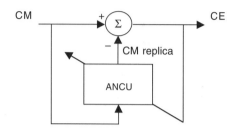

FIGURE 14.8
CM noise suppression.

rithm (Chapter 1) and its variants comply with the above requirements and have been traditionally employed in similar cases.[11]

Algorithm Considerations

As mentioned in Chapter 1, the implementation complexity of LMS is on the order of $2N$ multiplications and additions, where N is the number of taps used in the FIR filter. If the same number of taps is used for both the CM and DM noise suppression, then this complexity rises to $4N$, which is high enough to set a maximum limit on the number of taps that can be used (due to the DSP operation). It also should be mentioned that this complexity does not take into account the processing required for the overall control of the system, such as power normalization, voice/data activity detection, and monitoring.

A significant simplification of the replication procedure can be achieved if the periodicity of the involved signals is taken into account. This is the basic idea behind the recently introduced adaptive phase-locked buffer (APLB) algorithm, which is described in full detail in Reference 12. A brief description of the APLB algorithm is given next.

APLB Algorithm

Consider the noise cancellation configuration shown in Figure 14.7. Assuming that the DM samples do not change significantly from period to period, one cycle of the incoming DM waveform can be reconstructed by the ANCU using previous DM cycles. In this case, the ANCU is simply a buffer (as shown in Figure 14.9) that contains samples of one cycle of the predicted DM noise waveform. These samples are subtracted from the incoming noise waveform to provide interference suppression.

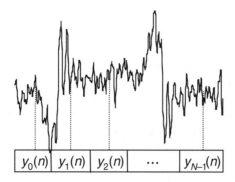

$$y_0(n) \quad y_1(n) \quad y_2(n) \quad \cdots \quad y_{N-1}(n)$$

FIGURE 14.9
The APLB Method.

To account for amplitude variations of the DM signal, the buffer values are continuously updated using the DE samples. Assume that the incoming DM noise and the APLB are phase locked and in phase. Let the APLB have N taps (buffer locations), which span one cycle of the DM noise. Also, let $y_m(n)$ be the APLB sample stored in tap m during the period n of the DM noise.

After each tap is sequentially injected onto the line, its effect on the residual DE signal is stored and used to update the relative tap value for the next period. During one noise period, all the taps in the buffer are updated once according to the equation:

$$y_m(n + 1) = y_m(n) + g\mathrm{DE}_m(n) \quad m = 1 \text{ to } N \tag{14.2}$$

where $\mathrm{DE}_m(n)$ is the value of the residual DE signal, which corresponds, to the tap m during period n. The parameter g is a weighting factor ($0 \le g \le 1$), used to smooth the DE fluctuations.

The operation of the APLB technique depends heavily on the in-phase locking mentioned above. This can be achieved by processing the strong CM signal, with a specially designed digital phase-locked loop (DPLL).[12] Once the DPLL is phase locked with the CM signal, the APLB waveform can easily be brought to an in-phase lock with the incoming DM component of the interference.

The CM noise suppression can be achieved in a similar manner. A second APLB is used to store the antiphase CM samples. The APLB values are updated with the CE values in a similar manner to Equation (14.2), and locking is achieved using the same DPLL.

As can be seen from Equation (14.2), the updating part of the APLB method requires one multiplication and one addition per tap and the DPLL operation requires only a few multiplications, additions, and logical operations. This is a very small amount of processing compared with the $2N + 1$ multiplications and $2N$ additions that are required to implement an LMS filter with N taps. The computational savings in the APLB approach stem from the fact that the DM noise replica is directly estimated and updated, eliminating the need to estimate/update an underlying model, as in the case of the LMS algorithm.

Discussion

The APLB method described above is similar to the waveform synthesis (WS) method derived by Chaplin et al. in 1979 (see Reference 11, pp. 103–110). In the WS method, a reference input derived from the noise source is phase locked with the generated waveform. By doing so, only the harmonics of the fundamental frequency are canceled, leaving unaffected the frequency bands between the harmonics. Assuming a perfect lock between the reference input and the synthesized waveform, the APLB and WS methods can be analyzed as if an FIR filter with N taps were excited by an impulse train of period N samples:[13]

$$x(n) = \sum_{k=-\infty}^{k=\infty} \delta(n - kN) \qquad (14.3)$$

where $\delta(\cdot)$ is the Kronecker delta function. In this case (the order of the adaptive filter is equal to the period of the impulse train), the LMS weight updating reduces to the simplified APLB update given by Equation (14.2).

The analysis shows that the transfer function between the input $DM(n)$ and the error $DE(n)$ can be represented by a linear time-invariant comb filter with notches at each harmonic of the interference.[13] The DPLL tracks the variations of the fundamental frequency of the interference and the notches are accordingly adjusted. The overall scheme acts like an adaptive comb filter, implemented in a parallel configuration, and has none of the disadvantages of the fixed comb filtering mentioned in "Problem Modeling."

Practical Noise Canceller

A prototype system was designed by British Telecommunications (BT) for the implementation of various noise cancellation algorithms. This is shown in Figure 14.10.

The analog stages 1 and 2 detect the single-ended voltages on each wire of the input and output pairs and form the CM, CE, and DE signals. These signals are then digitized and sent to a DSP chip. The DSP implements the noise cancellation algorithms and creates the antiphase waveforms. These are then converted to analog signals, which are injected on the line through the use of transformers.

Overall the device can be seen to be highly transparent to telephony signals, offering only a small series impedance and high parallel impedance.

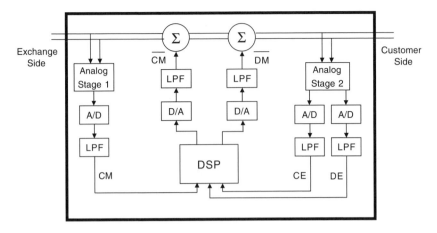

FIGURE 14.10

The adaptive noise cancellation unit (ANCU). LPF, low-pass filter; A/D, analog-to-digital converter; D/A, digital-to-analog converter; DSP, digital signal processor; Σ, injection transformer.

To reduce the cost and power consumption, a fixed-point DSP was chosen for the algorithm implementation.

Experimental Set-Up

In order to test the effectiveness of the designed noise canceller, the configuration shown in Figure 14.11 was used. A telephone exchange emulator was used to set up a call between two handsets to emulate a normal telephone connection between two subscribers. An artificial transmission line with length set to 6 km was used to represent a typical twisted pair connecting the two subscribers. The interference coupling mechanism was modeled as an additive process, and two transformers were used for the CM and DM noise injection.

Real CM and DM noise recordings were converted in formats suitable for CD reproduction. A stereo track was used with the CM recording on the left channel and the DM recording on the right channel. The volume of the CD player was adjusted so that the peak-to-peak level of the DM signal was approximately 300 mV, a value similar to the one found in practical situations. Since the practical CM peak-to-peak values are on the order of 50 V, an audio amplifier was used to amplify the CM output from the CD player.

The ANCU was connected in series with the pair and its output was then connected to the customer's handset. The effectiveness of the noise cancellation was assessed using an oscilloscope, a spectrum analyzer, and a typical telephone handset.

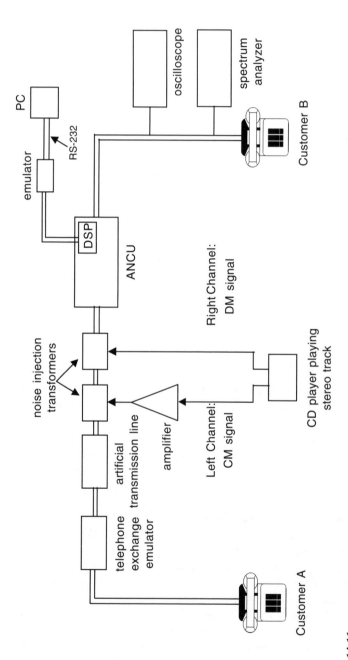

FIGURE 14.11
The experimental set up used to test the noise cancellation unit.

Results

Figures 14.12 and 14.13 are photographs of a digital oscilloscope showing the CM and DM noise signals appearing at the input of the ANCU. The peak-to-peak levels of the CM and DM signals were 32 V and 234 mV, respectively. The APLB algorithm implemented on the DSP chip used 160 taps, and the rest of the parameters were adjusted so that convergence was achieved within approximately 1 s. Figures 14.14 and 14.15 are photographs that were taken after a few seconds of operation.

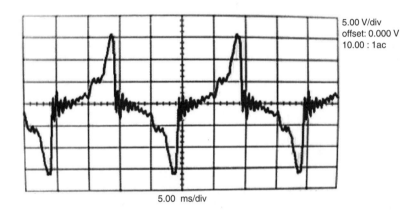

5.00 V/div
offset: 0.000 V
10.00 : 1ac

5.00 ms/div

FIGURES 14.12
CM signal before suppression.

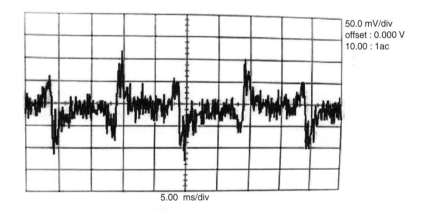

50.0 mV/div
offset : 0.000 V
10.00 : 1ac

5.00 ms/div

FIGURES 14.13
DM signal before suppression.

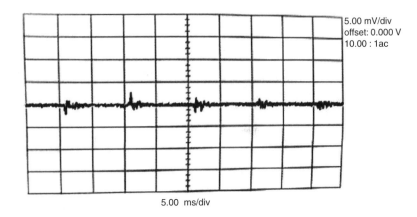

5.00 mV/div
offset: 0.000 V
10.00 : 1ac

5.00 ms/div

FIGURE 14.14
CM signal after suppression.

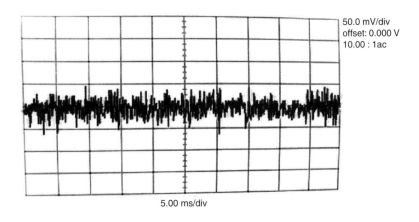

50.0 mV/div
offset: 0.000 V
10.00 : 1ac

5.00 ms/div

FIGURE 14.15
DM signal after suppression.

As can be seen, the CM amplitude is reduced significantly, and only the "ringing" parts of the waveform appear on the residual signal.

On the other hand, the residual DM signal consists mainly of high-frequency components, as can be seen in Figure 14.15. The periodic pattern has been completely suppressed, an indication of very good attenuation of low-frequency harmonics. The cancellation bandwidth extends up to 2 kHz, and the attenuation is better for lower frequency harmonics. The first five harmonics are suppressed by approximately 20 dB, whereas the suppression of higher order harmonics is ~ 10 dB on average. As mentioned in "APLB Algorithm," the performance of the APLB method depends on the periodicity of the waveforms and the synchronization between the incoming signals and the antiphase (synthesized) ones. The DM signal can be decomposed into two parts. The first one corresponds to the lower-order harmonics and is the one that gives to the DM signal the periodic pattern, which is obvious in Figure 14.5(b). This part of the DM signal is the one that is easily replicated by the APLB. The second part of the DM signal constitutes higher frequency components, which exhibit random fluctuations and cannot be easily replicated by the proposed method. However, taking into account the psophometric weighting characteristic shown in Figure 14.4, the achieved suppression in the region of approximately 1 kHz is adequate to reduce significantly the subjective effect of the interference.

By increasing the number of taps in the APLB, the resolution of the replication procedure can be improved at the expense of extra computations. If the number of taps is doubled, the cancellation bandwidth increases from 4 to 8 kHz, assuming a perfect DPLL operation. However, in the fixed-point implementation of the approach, a trade-off exists between the number of taps and the performance of the DPLL. As the number of taps is increased, the performance of the DPLL is reduced in terms of stability and jitter. This can be attributed to the simplified building blocks used for the DPLL.

Conclusions

This chapter provided an introduction to the problem of interference in telephone circuits. The coupling mechanisms for two basic types of low-frequency interference were described and emphasis was given to the power-line interference problem, which is one of the most common types of noise appearing on a telephone network. The existing mitigating methods were described, revealing the need for a stand-alone, cost-effective solution.

It was shown that the problem can be modeled by an adaptive noise cancellation configuration. Assuming that the CM and DM components of the interference are correlated, the strong CM signal can be used as a reference by an adaptive filter, which models the CM-to-DM transfer function. The output of the adaptive filter, which is a replica of the DM signal, is then inverted and injected on the line and the residual noise is used to improve the suppression. The overall configuration can be regarded as an adaptive comb filter, with notches at each harmonic of the interference.

A novel algorithm was described that was specifically designed to exploit the periodicity of the power line interference. The basic advantage of the APLB method is the small computational complexity required for its implementation, allowing more processing power to be used for other functions, such as voice activity detection, power normalization, and stability monitoring. These secondary operations can have a significant impact on the efficiency of the system and careful implementation is required for a robust performance.

To demonstrate the feasibility of the proposed solution, a prototype noise cancellation unit was built that can be used for the implementation of various adaptive algorithms. The unit was part of an experimental configuration that was replicating a typical interference scenario. The results obtained indicate that the proposed solution can achieve significant noise suppression, and is a clear demonstration of how advanced signal processing techniques can provide solutions to a real-world noise problem.

References

1. Ruddock, B., Low frequency interference to the network: mechanisms, measurements and mitigation, *N&S Field Engineering Unit*, Issue 4, British Telecommunications, 1999.
2. Foster, K.T. et al., Realising the potential of access networks using DSL, *Br. Telecom. Tech. J.*, 16, 34, 1998.
3. Galagher Power Fence Systems, *8th International Power Fence Manual*, Gallagher RSM Pty Ltd., Sydney, Australia, 1989.
4. Keratiotis, G., Adaptive Algorithms for Real-Time Noise Cancellation, Ph.D. thesis, University of Essex, Essex, U.K., 2001.
5. Telecom Australia, *Electric Fence Interference on Telephone Lines: A Guide to Mitigation of Induced Noise*, Issue 1, 1984.
6. Cook, J.W., Proposal for project to develop prototype balance-about-earth improvement system for local access copper pairs, British Telecommunications Laboratories, Copper, Radio and Satellite Systems, 1991.
7. Hierman, D.N., Time variations and harmonic content of inductive interference in urban/suburban and residential/rural telephone plants, *IEEE Trans. Commun.*, Vol. COM-23, 1975, 1484.
8. Gundrum, R., *Power Line Interference; Problems and Solutions*, 14, ABC of the Telephone, abc TeleTraining, Geneva, IL, 1988.
9. Adams, F. et al., Today's access technologies, *Br. Telecom. Tech. J.*, 16, 21, 1998.
10. Widrow, B. et al., Adaptive noise cancelling: principles and applications, *IEEE Proc.*, 63, 1692, 1975.
11. Kuo, S. and Morgan, D.R., *Active Noise Control Systems: Algorithms and DSP Implementations*, John Wiley & Sons, New York, 1996.
12. Keratiotis, G. et al., A novel method for periodic interference suppression on local telephone loops, *IEEE Trans. Circuits and Syst.*, 47, 1096, 2000.
13. Elliott, S.J. and Darlington, P., Adaptive cancellation of periodic, synchronously sampled interference, *IEEE Trans. Acoust., Speech, Signal Processing*, ASSP-33, 1985, 715.

15

An Adaptive Beamforming Perspective on Convolutive Blind Source Separation

Lucas Parra and Craig Fancourt

CONTENTS

Introduction

Microphones in an acoustic environment typically capture a mixture of several sources. The goal of convolutive blind source separation (BSS) is to filter the signals from a microphone array to extract the original sources while reducing interfering signals. Due to the spatial variability of a room response, different microphones receive different convolved versions of each source.

Separation, therefore, also requires a convolutive filtering of the sensor signals, effectively resulting, for each source, in a spatially selective filter or "beam." As opposed to conventional geometric or adaptive beamforming, in BSS no assumptions on array geometry or source location are made. Instead, it is only assumed that the desired sources are statistically independent. Convolutive blind source separation can therefore be understood as multiple adaptive beamformers that generate statistically independent outputs, or more simply, outputs with minimal crosstalk.

Considerable progress has been made in formulating sufficient conditions on the source signals and deriving corresponding optimization criteria. Strict independence criteria involve higher-order statistics (HOS) of the signals. Unfortunately, HOS are difficult to estimate and lead to complex and computationally demanding algorithms. An alternative to HOS is to constrain the crosstalk minimization to a second-order criteria and instead exploit the nonstationarity of the signals.[1,2] Estimating second-order statistics is numerically more robust and the criteria lead to simpler algorithms. Most results reported in the literature on real-room recordings are based on second-order methods, whereas higher-order separation algorithms are often only demonstrated on simulated data.

Aside from this, the independence criteria exhibit a number of ambiguities: (1) the recovered sources are only determined up to an arbitrary convolution, (2) more microphones than sources results in underconstrained filter coefficients, and (3) frequency bins may not be assigned consistently to the correct channels. We propose to reduce the inherent ambiguities of convolutive BSS by introducing geometric constraints similar to those used in the linearly constraint minimum variance (LCMV) algorithm and generalized sidelobe canceler (GSC).[3] We have termed the resulting algorithms geometric source separation (GSS)[4] and the generalized sidelobe decorrelator (GSD),[5] respectively. Efficient frequency domain on-line and off-line implementations will be outlined. Results on noise reduction for speech recognition in different real-room environments and applications will be given.

Convolutive Blind Source Separation

Consider M uncorrelated sources, $\mathbf{s}(t) \in \mathbb{R}^M$, originating from different spatial locations and $N > M$ sensors detecting signals $\mathbf{x}(t) \in \mathbb{R}^N$. In a multipath environment, each source j couples with sensor i through a linear transfer function $A_{ij}(\tau)$, such that $x_i(t) = \sum_{j=1}^{M} \sum_{\tau=0}^{P-1} A_{ij}(\tau) s_j(t - \tau)$. Using matrix notation and denoting the convolutions by $*$, we can write this briefly as $\mathbf{x}(t) = A(t) * \mathbf{s}(t)$, or after applying the discrete-time Fourier transform (DTFT),

$$x(\omega) = A(\omega)s(\omega) \tag{15.1}$$

The task of convolutive source separation is to find filters $W_{ij}(\tau)$ that invert the effect of the convolutive mixing $A(\tau)$. One generates model sources $\mathbf{y}(\omega)$

$$y(\omega) = W(\omega)x(\omega) \tag{15.2}$$

that correspond to the original sources $\mathbf{s}(t)$. Although any linear system is compatible with Equation (15.2), we restrict ourselves to finite impulse response (FIR) filters since this allows the algorithms to be efficiently implemented in the frequency domain.

Higher-Order Methods vs. Second-Order Nonstationarity

Different criteria for convolutive separation have been proposed.[1,2,6-10] All criteria can be derived from the assumption of statistical independence of the unknown signals. However, typically only pairwise independence of the model sources is used. Pairwise independence implies that all cross moments factor, yielding a set of necessary conditions for the model sources

$$\forall t, n, m, \tau, i \neq j: \quad E\left[y_i^n(t)y_i^m(t+\tau)\right] = E\left[y_i^n(t)\right] E\left[y_j^m(t+\tau)\right] \tag{15.3}$$

$E[\,\cdot\,]$ represents the ensemble average and will in practice be replaced with a sample average over a given time window surrounding time t. Convolutive separation requires these conditions to be satisfied for multiple delays τ, corresponding to the delays of the filter taps of $W(\tau)$. For stationary signals, multiple n, m, i.e., higher-order criteria, are required. For nonstationary signals, multiple t with $n = m = 1$ are sufficient.[1,2,11] In this case, conditions (Equation [15.3]) state that cross-correlation matrices $R_{yy}(\tau, t) = E[\mathbf{y}(t)\mathbf{y}^T(t+\tau)]$ have to be diagonal at all times.

Separation Based on Second-Order Nonstationarity

Joint diagonalization of $R_{yy}(\tau, t)$ has to find filters $W(\tau)$ that decorrelate model sources $\mathbf{y}(t)$ at multiple t. This can be implemented efficiently in the frequency domain[2] using the Fourier transform of the cross correlations — the cross-power spectra. Currently, we obtain the best results with a diagonalization criteria based on the coherence function,[12] defined as

$$C_{y_iy_j}(\omega,t) = \frac{R_{y_iy_i}(\omega,t)}{\sqrt{R_{y_iy_i}(\omega,t)R_{y_jy_j}(\omega,t)}} \tag{15.4}$$

where $R_{y_iy_j}(\omega,t)$ is the cross-power spectra between outputs y_i and y_j at frequency ω and time t. In matrix notation this can be written as

$$C_{yy}(\omega,t) = \Lambda_{yy}^{-1/2}(\omega,t)R_{yy}(\omega,t)\Lambda_{yy}^{-1/2}(\omega,t) \qquad (15.5)$$

with $\Lambda_{yy}(\omega,\ t) = \mathrm{diag}\ R_{yy}(\omega,\ t)$. The squared coherence function is real and constrained to lie between 0 and 1 for all frequencies. The coherence function matrix $C_{yy}(\omega,\ t)$ is identically equal to one on the diagonal. Its off-diagonal elements vanish only if $R_{yy}(\omega,\ t)$ is diagonal, and so we can use the following diagonalization criteria

$$J(W) = \sum_t \sum_\omega \left\| C_{yy}(\omega,t) \right\|^2 \qquad (15.6)$$

with the Frobenius norm, $\|C\|^2 = Tr[C^H C]$, representing the square sum of all the elements in the matrix C. The minimization of Equation (15.6) can be solved using gradient descent methods. The advantage of the coherence function criteria is that the normalization guarantees uniform convergence speed irrespective of the power present in any given frequency bin. The optimization of Equation (15.6) requires multiple estimates of the cross-power spectra estimated at different times, t. Parra and Spence[2] did this using an off-line algorithm that first estimates the cross-power spectra of the microphones over different time windows, $R_{xx}(\omega,\ t)$, and in a second step computes the simultaneously diagonalizing filters $W(\omega)$. The approximation of linear and circular convolution is used there, $R_{yy}(\omega,\ t) \approx W(\omega)R_{xx}(\omega,\ t)W(\omega)^H$, which is valid if the filters are short in comparison to the length of the discrete Fourier transform (DFT).

Online Decorrelation

In attempting to convert the offline algorithm into an online algorithm, we are faced with the problem of designing an algorithm that *requires* nonstationary signals for convergence. The reason for this is that what we do with each new measurement depends on whether it is part of the previous stationary regime, or represents a transition to a new stationary regime. In the first case, the new data should be used to improve the estimate of the current covariance, implying the use of a long effective memory. In the second case, the data represent the beginning of new covariance matrix for simultaneous diagonalization with previous covariance matrices, implying a short memory is appropriate. Therefore, in addition to the conventional trade-off between convergence speed and misadjustment, we now have a trade-off between estimation accuracy and novel information when measuring correlation.

Note that there are actually two sums over time in Equation (15.6). First, there is the explicit summation over multiple coherence matrices estimated at different times. There is also an implicit summation over the block of time necessary to estimate each coherence matrix. The key insight is that these two sums are not interchangable because the criteria are nonlinear in the power estimates. Often online second-order decorrelation has proposed a stochastic optimization method whereby the sums over time are entirely removed. In doing so, however, nonstationarity is not properly captured and the algorithms reduce to simple decorrelation, which is not sufficient for separation. Therefore, we propose to preserve the time-averaging process by recursively estimating the cross-power spectra to capture short-term nonstationarity[12]

$$R_{yy}(\omega,t) = \gamma R_{yy}(\omega,t-T) + (1-\gamma)y(\omega,t)y^H(\omega,t) \qquad (15.7)$$

where γ is a forgetting factor, constrained to $0 < \gamma < 1$ for stability, and T is a block processing time (frame rate) that represents the time it takes to estimate $y(\omega, t)$. The forgetting factor and block processing time combine to make the effective memory of the estimator to be $T/(1 - \gamma)$.

We consider the sum in Equation (15.6) as an estimator of the instantaneous cost, $\Sigma_\omega \|C_{yy}(\omega,t)\|^2$. Stochastic gradient descent uses the instantaneous cost for the weight updates. We take the derivative with respect to the complex weights in the frequency domain and update the weights at the end of each time block

$$\Delta W = -\mu \, (\Lambda_{yy}^{-1} R_{yy} \Lambda_{yy}^{-1} - \text{diag}[\Lambda_{yy}^{-2} R_{yy} \Lambda_{yy}^{-1} R_{yy}])R_{yx} \qquad (15.8)$$

where μ is the learning rate and R_{yx} is a matrix of cross-power spectra between the outputs and the inputs:

$$R_{yx}(\omega,t) = \gamma R_{yx}(\omega,t-T) + (1-\gamma)y(\omega,t)x^H(\omega,t) \qquad (15.9)$$

The online blind source separation algorithm consists of Equation (15.2) and Equations (15.7) through (15.9) and is entirely compatible with the overlap-save method of frequency domain adaptive filtering.[13] The overlap-save method implements linear convolution in the frequency domain with the DFT, or its efficient counterpart, the fast Fourier transform (FFT). However, since the DFT corresponds to circular convolution in the time domain, the filters must be padded with zeros, in turn requiring the use of a larger input buffer. As a result, only the latter part of the output in the time domain is valid. In the context of the present algorithm, it is thus incorrect to directly use the complex output (Equation [15.2]) in updating the cross-power spectral densities in Equations (15.7) and (15.9). Rather, they must first be transformed into the time domain (also required to obtain the system output), and the invalid parts zeroed prior to transforming back into the frequency domain for use in Equations (15.7) and (15.9). Note that this is not required for x, since the input buffer is always filled with valid input samples prior to transforming into the frequency domain.

The computational complexity of the algorithm scales linearly in the number of inputs and quadratically in the number of outputs. Although other frame rates relative to the filter size can be used, a 50% overlap is the most computationally efficient. For a two input–two output problem at a sampling rate of 8 kHz with 512 taps, the algorithm runs in approximately one tenth real-time on a 866-MHz Pentium III. It is thus entirely suitable for real-time operation in many-input–many-output problems.

Combining Source Separation with Beamforming

This section discusses the ambiguities of convolutive blind source separation. We will review how geometric information is utilized in conventional adaptive beamforming and suggest that second-order BSS can readily be combined with adaptive beamforming methods, because they both operate on the power spectra of the signals.

Ambiguities of Independence Criteria

Regardless of the independence criteria, there remains an ambiguity of permutation and scaling in the separating filters. In the convolutive case the scaling ambiguity applies to each frequency bin, resulting in a convolutive ambiguity for each source signal. This expresses the fact that filtered versions of independent signals remain independent. Furthermore, when defining a frequency domain independence criteria such as

$$\forall n, m, \omega, i \neq j: \quad E[y_i^n(\omega)y_j^m(\omega)] = E[y_i^n(\omega)]E[y_j^m(\omega)] \qquad (15.10)$$

there is a permutation ambiguity for each frequency. The criteria (Equation [15.10]) are equally well satisfied with arbitrary scaling and assignment of indices i, j to the model sources, i.e.,

$$W(\omega)A(\omega) = P(\omega)S(\omega) \qquad (15.11)$$

where $P(\omega)$ represents an arbitrary permutation matrix and $S(\omega)$ an arbitrary diagonal scaling matrix per frequency. The most immediate problem with this is that contributions of a given source may not be consistently assigned to a single model source across different frequency bins.[2,8,14,15] Parra and Alvino[4] argue that the permutation problem (Equation [15.10]) also exists in the time domain criteria (Equation [15.3]).

In practice, one may want to use a larger number of microphones to improve spatial resolution or reduce aliasing. Aside from the permutation and scaling ambiguity, Equation (15.11) suggests that for a given $A(\omega)$ there is an $N - M$ dimensional linear space of solutions $W(\omega)$. In effect, this indicates that there are additional degrees of freedom in terms of shaping a beam pattern represented by the separating filters $W(\omega)$.

Linear Constraints in Geometric Beamforming

To disambiguate the permutation, convolution, and underdetermined filter coefficients, one can use geometric information. In conventional geometric and adaptive beamforming, information such as microphone position and source location are often utilized. A good review of these methods is given in Reference 3. We want to emphasize that geometric assumptions can be incorporated and implemented as linear constraints to the filter coefficients.

If the source location, array geometry, and microphone response characteristics are known, then we can specify an *array response vector*, $d(\omega, q) \in \mathbb{C}^N$, which represents the complex response from the source at location q to the outputs of the N sensors. Then, for a given beamforming filter, $w(\omega)$, the total system response is given by

$$r(\omega, q) = w(\omega)d(\omega, q) \qquad (15.12)$$

For a linear array with omnidirectional microphones and a far-field source, the microphone response depends in good approximation only on the angle $\theta = \theta(q)$ between the source and the linear array

$$d(\omega, q) = d(\omega, \theta) = e^{-j\omega \frac{p_i}{c}\sin(\theta)} \qquad (15.13)$$

where p_i is the position of the ith microphone on the linear array and c is speed of sound.

Constraining the response to a particular orientation is simply expressed by the linear constraint, $r(\omega, \theta) = w(\omega) \, d(\omega, \theta) = const$. This concept is used in the linearly constrained minimum variance (LCMV) algorithm.[16]

Power vs. Cross-Power Criteria

Most adaptive beamforming algorithms rely on a power criteria of a single output. Sometimes power is minimized such as in noise or sidelobe canceling. There the aim is to minimize adaptively the response at the orientation of interfering signals.[3] Sometimes power is maximized such as in matched-filter approaches that seek to maximize the response of interest.[17] As outlined in "Separation Based on Second-Order Nonstationarity," blind source separation of nonstationary signals minimizes the off-diagonal elements of $R_{yy}(t, \omega)$ rather than the diagonal terms as in conventional adaptive beamforming. It can thus identify proper beams for each source despite the fact that multiple sources are simultaneously active. Strict one-channel power criteria have a serious crosstalk or leakage problem, especially in reverberant environments.

Geometric Source Separation

We propose to combine blind source separation and geometric beamforming by minimizing cross-power spectra for multiple times while enforcing constraints used in conventional adaptive beamforming. This can be done explicitly by adding a geometric constraint to the optimization criteria, resulting in an algorithm we call geometric source separation,[4] or implicitly by embedding the constraint in the system architecture, resulting in the generalized sidelobe decorrelator.[5] The former approach will be discussed in this section and the latter in the next section.

Geometric Constraints for Source Separation

To include geometric information, we will assume that the sources we are trying to recover are localized at angles $\theta = [\theta_1, ..., \theta_M]$ and at sufficient distance for a far-field approximation to apply. Following "Linear Constraints in Geometric Beamforming," the response of the M filters in W for the M directions in θ is given by $W(\omega) D(\omega, \theta)$, where $D(\omega, \theta) = [d(\omega, \theta_1), ... d(\omega, \theta_M)]$. In this section, we consider linear constraints such as

$$\text{C1:} \quad \text{diag}\,(W(\omega)D(\omega,\theta)) = 1 \tag{15.14}$$

$$\text{or C2:} \quad W(\omega)D(\omega,\theta) = I \tag{15.15}$$

Constraint (Equation [15.14]) restricts each filter $w_i(\omega)$ — the ith row vector in $W(\omega)$ — to have unit response in direction θ_i. Constraint (Equation [15.15]) enforces in addition that they have zero response in the direction of interfering signals θ_j, $i \neq j$.

Note that condition (Equation [15.15]) requires that $D(\omega, \theta)$ is invertible for the given set of angles. This is, however, not always possible. At the frequencies where the grating lobes* of a beam pattern cross the interfering angles, $D(\omega, \theta)$ is not invertible. It is, therefore, not reasonable to try to enforce Equation (15.15) as a hard constraint. Rather, as we confirmed in our experiments, it is beneficial to enforce Equation (15.15) as a soft constraint by adding a penalty term of the form $J_{C2}(\omega) = \|W(\omega)D(\omega,\theta) - I\|^2$ to the optimization criteria (Equation [15.6]). Note also that power or cross-power mini-

---

* Periodic replica of the main lobe due to limited spatial sampling.

mization will try to minimize the response at the interference angles. This will lead to an equivalent singularity at those frequencies. It is, therefore, beneficial to enforce condition (Equation [15.14]) also only as a soft constraint by using a penalty term of the form $J_{C1}(\omega) = \|\text{diag}(W(\omega)D(\omega,\theta)) - 1\|^2$.

Constraints as Penalty Terms

We implemented the linear constraints (Equations [15.14] and [15.15]) each as a soft constraint with a penalty term. We have further addressed the problem of noninvertibility discussed in "Geometric Constraints for Source Separation," by introducing a frequency dependent weighting of the penalty term. The idea is to eliminate the constraints from the optimization for those frequency bands for which $D(\omega, \theta)$ is not invertible. A rather straightforward metric for invertibility is the condition number. We, therefore, weight the penalty term with the inverse of the condition number of $\lambda(\omega) = $ cond$^{-1}(D(\omega, \theta))$, which converges to zero when $D(\omega, \theta)$ is not invertible and remains bounded otherwise, i.e., $0 \le \lambda(\omega) \le 1$. The total cost function including frequency dependent weighting of the geometric penalty term is given by

$$J(W) + \lambda \sum_{\omega} \lambda(\omega) J_{C1/2}(W(\omega)) \qquad (15.16)$$

In algorithm *GSS-C1* the penalty term J_{C1} will maximize the response of filters i in orientation θ_i. Note that the delay-sum beamformer ($w(\omega) = d(\omega, \theta)^H$) satisfies conditions C1 strictly. In algorithm *GSS-C2* the penalty term J_{C2}, in addition, will minimize the response for the orientations of the interfering sources. The filter structure that guarantees constraints C2 strictly can be computed with a least squares approach as the pseudo-inverse of $D^H(\omega, \theta)$, or including a regularization term βI for the noninvertibility problem the solution is given by $W(\omega) = D^H(\omega, \theta) (D(\omega, \theta) D^H(\omega, \theta) + \beta I)^{-1}$. We denote this solution by *LS-C2*. All GSS algorithms reported here minimize cross power using a straightforward gradient descent algorithm.[2]

Performance Evaluation and Discussion

Examples of typical response patterns for the GSS algorithms are shown in Figure 15.1, which shows the beam patterns of the filter weights for a linear array of four microphones with an aperture of 70 cm. There were two sources located at 0 and $-40°$ broadside to the array.

Algorithms *GSS-C1* and *GSS-C2* place a zero at the angles of interfering sources while maintaining a main lobe in the directions of the corresponding source. For conflicting frequency bands, where a grating lobe coincides with the location of an interfering source, multiple cross-power minimization reduces the main lobe. Qualitatively, the results for the data-independent

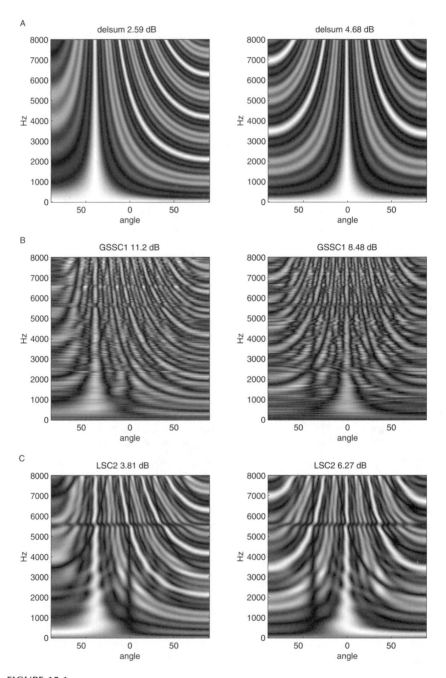

FIGURE 15.1
Response for geometrically constrained source separation. Algorithms *GSS-C1* and *GSS-C2* minimize (Equation [15.16]) with constraints C1 and C2, respectively. *del-sum* and *LS-C2* satisfy the respective constraints explicitly and are shown for comparison.

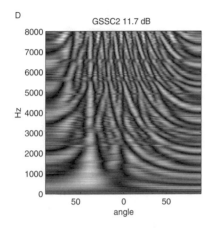
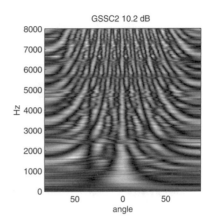

FIGURE 15.1
Continued

LS-C2 algorithm capture both main lobe and zeros at the correct angles. Its performance, however, is inferior to the data-adaptive algorithms.

A systematic performance evaluation of the algorithms for the case of two sources in a moderately reverberant room (T_{30} = 50 ms) is presented in Figure 15.2. We varied the locations of two speakers who were always at least 2 m from the array. The number of microphones was varied (two to eight), but the array aperture was kept at 70 cm. The top row shows the results for some known beamforming algorithms (*del-sum, LS-C2, LCMV*).

The criteria (Equation [15.6]) represent a nonconvex optimization problem. The results for the optimization procedure, therefore, strongly depend on the initial conditions. For comparison, the center row in Figure 15.2 presents the results for unconstrained multiple cross-power minimization with different initializations of the filter structure. Initializations that have been considered are unit filters ($W(\omega) = I$), delay-sum beamformer (*del-sum*), and least squares (*LS-C2*). The results for unconstrained optimization with the different initializations are labeled *BSS, GSS-I2*, and *GSS-I1*, respectively.

The last row shows the results for the geometrically constrained separation algorithms (*GSS-C1', GSS-C1, GSS-C2*). Algorithm *GSS-C1'* is the same as *GSS-C1* only with constant penalty term λ. Within each row the algorithms are sorted by average performance. Comparison of the results for *GSS-C1'* and *GSS-C1* show the advantage of the frequency-dependent weighting of the penalty term. Due to the limited angular resolution all algorithms perform poorly when the sources are too close.

We now present results obtained for the separation of three sources. Note that the permutation problem discussed in "Ambiguities of Independence Criteria" becomes worse as the number of sources increases. We show in Figure 15.3 the performance of separating two speakers and babble noise using a linear array of eight microphones. The performance mirrors mostly the results obtained for the separation for two sources.

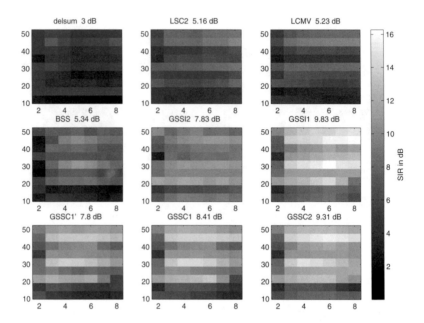

FIGURE 15.2
Performance comparison of the proposed algorithms and geometric beamforming for two sources. SIR performance in decibels is encoded in gray scale as a function of number of microphones (horizontal axis) and angular separation (vertical axis: 12°, 18°, 19°, 25°, 33°, 37°, 38°, 41°, 50°). SIR performance averaged over all positions and number of microphones is also given. (Signal-to-interference ratio (SIR) is used as a separation metric, which measures the ratio of power (dB) in the enhanced channel to the rejection channel during periods when only one speaker is active.)

In these experiments the cross-power spectra were estimated at five time instances with a time window of about 3 s each, such that a total of about 15 s of data is analyzed. In all experiments we used a linear array of cardioid condenser microphones. The user locations were identified acoustically.[4]

Generalized Sidelobe Decorrelator

As we mentioned previously, one possibility for enhancing a point source while suppressing noise is the linearly constrained minimum variance (LCMV) algorithm, which adaptively filters the sensor signals so as to minimize power, subject to a constraint that a delay-sum beam points in the direction of the source of interest. An alternative but equivalent approach is the generalized sidelobe canceller (GSC),[18] shown in Figure 15.4(a). It also implements power minimization criteria on the filtered sensor signals. However, unlike the LCMV, the requirement that a beam points in the direction of interest is enforced in the architecture rather than the criteria. Specifically, the GSC utilizes a delay-sum beam through the use of steering delays

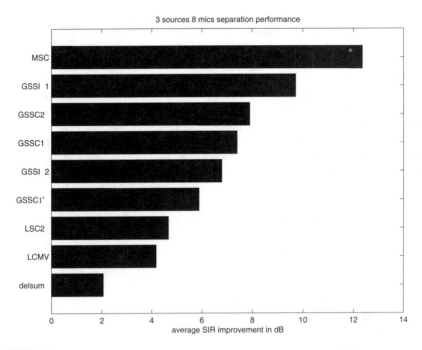

FIGURE 15.3
Performance for the separation of three sources using eight microphones. SIR improvement averaged over three configurations with angles –78, –41, 0°; –60, 0, 60°; and –43, 0, 36°. The initial average SIR is about –3 dB.

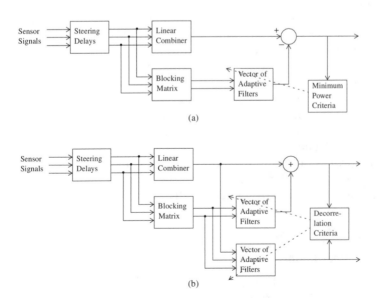

FIGURE 15.4
(a) Generalized sidelobe canceller; (b) generalized sidelobe decorrelator.

followed by a linear combiner. The linear combiner is a window that can be designed to vary the trade-off between main lobe width and sidelobe energy. After the steering delays but prior to the linear combiner, the signals are all in phase. This is exploited to form beams orthogonal to the primary beam through the use of a "blocking matrix."[3] Each row of the blocking matrix is constrained to sum to zero to ensure that the resulting secondary beams will all have a null in the direction of the primary beam. During adaptive power minimization, the secondary beams are adapted out of the primary beam but are prevented by the blocking matrix from canceling any signal that exclusively resides in the primary beam. The GSC approach has the advantage that the resulting optimization can be carried out using unconstrained power minimization, such as the least mean squares (LMS) algorithm. Unlike the LCMV, the constraint is always enforced and no extra steps have to be taken to ensure that the filter weights do not stray from the constraint over time due to finite precision effects.

However, although the GSC exploits the available prior geometric information, it does not exploit the independence prior, and is thus subject to the leakage problem associated with power minimization. That is, any leakage of the primary source into the secondary beams will result in cancellation of the primary source and a degradation of the signal-to-noise ratio (SNR) improvement. This leakage can be due to any of several factors, including (1) array calibration errors; (2) primary source location error; (3) a main beam that is narrower than the primary source, caused by a large array aperture; (4) spatial aliasing lobes, caused by an insufficiently spaced sensor array; (5) reverberation, caused by reflections of the primary source coming from directions outside the primary beam.

To overcome these deficiencies, we combine aspects of the generalized sidelobe canceller and blind source separation to create an algorithm we call the generalized sidelobe decorrelator (GSD),[5] shown in Figure 15.4(b). Like the GSC, it consists of steering delays that place all the sensor signals in-phase, a linear combiner that forms the primary beam, and a blocking matrix that forms secondary orthogonal beams. However, unlike the GSC, instead of adopting power minimization criteria that adapt the secondary beams out of the primary beam, we adopt cross-power minimization criteria, as described in "Online Decorrelation," that decorrelates the secondary beams from the primary beam. This allows for removing leakage of the primary source into the secondary beams, while the blocking matrix guarantees the integrity of the primary beam independent of whether the sources are continually active.

Results in Real Rooms

We conducted an acoustic experiment designed to demonstrate the superior performance of the algorithm for noise reduction. A two-dimensional rectangular sensor array of dimension 10×7 cm was formed, corresponding to

TABLE 15.1

Real-Room Experiment

Algorithm	SNR (dB)	CER (%)
None	1.2	77.6
Fixed delay-sum beam	1.3	19.4
Generalized sidelobe canceller	3.0	73.9
Blind source separation	3.6	100.0
Generalized sidelobe decorrelator	4.6	5.4

the dimensions of a personal digital assistant (PDA), using inexpensive omnidirectional lapel microphones (Audio-Technica ATR35S).

The array was located in a room of dimension $3.0 \times 3.6 \times 2.3$ m. A loudspeaker was placed 0.5 m directly in front of the array, which was used to replay a quiet recording of a male speaking 300 short commands over a period of 20 min, with a pause of 2 to 3 s between commands. The recording was automatically segmented into speech/nonspeech for the purpose of measuring SNR, and the speaker had previously trained an automatic speech recognition system for the purpose of measuring speaker-dependent command error rate (CER). The recognizer and all algorithms operated at 11.025 kHz.

Also in the room but in the corner and facing the wall 2.5 m from the array, a loudspeaker played babble (the sounds of many voices). Outside the room, another loudspeaker played a recording of street noises. The nominal SNR at the microphones was 1.2 dB, which corresponded to a CER of 77.6%. We then applied four online adaptive algorithms to the array signals, each of which used FIR filter sizes of 512 taps. The results are shown in Table 15.1.

Because the source was directly in front of the array, the fixed delay-sum beam could be obtained by a simple averaging of the four sensors. Although the fixed beam does not provide much SNR improvement, it does provide significant CER improvement, primarily because it does not distort the speech.

Next, we implemented the GSC using a "Walsh" blocking matrix (see Reference 18) to form three secondary beams orthogonal to the primary delay-sum beam. The secondary beams were adapted out of the primary beam using the frequency domain LMS algorithm. Although there is improvement in the SNR, there is degradation in the CER relative to the delay-sum beam, most likely due to spectral distortion of the speech.

Next, we applied BSS on the four raw input signals using the algorithm of "Online Decorrelation" with two outputs. Although BSS provides a small SNR improvement over GSC, the algorithm completely destroys the recognition performance. Part of the problem is that BSS requires that the sources be simultaneously active, and thus the filters start to degrade during the silent periods between commands. In addition, the frequency-domain permutation problem (see "Ambiguities of Independence Criteria") can distort the speech spectrum.

Finally, we applied our new hybrid GSD by performing BSS on the fixed delay-sum beam and blocking matrix outputs taps, and obtained very encouraging results. In addition to obtaining the largest SNR improvement of any of the algorithms, the CER was a very respectable 5.4%, approaching the single microphone CER of 2.0% in a quiet environment.

Conclusions

This chapter emphasizes the importance of second-order criteria and the use of prior geometric information to solve the problem of separating multiple sources in an acoustic environment. It combines notions from adaptive beamforming and blind source separation resulting in semiblind algorithms where at least microphone locations are known. The assumption is made that sources are reasonably well localized and that user location can be determined acoustically. The algorithms overcome the crosstalk problems of conventional adaptive beamforming and the ambiguity problems of convolutive blind source separation.

References

1. Weinstein, E., Feder, M., and Oppenheim, A.V., Multi-channel signal separation by decorrelation, *IEEE Trans. Speech Audio Processing*, 1(4), 405–413, 1993.
2. Parra, L. and Spence, C., Convolutive blind source separation of nonstationary sources, *IEEE Trans. on Speech and Audio Processing*, May 2000, 320–327.
3. Van Veen, B. and Buckley, K., Beamforming techniques for spatial filtering, in *Digital Signal Processing Handbook*, CRC Press, Boca Raton, FL, 1997, chap. 61.
4. Parra, L. and Alvino, C., Geometric source separation: merging convolutive source separation with geometric beamforming, in *IEEE International Workshop on Neural Networks and Signal Processing*, 2001, 273–282.
5. Fancourt, C. and Parra, L., The generalized sidelobe decorrelator, in *Proc. IEEE Workshop on Applications of Signal Processing to Audio and Acoustics*, 2001.
6. Thi, H.-L.N. and Jutten, C., Blind source separation for convolutive mixtures, *Signal Processing*, 45(2), 209–229, 1995.
7. Yellin, D. and Weinstein, E., Multichannel signal separation: methods and analysis, *IEEE Trans. Signal Processing*, 44(1), 106–118, 1996.
8. Shamsunder, S. and Giannakis, G., Multichannel blind signal separation and reconstruction, *IEEE Trans. Speech Audio Processing*, 5(6), 515–528, 1997.
9. Ehlers, F. and Schuster, H.G., Blind separation for convolutive mixtures and an application in automatic speech recognition in a noisy environment, *IEEE Trans. Signal Processing*, 45(10), 2608–2612, 1997.
10. Sahlin, H. and Broman, H., Separation of real-world signals, *Signal Processing*, 64, 103–104, 1998.
11. Kawamoto, M., A method of blind separation for convolved nonstationary signals, *Neurocomputing*, 22(1–3), 157–171, 1998.

12. Fancourt, C. and Parra, L., The coherence function in blind source separation of convolutive mixtures of nonstationary signals, in *Proc. IEEE Workshop on Neural Networks for Signal Processing*, 2001, 303–312.
13. Haykin, S., *Adaptive Filter Theory*, Prentice-Hall, Englewood Cliffs, NJ, 1996.
14. Capdevielle, V., Serviere, C., and Lacoume, J.L., Blind separation of wide-band sources in the frequency domain, in *Proc. ICASSP*, 1995, 2080–2083.
15. Diamantaras, K., Petropulu, A., and Chen, B., Blind two-input-two-output FIR channel identification based on frequency domain second-order statistics, *IEEE Trans. on Signal Processing*, 48(2), 534–542, 2000.
16. Frost, E., An algorithm for linearly constrained adaptive array processing, *Proc. IEEE*, 60(8), 926–935, 1972.
17. Affes, S. and Grenier, Y., A signal subspace tracking algorithm for microphone array processing of speech, *IEEE Trans. on Speech and Audio Processing*, 5, 425–437, 1997.
18. Griffiths, L.J. and Jim, C.W., An alternative approach to linearly constrained adaptive beamforming, *IEEE Trans. Antennas Propagation*, AP-30(1), 27–34, 1982.

16

Use of DSP Techniques to Enhance the Performance of Hearing Aids in Noise

Douglas Chabries and Victor Bray

CONTENTS

Introduction

The major complaint of individuals with hearing impairment, regardless of whether they wear hearing aids, is the reduced ability to understand speech in a noisy environment. Hearing in the presence of noise is a complex problem that will be discussed for several classes of interference. The

0-8493-0949-2/02/$0.00+$1.50
© 2002 by CRC Press LLC

engineering challenge is to provide audibility of speech in order to overcome the hearing loss and simultaneously reduce background noise, thereby enhancing the signal-to-noise ratio (SNR) for the hearing-impaired person.

Digital signal processing (DSP) holds significant promise of precise and accurate compensation for hearing loss. Hearing aid signal processing purposefully distorts the speech signal in the frequency domain through frequency shaping and distorts the signal in the amplitude domain through compression. Although these signal alterations are desirable, it is also important to minimize distortion that occurs in hearing aids at high amplification levels. Digital systems can also reduce or eliminate feedback (acoustic squeal), which is a normal consequence of an open-loop amplification system. In addition, SNR enhancement algorithms are possible and may be categorized in several classes that include binaural processing, microphone technologies, and noise reduction algorithms.

Hearing Loss

The degree of hearing loss often varies with frequency. A typical pattern is increased hearing loss for high-frequency signals rather than for low-frequency signals. With speech, this results in a loss of sensitivity for consonants but not vowels. The result is the hearing-impaired person is aware of the talker's voice but has difficulty understanding the exact words that are being spoken.

Hearing aids compensate for this frequency-domain problem using frequency-shaping amplification, whereby more electronic gain is applied in the regions of greater hearing loss. For sensorineural hearing impairments, the gain correction applied is typically up to one half the loss for moderate impairments, and up to two thirds the amount of loss for severe impairments.[1,2] Emphasis in frequency shaping is in the region of 750 to 4000 Hz, which provides approximately 75% of the cues for speech understanding.[3]

Sensorineural hearing loss results from damage to the cochlea, or inner ear. When due to cochlear pathology, The loss is characterized by recruitment, which is a reduction in sensitivity to low- and moderate-intensity auditory signals but near-normal sensitivity for high-intensity signals. Modern hearing aids compensate for this with wide dynamic range compression (WDRC) amplification, whereby more gain is applied to low-intensity signals than to high-intensity signals. For moderate-to-severe hearing loss, the differential gain correction for low-intensity signals can be twice that for high-intensity signals.[4]

SNR Loss

Not only does sensorineural hearing loss produce a loss of sensitivity for auditory signals, as measured in quiet, but also a further impairment in processing ability, as measured in noise. Some of the impairment occurs in

FIGURE 16.1
Basic configuration of a DSP hearing aid receiving a signal as a function of time [s(t)] and processing it as a signal as a function of the number of samples [s(n)].

the frequency domain, where a sound at one frequency masks the audibility of the target sound at an adjacent frequency. Additional impairment occurs in the temporal domain, where a sound immediately preceding or following the target sound can also produce a masking effect.

The hearing loss for speech intelligibility in noise is measured as the SNR required for the listener to maintain understanding of conversational speech in the presence of noise. As hearing loss in quiet increases, there is a general trend for the SNR impairment to increase.[5] The degradation in SNR occurs at a rate of about 1.0 to 1.5 dB SNR for each 10 dB of hearing loss in quiet.[6] The implication of the SNR loss is that persons wearing hearing aids must be provided with a more positive SNR than normal-hearing people in order to understand speech in the presence of noise in the same listening environment.

The DSP Hearing Aid

In this chapter, DSP will be presented in the context of a digital hearing aid, which comprises a microphone, an analog-to-digital converter, a digital signal processor, a digital-to-analog converter, and a receiver, as shown in Figure 16.1.

Many hearing aids exist that provide digital control of analog circuitry. The true DSP hearing aid should not be confused with digitally controlled analog aids. To date, these digitally controlled analog devices have not been effective in noise suppression and have occasionally suffered from problems associated with distortion and inadequate filtering ability. This discussion will focus on the DSP techniques that are now emerging and hold promise of increasing the SNR in a variety of noisy environments.

The most recent all-digital hearing aids contain between 400,000 and 1,300,000 transistors on a silicon chip small enough to fit inside the ear canal. These digital hearing aids operate nominally at 1 V with a battery drain of about 1 mA, using approximately 40% of the power budget for digital signal processing and 60% for analog-to-digital and digital-to-analog conversion tasks. Battery life for the DSP hearing aid is 1 to 2 weeks for a small 1.3 V battery.

Modern digital circuitry allows such algorithms as feedback suppression, digital fine-tuning of directional microphones, and spectral subtraction to be implemented in current DSP hearing aids. As integrated circuit densities increase, it is reasonable to expect that algorithmic complexities can be added

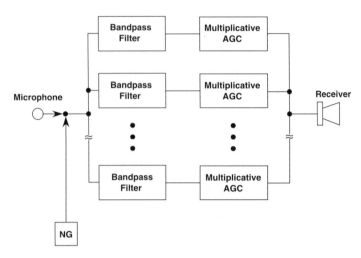

FIGURE 16.2
A multichannel DSP hearing aid with a separate automatic gain control (AGC) for each compression channel.

as the normal advances in integrated circuit capabilities progress on a time-table consistent with Moore's Law.*

An approach which offers promise when applied to hearing aids is based on multichannel/multiband signal processing. A technique shown in Figure 16.2 operates on the output of the microphone and an internal noise generator (NG) and divides the acoustic spectrum into multiple bands.[7-9] The bands are preferably chosen to mimic the bandwidth of the critical bands of hearing, although in practice the bands may cover a broader region to simplify the processing. Each band is then processed through an automatic gain control (AGC) that adjusts the applied gain such that signal levels between threshold and the desired maximum intensity are mapped to evoke the corresponding desired perceptions of loudness for the hearing-impaired individual.

A simple method of achieving this result is to extract the amplitude envelope of the acoustic signal in each band of the multiband system, measure the intensity, and map the signal to the desired output level. One method of envelope extraction and loudness mapping is shown in Figure 16.3. In this figure, the value of the constant e_{max} is chosen to be the maximum envelope amplitude for which gain is desired, hence for all envelope amplitudes larger than e_{max}, attenuation rather than gain is the result. The parameter K is chosen to provide the appropriate sensation for acoustic intensities near the hearing

---

* In the April 19, 1965, *Electronics* magazine 35th anniversary issue was an obscure article entitled, "Cramming more components onto integrated circuits," by Gordon E. Moore, Director, Research and Development Laboratories, Fairchild Semiconductor. In this article, the following quote, observed to be accurate now over 36 years, has become referred to as Moore's Law: "The complexity for minimum component costs has increased at a rate of roughly a factor of two per year. Certainly over the short term this rate can be expected to continue, if not to increase. Over the longer term, the rate of increase is a bit more uncertain, although there is no reason to believe it will remain nearly constant for at least 10 years."

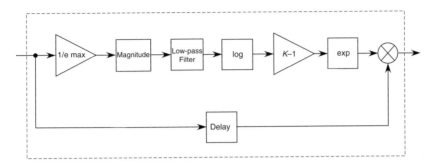

FIGURE 16.3
A multiplicative AGC used in an all-DSP hearing aid.

threshold. It is important that the low-pass filter used to smooth the envelope be selected accordingly to vary with the critical bandwidth associated with the band being processed.

Feedback Reduction

A frequent complaint about hearing aids is the annoying acoustic feedback that occurs at higher amplification levels. Often the squeal is not audible to the hearing aid user, but becomes an annoyance to those nearby. In addition to being annoying, the feedback can drive the system to its maximum operating limits introducing distortion and quickly draining battery energy. Digital feedback compensation has been proposed by Best,[10] applying a least mean square error (LMS) adaptive filter to provide feedback suppression. Best reported gains of approximately 10 dB for his implementation.

Adaptive feedback cancellation that varies with the dynamics of the environment is very attractive because it reflects typical user environments. Recently developed implementations of adaptive feedback cancellation have been proposed with an eye toward implementation in digital hearing aids.[11,12]

Joson et al.[13] have proposed an adaptive canceller that uses frequency compression as a preprocessor to the hearing aid signal. The frequency compression is provided to decorrelate the output signal from the input signal as sound from the hearing aid receiver leaks back into the hearing aid microphone. The use of this frequency compression for feedback cancellation indicates a significant improvement and additional gain is achievable. Joson et al. report a feedback margin increase of about 18 dB for one set of operating conditions.

The frequency compression implementation of feedback suppression offers the advantage of preserving features of the tonal quality of speech, thereby suppressing feedback while preserving desired input signals to the hearing aid. Hearing aid users can expect new feedback algorithms to be implemented shortly because digital techniques to achieve feedback reduction can now be implemented within current hearing aid power and computation constraints. As a note, the spectral subtraction noise suppression techniques also provide, to a limited extent, a reduction in feedback, thereby allowing for modest additional gain in digital hearing aids.

Treatment for SNR Loss

Treatment for SNR loss, or the impairment of processing ability in the presence of background noise, is an area of great potential for DSP applications in hearing aids. Techniques to improve speech understanding in noise include binaurally integrated amplification, directional microphone technology, and digital noise reduction algorithms.

Binaural Amplification

Binaural fitting of hearing aids provides many advantages for complex listening tasks, including better localization in the horizontal plane and increased speech understanding in noisy situations through the physical factor of head shadow effects and the physiological factor of binaural squelch.[14] A binaural fitting approach, for symmetric or asymmetric hearing losses, is first to adjust the frequency-shaped gain response of the multiband hearing aid in the patient's ear with the best hearing. Once the fitting of the first aid is completed, a narrowband noise confined in bandwidth to one of the multiple bands of the hearing aid is presented to the patient via headphones or in a soundfield. With an aid in each ear — the first aid having already been fitted — the gain of the second aid is adjusted in the band corresponding to the narrowband noise to provide the patient with the sensation that the sound source is directly in front or centered in the head (i.e., the patient hears the sound equally in both ears). The process is repeated until all bands of the multiband aid are binaurally balanced. When the sound source is placed directly in front of the individual and the hearing in both ears is balanced, the hearing aid wearers will perceive the sound to be equally loud in both ears. Then as the sound source is moved to other positions, most hearing aid wearers will be able to localize correctly. This binaural restoration is generally not possible with analog aids.

Microphones

Directional Microphones

Modern first-order directional microphones, in contrast to more complex arrays of microphones, are currently available for implementation with either analog or digital hearing aids. A common implementation in digital hearing aids is to use two omnidirectional microphones. For the omnidirectional mode, output is only processed from the front microphone. For the directional mode, the output from the back microphone is delayed and subtracted from the output of the front microphone, and the resulting signal serves as the input to the DSP.

Directionality is optimized when the two microphones are equal in sensitivity, and a digital equalization filter can be used to fine-tune the response of one microphone to match the other. Directional efficiency, as measured by polar plot, is controlled by the time delay between the back and front microphone. Digital circuits are used to alter the time delay so that the pattern can be changed among cardioid, hypercardioid, or supercardioid patterns. The changes may either be static, i.e., always set to hypercardioid pattern, or dynamic, i.e., changing between cardioid and hypercardioid patterns as the location of environmental noises change.

Directional microphones have consistently produced significant gains in laboratory measures of word recognition tests in noise, provided there is some spatial separation between the speech signal and the masking noise. In the idealized situation, where the signal is from the front and the masker is located from behind and/or near the polar plot null, typically 6 dB of improvement in SNR can be achieved.[15] However, many user environments include parties, cafeterias, public transportation, and conferences. In these environments, both the speech from a desired talker and competing signals bounce off walls, ceilings, floors, and other objects to produce a spatially diffuse reverberatory environment thereby complicating or reducing the directional gains. In laboratory simulations of these situations, where the signal is from the front and the masker is diffuse, either by using multiple loudspeakers or introducing reverberation, the measured benefit is reduced to about 3 dB of improvement in SNR.[16]

Array Microphones

An assembly of microphones with spatial separation can be used in an array to provide spatial processing gains. DSP can be used to distinguish a variety of noise and speech signals based on temporal cues, frequency characteristics, and the spatial character of the noise. Algorithms have been proposed which adaptively adjust the spatial sensitivity of a hearing system to focus in a particular direction while minimizing competing sources from other directions.[17-20]

The advantages of these adaptive array systems do not come without cost. These complex systems produce second-order and higher polar responses and require the use of several microphone inputs and some kind of external mounting system such as the top rim of eyeglasses for a broad-fire approach or the temple of the eyeglasses for an end-fire approach. Further, there are dynamic trade-offs between the rate of spatial adaptation and the effectiveness of the array beam former. Although these systems can optimize spatial discrimination, they present cosmetic challenges and processing complexity that likely will restrict their use to specialized applications.

Digital Noise Reduction

Noise suppression is a topic that has occupied the attention of many research-ers during the past two decades. In the hearing aid literature, noise suppres-sion refers to reducing the noise level to provide an improvement in the intelligibility of speech in the presence of noise. To address this complaint several noise suppression algorithms have been proposed, but these have generally failed to provide an increase in intelligibility as measured in stan-dardized behavioral tests.

The ability to remove noise from speech is reliant upon differences between the characteristics of the speech and unwanted noise. Many of the noise suppression techniques have focused on differentiating a speech signal from a noise background, which is characterized as either white or pink noise. A hearing-impaired individual more typically finds that the noisy environ-ments they encounter are of competing conversations or speech babble. The frequency-amplitude characteristics of the speech babble closely resemble those of the desired speech spectrum, thereby complicating the task of noise suppression. Further, there is a tendency in the application of the DSP algo-rithm to remove the noise as completely as possible in order to maximize the SNR.

There are situations, however, when background noise can be most helpful or even critical and, therefore, must be preserved. For example, a hearing-impaired person who begins to step off a curb would like to hear the approach of a bus to warn that it is not safe to enter into the roadway. Or, in a public environment, such as a theater, it is critical to preserve a distant warning of "fire." To this end, several different DSP techniques have been proposed, each of which addresses the improvement in SNR to differing degrees.

The single-channel methods of speech enhancement generally attempt to minimize measures of mean square error to improve the ratio of speech signal power to noise power. If the speech and noise can be precisely char-acterized, then such an approach offers some promise. However, in most situations, the precise character of the speech and noise vary from moment to moment as encountered by a hearing-impaired listener. Further, when the characteristics of the noise and speech become sufficiently similar, then speech enhancement techniques become ineffective.

Adaptive Predictive Filtering

One of the early techniques using only the input from a single microphone for removing noise was proposed by Sambur.[21] In this application, an adap-tive filter was used as a linear predictor to estimate the current speech plus signal characteristics based upon previous samples of the same input. The microphone input was fed into an adaptive filter delayed by an amount of time equivalent to the pitch of the desired speech signal that was embedded in the noise. An LMS adaptive algorithm, proposed earlier by Widrow,[22] was employed to estimate the correlated portions of the signal that corresponded to speech and to deemphasize the noise signals. Gains in SNR of approximately 7 dB were reported; however, no measures showing gains in intelligibility for this algorithm have been reported.

Near the same time period, Graupe and Causey introduced an adaptive filtering technique based on a different approach.[23] Their approach used signal analysis to distinguish between speech and nonspeech signals and attempted to filter out the near-stationary noise. Listeners generally did not report a significant benefit, and the technique is no longer used.

Spectral Subtraction

A more commonly used technique known as spectral subtraction has been proposed.[24-27] Spectral subtraction (SS) uses a discrete Fourier transform (DFT) to obtain estimates of the noise spectrum, which are then subtracted from the spectral magnitude in those areas where noise is present. The use of this algorithm results in a background artifact, which sounds much like a babbling brook and has been reported as a warbling or residual noise. The elimination of this residual noise has been the object of significant research over the past decade.

Because problems occur when SS errors are made near the noise floor, researchers have suggested the use of a noise floor to limit the amount of suppression.[28] The imposition of a noise floor provides a lower limit on the gain and is typically imposed when the SNR ratio becomes 0 dB or less. In effect, this maintains the gain at or above a specified threshold for low-level signals and for noise inputs. Other techniques of SS have capitalized on the nature of the residual noise peaks.[29] These techniques have enjoyed some considerable success. Although the initial approaches to spectral subtraction utilize a DFT, others have proposed using the Karhunen–Loeve transform (KLT)[30,31] or wavelets to achieve a similar noise suppression effect, while simultaneously minimizing the residual noise energy.

Many of these subtraction techniques have been evaluated in a variety of environments. Several tests have indicated improved listener comfort for the processed signal using spectral subtraction. However, to date, intelligibility tests have failed to verify an increase in speech intelligibility.

In contrast to the foregoing methods, a simple, but effective approach to noise reduction has been employed in a commercially available DSP hearing aid, the Sonic Innovations (Salt Lake City, Utah) NATURA 2 SE. The technique uses an estimate of the output signal level from the logarithmic operation of Figure 16.3 when the signal is deemed to be void of a desired signal (normally, one looks for a long-term background level for this estimate). A separate estimate is made in each band of the multiband system. When the time-varying amplitude envelope of each of the multiband signals is near the level of the respective background noise estimate, a constant is subtracted from the output signal of the logarithmic operation.

The algorithm uses SS with a noise floor, but it operates not on the input signal as described previously but rather operates on the logarithm of the signal envelope magnitude in each of the nine parallel channels. The effect of subtraction in this logarithmic domain is quite different from the application of subtraction to an estimate of the signal power. Subtraction in the logarithmic domain is equivalent to division. Hence, the final signal obtained at the output of the multiplier in Figure 16.2 is scaled in amplitude, thereby reducing its level. The scaling in the log domain can be chosen to reduce background signals by 6, 12, or 18 dB. When amplitude envelope levels exceed the estimate, the envelope levels in each band are expanded so that the loudest signals are restored to their original levels. This approach to noise reduction does not increase the instantaneous SNR in any single band; however, the constantly adjusting gain levels in the multiple bands of the aid can selectively attenuate and enhance information across frequencies and over time.

The effect on SNR is complex, including rapid gain adjustments that are dependent upon the short-term SNR, but clinical results show an increase in speech intelligibility in noise using behavioral tests.[32] To the listener, background noises are attenuated but not eliminated. For complex noise backgrounds, such as speech babble that present a somewhat constant amplitude envelope, the babble is attenuated and foreground speech with more energy is expanded. A secondary benefit of this approach is that feedback is modestly ameliorated, since a constant feedback level is interpreted as background and the gain is therefore reduced.

The foregoing noise suppression signal processing algorithms have been developed to provide noise suppression with the use of a single microphone. These methods, however, are generally ineffective when the noise power in the background is at a level equal to or greater than the speech. For such cases, it is necessary to provide to the digital signal processing system an independent measure of the interfering signal, which can be used to remove the noise from the speech plus interference.

Noise Cancellation

Two other techniques are available to eliminate interference in special situations. The adaptive filter, as originally proposed by Widrow,[22] has been used effectively to reduce interference in the case where a secondary noise reference is available. SNR improvements of greater than 15 dB have been obtained from this technique with attendant gains in intelligibility of approximately 38% on some word lists.[33] Although the use of such a two-channel system provides very promising results, its application to a hearing aid is problematic and is perhaps limited to the use of an auditory trainer. In such applications a signal from a central point can be relayed to the hearing aid user for processing. To date, no such systems have been implemented or employed. Further, the number of filter taps used in these adaptive filters is typically very large, thereby making their application prohibitive in hearing aids. These filters do have the advantage, however, that they are effective against all classes of noise, ranging from speech babble to white noise, and work well in a number of dynamic noise environments.

Reverberation Cancellation

Reverberation refers to the presence of replicas of the original speech signal, slightly altered in amplitude and timbre, and arriving at the listener at different times corresponding to reflections from floors, ceilings, walls, and adjacent objects. The reverberation or echo properties tend to have a significant negative effect on the speech intelligibility. DSP algorithms have been investigated for reverberant environments.[34-37] However, these algorithms require that the signal be collected and processed over a duration of time roughly equivalent to the reverberation time constant. Since these time constants can be large, the digital storage and processing requirement would indicate that dereverberation is not to be envisioned in the near future as a candidate for application to digital hearing aids.

Combined Treatment Effects

The complex nature of these signal manipulation interactions is accentuated in the application of directional microphones to binaural hearing. By correctly using directional microphones, providing 3 to 4 dB of SNR benefit, and then balancing the binaural response, providing 2 to 3 dB of SNR benefit, the combination of binaural hearing and directional microphones have an additive effect.[38] Further effects of combining strategies can be seen in Figure 16.4, measuring the increase in speech intelligibility in the presence of noise with the digital hearing aid described previously in this chapter.

In Figure 16.4, five listening situations are plotted with respect to the SNR for understanding speech in noise using a modified version of the Hearing in Noise Test (HINT).[39] The HINT is a standardized test to measure speech intelligibility in noise, where the speech is sentences spoken by a male talker and the noise is steady state, filtered to match the long-term average spectrum of the speech. The noise level is fixed at 65 dBA and the level of the sentences is adjusted to find the 50% intelligibility point. Results are reported in SNR, with lower scores indicating better performance.

Subjects were 14 adults with bilateral, moderate to severe sensorineural hearing loss. In the unaided condition, they required a +3 dB SNR to maintain speech intelligibility in steady-state noise that is spectrally matched to the long-term average spectrum of the male talker. Using omnidirectional, binaural DSP aids employing multichannel compression, the improvement was about 2.4 dB SNR without noise reduction and 3.3 dB SNR with noise reduction. In the directional mode, the improvement was about 4.7 dB SNR

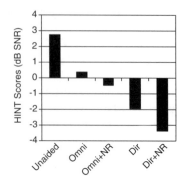

FIGURE 16.4
Speech intelligibility in noise measurements for 14 hearing-impaired listeners across five listening conditions.

without noise reduction and 6.2 dB SNR with noise reduction. An improvement of 1 dB SNR on the HINT corresponded to an approximately 8.5% improvement in speech intelligibility.[40]

Conclusions

About 10% of the U.S. population is reported to have a hearing loss, and the major complaint of hearing-impaired persons is difficulty understanding speech, especially in the presence of background noise. The DSP hearing aids of today offer many technical features to improve speech understanding that are not, and will not be, available in analog hearing aids. These features include compression circuits with as many as 20 or more bands/channels, allowing complex shaping of amplified sound in the frequency and amplitude domains; feedback reduction circuits, allowing increased amplification to be delivered to the ear without acoustic squeal; binaural integration algorithms to improve hearing and localization in complex listening environments; adaptive directional microphone systems that improve the SNR by exploiting spatial separations between speech and noise; and adaptive noise reduction algorithms that can improve the SNR in the presence of steady-state noises.

Performance measures show that many of these features have additive benefits, allowing hearing-impaired individuals to more than double their speech understanding in noise, thereby cutting their SNR deficit by more than half. The digital techniques available in hearing aids today, coupled with projected near-term advances, lead to the hope that hearing aid users will soon be able to understand speech in noise almost as well as normal-hearing individuals.

References

1. Byrne, D. and Dillon, H., The National Acoustic Laboratories' (NAL) new procedure for selecting the gain and frequency response of a hearing aid, *Ear and Hearing*, 7(4), 257–265, 1986.
2. Byrne, D., Parkinson, A., and Newell, P., Hearing aid gain and frequency response requirements for the severely/profoundly hearing impaired, *Ear and Hearing*, 11(1), 40–49, 1990.
3. American National Standards Institute, *American National Standard Methods for Calculation of the Speech Intelligibility Index* (S3.5-1997), American National Standards Institute, New York, 1997.
4. Dillon, H., NAL-NL1: A new prescriptive fitting procedure for non-linear hearing aids, *Hearing J.*, 52(4), 10–16, 1999.
5. Killion, M., Hearing aids: past, present, future: moving toward normal conversation in noise, *Br. J. Audiol.*, 21, 141–148, 1997.
6. Nilsson, M. et al., Single microphone noise reduction: New findings. Research poster presented at the American Academy of Audiology, April 2001.
7. Stockham, T.G., Jr. and Chabries, D.M., Hearing aid device incorporating digital signal processing techniques, U.S. Patent 5,500,902, March 1996.
8. Chabries, D.M. et al., Application of a human auditory model to loudness perception and hearing compensation, *Proc. IEEE ICASSP*, May, 1995.
9. Anderson, D.V., Harris, R.W., and Chabries, D.M., Evaluation of a hearing compensation algorithm, *Proc. IEEE ICASSP*, May 1995.
10. Best, L.C., Digital Suppression of Acoustic Feedback in Hearing Aids, Master's thesis, University of Wyoming, Laramie, 1985.
11. Kates, J.M., Feedback cancellation in hearing aids: Results from a computer simulation, *IEEE Trans. ASSP*, 39(3), 553–562, 1999.
12. Kates, J.M. and Melanson, J.L., Feedback cancellation improvements, U.S. Patent 6,219,247, April 2001.
13. Joson, H. et al., Adaptive feedback cancellation with frequency compression for hearing aids, *J. Acoust. Soc. Am.*, 94(6), 3248–3254, 1993.
14. Dillon, H., *Hearing Aids*, Thieme, New York, 2001.
15. Ricketts, T. and Dhar, S., Comparison of performance across three directional hearing aids, *J. Am. Acad. Audiol.*, 10(4), 180–189, 1999.
16. Ricketts, T., Impact of noise source configuration on directional hearing aid benefit and performance, *Ear and Hearing*, 21(3), 194–205, 2000.
17. Peterson, P.M. et al., Multimicrophone adaptive beamforming for interference reduction in hearing aids, *J. Rehab. Res. Dev.*, 24(4), 103–110, 1987.
18. Griffiths, L. J., An adaptive lattice structure for noise-canceling applications, *Proc. IEEE ICASSP*, 1978, 87–90.
19. Widrow, B. et al., Adaptive antenna systems, *Proc. IEEE*, 55, 2143, 1967.
20. Widrow, B. and Stearns, S.D., *Adaptive Signal Processing*, Prentice-Hall, Englewood Cliffs, NJ, 1985, 368–404.
21. Sambur, M.R., Adaptive noise canceling for speech signals, *IEEE Transactions ASSP*, 26(5), 419–423, 1978.
22. Widrow, B. et al., Adaptive noise canceling: Principles and applications, *Proc. IEEE*, 63(12), 1692–1716, 1975.

23. Graupe, D. and Causey, G.D., Method of and means for adaptively filtering near-stationary noise from speech, U.S. Patent 4025721, May 1977.
24. Boll, S., Suppression of acoustic noise in speech using spectral subtraction, *IEEE Trans. ASSP*, 27(2), 113–120, 1979.
25. Kushner, W. et al., The effects of subtractive-type algorithms on parameter estimation for improved recognition and coding in high noise environments, *Proc. IEEE ICASSP*, 1989, 211–214.
26. Crozier, P.M. et al., Speech enhancement employing spectral subtraction and linear predictive analysis, *Electronics Lett.*, 29(12), 1094–1095, 1993.
27. Nalazco Flores, J.A. and Young, S.J., Continuous speech recognition in noise using spectral subtraction and HMM adaptation, *ICASSP-94*, 1, 409–412, 1994.
28. Berouti, M., Schwartz, R., and Makhoul, J., Enhancement of speech corrupted by additive noise, *Proc. IEEE ICASSP*, 1979, 208–211.
29. Boll, S. et al., Improving the quality of the noisy speech signal, *Bell Systems Tech. J.*, 60, 1847–1859, 1981.
30. Ephraim, Y. and Van Trees, H.L., A spectrally-based signal subspace approach for speech enhancement, *ICASSP-95*, 1, 804–807, 1995.
31. Ephraim, Y. and Van Trees, H.L., A signal subspace approach for speech enhancement, *ICASSP-93*, 2, 355–358, 1993.
32. Bray, V. and Nilsson, M., Objective test results support benefits of a DSP noise reduction system, *Hearing Rev.*, 7(11), 60–65, 2000.
33. Brey, R.H. et al., Improvement in speech intelligibility in noise employing an adaptive filter with normal and hearing-impaired subjects, *J. Rehab. Res. Dev.*, 24(4), 75–86, 1987.
34. Lang, D.A., Methods of Adaptive Echo Removal, Master's thesis, Brigham Young University, Provo, UT, 1989.
35. Schafer, R.W., Echo removal by discrete linear filtering, Technical Report 466, MIT Research Laboratory of Electronics, Cambridge, MA, February 1969.
36. Neely, S.T. and Allen, J.B., Invertibility of a room impulse response, *J. Acoust. Soc. Am.*, 66(1), 165–169, July 1979.
37. Mourjopuoulos, J., Clarkson, P.M., and Hammond, J.K., Dereverberation of speech using optimism control, Cappellini, V. and Constantinides, A.G., eds., *Digital Signal Processing — 84*, Elsevier, North-Holland, Amsterdam, 1984.
38. Hawkins, D. and Yacullo, W., Signal-to-noise ratio advantage of binaural hearing aids and directional microphones under different levels of reverberation, *J. Speech Hear. Disord.*, 49, 278–286, 1984.
39. Nilsson, M., Soli, S., and Sullivan, J., Development of the Hearing In Noise Test for the measurement of speech reception thresholds in quiet and in noise, *J. Acoust. Soc. Am.*, 95(2), 1085–1099, 1994.
40. Soli, S. and Nilsson, M., Assessment of communication handicap with the HINT, *Hearing Instruments*, 45(2), 12–16, 1994.

Index

A

Absolute Category Rating Method (ACRM), 138–139
Absolute delay, 283
Acceptable timeliness, 104
Acoustic crosstalk cancellation, 331–334
Acoustic echo, *see also* Echo cancellation
  feedback fundamentals, 6
  hands-free telephony, 200
  sources of, 297
ACRM, *see* Absolute Category Rating Method (ACRM)
Active noise control, 309–322
Active testing, 300
Acute phonemes, 131
Adaptive beam forming, *see also* Beam forming
  alternative to, 191
  delay-and-sum, 183–185, 195
  echo cancellation, 192
  Griffiths-Jim combination, 187–188
Adaptive digital filtering
  additional techniques, 42–43
  basic fundamentals, 35
  gradient techniques, 40–41
  least mean squares, 41–42, 383
  mean square error optimization, 37–40
  structure and architectures, 35–37
Adaptive echo cancellation, 203–208
Adaptive feedback controllers, 318–321
Adaptive noise cancellation unit (ANCU), 351, 352, 355
Adaptive phase-locked buffer (APLB)
  algorithm, 352–354, 358–359
Adaptive predictive filtering, 386–387
ADC, *see* Analog-to-digital circuit (ADC)
Addition operation, 93–94
Additive vs. subtractive distortion, 284
Address generation, 95
AGC, *see* Automatic gain control (AGC)
Agricultural electrical fences, 346
Aircraft applications, 37, 305, 318
Algorithms, *see also* specific algorithm
  adaptive filtering, 43

address generation, 95
APLB, 352–354
Baum-Welch, 248, 253, 256
building blocks, 96–97
cooperative multitasking, 116
designing, 109–110
Euclidean, 193
expectation maximization (EM), 231, 234, 252, 255, 267
FXLMS, 312, 314–320
least mean squares (LMS), 34, 41–42, 203–204, 205, 207, 351–352, 374, 383, 386
linearly constrained minimum variance (LCMV), 362, 367, 372–374
minimum mean square error (MMSE), 171
PAMS, 302
perceptual analysis measurement system (PAMS), 293
preemptive multitasking, 111, 116
recursive, 173–175
recursive least squares (RLS), 43
root-mean square (RMS), 61
single-channel speech enhancement, 155–177
stochastic gradient, 41
time-sharing multitasking, 116
vector Taylor series (VTS), 231, 233–236, 237, 341–342
Viterbi, 248
Aliasing, spatial, 191
Allen, Neely and, studies, 193
Alvino, Parra and, studies, 366
Amplifier noise, 6
Amplitude clipping, 201
Analog circuitry
  analog-to-digital gateway, 49–50
  circuit complexity, 53
  multiband processor design, 66
  problems, 91–92
Analog/digital interfacing, 4–12
Analog techniques
  basic fundamentals, 47–48
  current-to-voltage conversion, 53–56
  filters, 49, 92